The Columbia Guide to
American Environmental History

D0764691

The Columbia Guides to American History and Cultures

The Columbia Guides to American History and Cultures

The Columbia Guide to the Cold War
Michael Kort

The Columbia Guide to American Women in the Nineteenth Century
Catherine Clinton and Christine Lunardini

The Columbia Guide to America in the 1960s
David Farber and Beth Bailey

The Columbia Guide to Asian American History
Gary Y. Okihiro

The Columbia Guide to
American Environmental History

Carolyn Merchant

COLUMBIA UNIVERSITY PRESS   NEW YORK

COLUMBIA UNIVERSITY PRESS
Publishers Since 1893
New York, Chichester, West Sussex

Library of Congress Cataloging-in-Publication Data
Merchant, Carolyn.
    The Columbia guide to American environmental history / Carolyn Merchant.
        p. cm.
    Includes bibliographical references and index.
    ISBN 10: 0–231–11232–7 (cloth : alk. paper)
    ISBN 13: 978–0–231–11232–1 (cloth : alk. paper)

    ISBN 10: 0–231–11233–5 (pbk. : alk. paper)
    ISBN 13: 978–0–231–11233–8 (pbk. : alk. paper)
    1. Human ecology—United States—History.
    2. Nature—Effect of human beings on—United States—History.
    3. Landscape changes—United States—History.
    4. United States—Environmental conditions.
    Series: The Columbia guides to American history and cultures.
    GF501 .M47 2002
    304.2 21

                                                                    2001– 056192

Casebound editions of Columbia University Press books
are printed on permanent and durable acid-free paper.
Printed in the United States of America

c 10 9 8 7 6 5 4 3 2
p 10 9 8 7 6 5 4 3 2

Contents

Acknowledgments

Writing the *Columbia Guide to American Environmental History* has afforded me the unique opportunity to review and synthesize developments and resources in this relatively new and dynamic field. Environmental history emerged out of 1960s concerns over the impacts of pesticides, population, urbanization, and technologies on the environment. Propelled by popular interest in the state of the environment following Earth Day 1970, many people began rethinking the relationships between the environment and academic fields such as history, ethics, political science, and economics. The American Society for Environmental History was founded in 1977, a year after the journal *Environmental Review* (subsequently called *Environmental History Review*), which merged with *Forest History* in 1996 to become the journal *Environmental History*. A second journal published in England, entitled *Environment and History*, was launched in 1995. As the field has continued to grow, articles pertaining to the history of the environment have appeared in many journals and magazines. The *Columbia Guide to Environmental History* presents a survey of the field that includes an overview of topics and themes, a compendium of persons, concepts, and laws, a chronology of major historical events, and a guide to additional resources. No book of this type can ever be complete in light of the many topics and resources that continue to emerge in the environmental history field, and difficult decisions have led to what is included in this volume.

Although the final product published here is my own, I would like to thank the many persons who have made substantial contributions to the outcome. I owe much to the inspiration and careful guidance of my editor, James Warren of Columbia University Press, editor in chief of the Columbia Guide series, and to Joshua Lupkin, Nickolas Frankovich, and Leslie Bialler for editorial assistance.

The final draft of the book was prepared at the National Humanities Center and I am grateful to the John D. and Catherine T. MacArthur Foundation for supporting my work there as a fellow in the ecological humanities. Jessica Teisch compiled the bibliography, incorporating materials she had prepared for the National Humanities Center with funding from the National Endowment for the Humanities. Under a grant from the Committee on Research of the University of California at Berkeley, Alison Lozner compiled materials for the topical compendium, bibliographical essay, and the visual and electronic resources sections. Valerie Peters and Joshua Volz contributed many of the details in the chronology and William Yaryan, Earth Trattner, and Yvan Chantery assisted with materials included in the historical overview of topics and themes and in the sections devoted to visual and electronic resources. Celeste Newbrough prepared the index with funding from the University of California at Berkeley's Committee on Research.

I would like to thank Shepard Krech III and Timothy Silver, colleagues for three different summers in the Nature Transformed seminar at the National Humanities Center, for their ideas, careful reading, and many suggestions for the historical narrative. I would also like to thank my students and teaching assistants in my course American Environmental and Cultural History, taught since 1979, for their ideas and contributions to the intellectual content and methodology of the course and the field. I am indebted to my partner and husband, Charles Sellers, whose inspiration and intellectual contributions over many years have played a major role in my interpretation of American history and to whom I am deeply grateful for reading, editing, crafting, and rethinking portions of the historical narrative. Final responsibility for the content of the book is, of course, my own.

CM
Berkeley, California

Introduction

The Columbia Guide to American Environmental History introduces the many dimensions of human interaction with nature over time. As people have lived and spread out over the planet, they have modified its forests, plains, and deserts. Those changes in turn have affected the ways in which people organize their social and religious systems. The *Guide* offers the reader a brief history of that interaction as it took place in North America; a mini-encyclopedia of concepts, laws, agencies, and people pertinent to the field; a timeline of important events; and a set of print, visual, and electronic resources for further reading and research.

Environmental history is both one of the oldest and newest fields within human history. All cultures have oral and written traditions that explain human origins and encounters with the natural world through stories about local landscapes and ways to perpetuate life from the land. Many cultures developed these early ideas into elaborate oral and written traditions, and finally into modern scientific approaches to explaining and managing the vicissitudes of nature. Religion, science, art, and literature provided ideas as they evolved over time, while records such as calendars, diaries, account books, treatises, and museum collections give access to human practices that modified the landscape.

Environmental history comprises a set of approaches to doing history that brings nature into the story. Natural conditions such as climate, rainfall, terrain, vegetation, and animal life create possibilities for the quality of human life. Human systems of producing and reproducing life over time entail technologies, economies, governance, and social structures. Such systems include gathering, hunting, and fishing; agriculture; and industrialization. These human systems, however, result in a transformed nature, and the character of that transformation is a major theme for environmental history. And, as nature changes, people's ideas of what it ought to look like also change. For some a vanishing "wilderness" has positive value; for others it is a tragic loss. The character of such evolving ideas and how to implement or reverse landscape changes is another major topic for environmental historians.

Environmental historians ask the following kinds of questions:

- How did American Indians use, manage, and conserve the land?
- What ideas, animals, plants, diseases, and systems of producing necessities were introduced by European, African, Asian, and other immigrants? How did they change the land?
- How did various racial and ethnic groups interact in transforming the various regions of North America?
- What resources and commodities were important in various regions and to various groups of people?
- What kinds of practices, regulations, and laws were used from earliest times to the present to manage the land? (E.g., timber cutting, firewood, fires, water, range, etc.)
- What was the impact of the rise of cities on the surrounding country? What contributions did cities make to resource use?
- What environmental problems were created by urbanization? (E.g., air and water pollution, disease, hazardous wastes, noise, etc.)
- What were the sources of opposition to development? (E.g., the wilderness preservation, resource conservation, and environmental movements.)
- What were some of the ideas driving environmental change? These might include: religion — replenish and subdue the land; stewardship or reverence for nature; manifest destiny; wise use of nature.
- How did past ideas about nature, such as wilderness, the Jeffersonian ideal of farms on the land, concepts of land reclamation, ideas of hydraulic societies, and the aesthetic appreciation of nature help to propel change?

Because of its vast scope, environmental history is very complex. Among the most complicated aspects of the field are the very meanings of terms such as nature, environment, wilderness, garden, conservation, and ecology. Not only do such terms mean different things in different eras, they mean different things to contemporary historians. A concept such as wilderness, for example, was synonymous with home for Indians, anathema to Puritans, the basis for national pride to romantics, and a way to retain masculine, frontier virtues to turn-of-the-century urbanites. For contemporary historians wilderness may be a complex idea that has no foundation other than its changing historical meanings, or it may mean a real, forested landscape that has evolved over millions of years and exists today only in isolated remnants. Debates such as these have great practical consequences. How wilderness is defined lies at the core of development policies. If it is an evolved reality that can be documented through evolutionary and ecological science and areas of pristine "wilderness" can be identified, then laws to preserve these remnants can be passed, implemented, and adjudicated through the courts. If, on the other hand, what wilderness means is an ephemeral semantic debate having different meanings in different eras, then some will argue that no particular place has any greater claim to preservation or development than any other. Environmental history therefore lies at the core of current policy choices.

Environmental historians approach their field from a variety of perspectives. One approach is to focus on biological interactions between humans and the natural world. Animals, plants, pathogens, and people form an ecological complex in any one place that can be sustained or disrupted. When Europeans settled in North America and other temperate regions of the world, they introduced diseases, such as smallpox, measles, and bubonic plague; livestock, such as horses, cattle, and sheep; European grains such as wheat, rye, barley, and oats, along with varmints, such as rats; and weeds, such as plantain and dandelions. These ecological introductions, especially diseases, devastated the lives of native peoples. While some of the introductions, such as the horse, gave some Indians a temporary advantage over Europeans, the introduced ecological complex as a whole altered the landscape in ways that benefited the settlers and disrupted Indian lifeways.

A second way to think about environmental history is in terms of a series of levels of human interactions with nature, such as ecology, production, reproduction, and ideas. On the first level is nature itself. Nature's own history can be described in terms of the evolution of the geology and biology of a given place; the ecological succession of plants and animals found there; and the variations in temperature and climate that create the potential for human systems of production. At the second level, human forms of production also vary over time. North American Indians evolved complex systems of gathering, hunting, fishing, and horticulture, combined with trading across tribal boundaries. European settlers who arrived on the continent in the sixteenth and seventeenth centuries developed sophisticated technologies such as ships, gunpowder, iron tools, clothing, and agricultural systems that created complicated, often uneven systems of interaction among Europeans, Indians, and nature. At the third level is reproduction. This includes biological and social forms of reproducing human and non-human life, as well as means of reproducing human social and political life over time. Finally, on the fourth level, are ideas, such as narrative, science, religion, and ethics that explain nature, the human place within it, and means of behaving in relation to it.

A third approach to doing environmental history is in terms of environmental politics and transformations in political and economic power. The history of the conservation and preservation movements of the late nineteenth and early twentieth centuries, for example, can be delineated in terms of political struggles within a presidential administration, the role of citizen movements in pressing for the preservation of natural areas, and the creation of government and state agencies to manage and conserve natural resources.

A fourth approach to the field is to focus on the history of ideas about nature. Histories of a philosophical idea such as wilderness, a scientific idea such as ecology, or an aesthetic idea such as natural beauty form the topics of numerous books about the nature of nature in North America. These works examine the ideas and creative products of artists, nature writers, science writers, explorers, and travelers for clues as to how people felt about nature and how their feelings led to actions with respect to its visual or economic resources. Such intellectual histories help us to understand how changing ideas about nature and beauty can be influential in creating the environments we see around us today.

A fifth way to do environmental history is in terms of narrative. One can argue that all peoples interpret their world through stories, whether the origin stories of Native Americans or Europeans, the stories of various fields of science as they progress over time, or morality stories that tell us how to behave in the world. Environmental historians often contrast their histories with histories of progress and enlightenment, inasmuch as developments for human well-being — such as industrialization — often result in the degradation of the environment through pollution and depletion. Yet all environmental history does not necessarily view history as a decline from a pristine environment that was irrevocably and negatively transformed when humans entered it. Environmental historians write narratives that are both progressive and declensionist, comic and tragic, intricate and bold. Nevertheless, the stories have a message. They explain the consequences of various past interactions with the natural world and warn us of potential problems as we form policies and make decisions that affect our lives and those of our children. Knowing and doing environmental history is therefore critical to the continuance of life on earth, whether that life be human or that of the other animals and plants that occupy the landscapes in which we dwell.

Part I of the *Columbia Guide to American Environmental History* is a historical overview of topics and themes in environmental history. The overview does not attempt to treat comprehensively every event, but rather to explore some of the themes that environmental historians have used to interpret the field. It is presented chronologically, beginning with the natural environment and proceeding through Native American lifeways to European settlement, the formation of the nation and industrialization, to the conservation and environmental movements of the present day. The dates associated with each section are somewhat arbitrary, and hence risk simplifying complex, interlocking events. The purpose of the overview, however, is to offer a framework within which further reading and research can be interpreted, elaborated, or contested. Many of the theories about doing environmental history introduced above appear in the historical overview. The reader is encouraged to approach this section critically and creatively — to think about how the story could have been written differently and what other topics would help to round out the presentation. Then, by using the materials in the Resource Guide (see part IV below), readers may become environmental historians on their own.

Part II of the guide is a topical compendium of agencies, concepts, laws, and people. It is arranged alphabetically within each category, but each entry can be accessed directly through the index. Here the reader can find out quickly who influential writers, ecologists, conservationists, and artists were; what the major thrust of a law is; or how an agency or environmental group is organized. Concepts relevant to the history of the environment are elaborated together with controversies and changes in their historical meanings. This section assists the reader in understanding these variations and controversies in meanings, as well as giving brief developments of major ideas and actors critical to the history of the environmental change.

Part III of the guide is a chronology of major events in environmental history. It is presented as an environmental history timeline, into which major political events — such as the American Revolution, the Constitution, the Civil War, and the two world wars — are inserted for reference and context. The timeline shows events in sequence — events that in some cases may have been formative and causal to each other or, in other cases, apparently unrelated — so that relationships among events can be perceived in new ways. Many of these environmental events are also elaborated in the historical narrative and the topical compendium, to which the reader can turn for additional details.

Part IV is a resource guide. As such, it offers visual, electronic, and print resources for further reading and research. The visual resource guide is a list of videos and films with brief synopses of their content, along with information on how to obtain them. They are arranged to dovetail with topics in the historical narrative. The visual materials can be used with courses on environmental history or as alternative means of understanding environmental change over time. The importance of electronic materials such as CD-ROMs and the Internet makes it imperative to include materials currently available in digital format. These resources, however, are the most ephemeral of all the materials in the resource guide. Some Web sites and links vanish without warning; others appear or are updated and changed in name; many become outdated as technology changes. They are included, however, to give a sense of the possibilities for doing environmental history and offering on-line courses in the field.

The bibliography of printed materials is arranged topically, and hence reinforces information in the historical narrative. It is introduced by a bibliographical essay that discusses many of the sources relevant to environmental history. The reader can gain a sense of what types of books and other printed materials constitute environmental history's primary sources. The bibliographical essay also discusses some of the most important books in environmental history in order to orient the reader to the field's historical evolution. The bibliography that follows the introductory essay is meant to be used for further reading and research on doing the environmental history of one's own region or for pursuing one's favorite topic or time period.

In summary, the overall goal of the *Columbia Guide to American Environmental History* is to provide a concise "first-stop" reference book on the history of the North American environment for high school and college students, teachers, researchers, and readers. The book places the subject of environmental history in the context of recent scholarship, introduces key questions about the topic, and includes brief sketches of significant persons, events, and themes. It provides an accessible overview of environmental history; a mini-encyclopedia of concepts, legislation and agencies, and people; a chronology of events and their significance; and a bibliographical introduction to printed materials, as well as to lists of films, videos, CD-ROMs, and on-line resources.

The Columbia Guide to
American Environmental History

Part I

Historical Overview: Topics and Themes

1   The American Environment and Native-European Encounters, 1000–1875

The North American environment contains rich natural resources that, over time, have supported a succession of modes of living on the land. A core topic in environmental history is how different peoples at different times have used, perceived, managed, and conserved their environments. Native Americans developed several forms of land use appropriate to the resources of different regions of the United States. This chapter compares three patterns of Native American subsistence and the processes by which European settlers colonized particular North American landscapes: southwestern horticulture; northeastern hunting and gathering; and the Great Plains buffalo and horse cultures.

The Physical Environment and Natural Resources

The physical environment that constitutes the present United States can be characterized in terms of its area, location, climate, rainfall, and topography. Its total area is approximately 3.6 million square miles. The 48 contiguous states extend from the 24th parallel at the tip of Florida to the 49th parallel at their northern border with Canada, falling between 66 and 125 degrees west longitude. Alaska lies between 54 and 72 degrees north latitude and 130 degrees east longitude, with the westernmost point of the Aleutian Islands being in the Eastern Hemisphere at 172 degrees east longitude. The tropical Hawaiian Islands are situated between 19 and 25 north latitude and 155 and 176 west longitude.

The climate of the 48 mainland states is temperate, having cold winters and hot summers. The states east of the 100th meridian are characterized as humid, with 20–60 inches on average of annual rainfall distributed throughout the year; those west of the 100th meridian have an average between 5 and 20 inches, distributed mainly in the winter months. Topographically, the country extends westward from the Atlantic coastal plain at sea level upward to the Piedmont Plateau, between 500 and 1,000 feet in elevation, and the Appalachian Mountains, which rise to higher than 6,000 feet. Beyond the Mississippi River the elevation again rises gradually upward to about 4,000 feet, where the high plains approach the Rocky Mountains. The Rockies, Sierra Nevadas, and Cascades rise westward of the plains from around 4,000 to higher than 10,000 feet, interrupted by the intermontane Great Basin, between 4,000 and 6,000 feet above sea level. Westward of the Sierras, Cascades, and Coast Ranges, peaks in Alaska rise to 20,000 feet and in Hawaii to over 13,000 feet.

The country contains the rich natural resources required for agricultural and commercial systems. It has extensive forests in its eastern and western regions for fuel and building materials and extremely fertile soils along the eastern coastal plains, the Mississippi Valley, midwestern prairies, and Pacific coast valleys. Ample rainfall and water for agriculture are supplied from the snowmelts of the Appalachian Mountains to water the fertile soils of the eastern states. Snowmelt from the Rocky Mountains and Cascades fill the western river systems, while aquifers now supply water to the Great Plains, creating possibilities for irrigated agriculture in the arid regions of the West. The river systems and lowlands of the eastern coastal plains and Mississippi Valley make extended inland transportation and commerce feasible. The vast mineral treasury of the Appalachians and the western mountains, with their deposits of gold, silver, copper, iron, and coal, supply the energy and minerals needed for thriving industries. These natural features, resources, and climate patterns were utilized first by Native Americans and then by colonists from other continents to create subsistence- and market-oriented societies.

Native Americans and the Land

Indians occupied almost all of North America on the eve of European colonization. Their primary mode of subsistence was hunting, gathering, and fishing, combined — in the Southwest, Mississippi Valley, and eastern woodlands — with horticulture. Indian ecological relations with the land and the transformation of those relations under European settlement can be illustrated by cases from the Southwest where horticulture predominated; the hunting cultures of the Northeast; and the buffalo cultures of the Great Plains. In each case the process of transformation varied, but several factors predominated: the introduction of pathogens and other biota; trading in furs and hides; and the addition of cultural factors such as Christian religion and alphanumeric literacy.

Estimates of the precolonial Indian population of North America have ranged from 1 to 18 million, depending on methods used to extrapolate backward from existing records of population densities and the devastation of diseases, with recent assessments in the range of 4 to 7 million. There is also considerable debate about the process of the peopling of the Americas. Indian origin stories tell of the continued presence of native peoples on the continent from time immemorial, of an original emergence of people out of the earth in distant times, or of arrival from other lands after a long journey. The theory accepted by most Westerners, including many Indians, however, has been that of arrival over the Bering land bridge, or the Beringia theory. Between 80,000 and 23,000 years ago sea levels dropped, opening up possibilities for people to migrate across ice-sheets, and were even lower 23,000 to 10,000 years ago, when a broad land mass some 1,000 miles wide appeared. The Beringia theory holds that people from Asia were able to migrate north into the area that is now Russia and then across the Bering land bridge, thus entering the Americas. These earliest bands, which arrived at least 13,000 years ago (and possibly, some believe, as early as 35,000 years ago), were able to survive on food resources found at latitudes above 60 degrees and to live in very cold temperatures. Much of that survival was the result of following migrating herds of game.

Between 13,000 and 11,000 years ago, the vast ice-sheets that covered present-day Canada and the Great Lakes receded, opening a corridor southward from Alaska through the Yukon and down along the slopes of the Canadian Rockies. A southward migration of people occurred through this Rocky Mountain Trench. Because of the relative scarcity of animal or plant life in that corridor, bands of people moved down it and then fanned out into the area that now constitutes the United States. A second theory holds that the migrations southward occurred primarily along the Pacific coast, where fish and wildlife were more abundant and sea levels were much lower than at present. Additionally, a third migration across the Bering land bridge into Alaska, between 11,000 and 8000 B.P., was made by Eskimo Aleut peoples.

People who migrated southward and eastward into the present-day United States also continued into Central and South America. Some recently discovered sites in Monte Verde, Chile, are perhaps 15,000 years old, although radiocarbon methods of dating the remains are subject to further verification, as is the question as to whether those sites were settled in the earliest Beringia migration or by migrations across the ocean. Other early sites, discovered in the eastern United States and in the Pacific Northwest, remain subject to additional verification as well.

The Beringia theory, however, supplies an additional explanation for later environmental transformations. The land bridge, or the Bering Strait itself, protected migrants from diseases associated with Europe and Asia because pathogens were unable to survive in the very cold temperatures. Although Amerindians did have a number of diseases (such as colds, intestinal parasites, and yaws), they lacked immunity to many diseases associated with domesticated animals and plants (such as smallpox, measles, bubonic plague, and scarlet fever) that the Europeans brought with them, and which rapidly decimated Indian populations. The timing and effects of epidemics varied according to type of disease, mode of transmission, rapidity of the spread of contagion, density of population, and types of contacts with Europeans. Debate centers on the validity of archaeological and historical records and the methods of counting and estimating population sizes in determining whether and when a disease reached a particular tribe. The vastly reduced populations gave European colonists an advantage over the weakened tribes. How did the early migrants from Asia to the Americas attain subsistence, and how did they transform the North American environment?

Pueblo Indians and the Southwest

In the American Southwest, a long history of environmental adaptation by Amerindians preceded the arrival of the Spaniards in the sixteenth century A.D. Early peoples changed and evolved over time in relationship to changing environmental conditions. Paleo-Indians (meaning ancient or very ancient) lived in the Southwest from about 11,500 to 8,000 B.P. (before present) by hunting mammoth and bison. The Clovis culture derives its name from the long pointed arrowheads, found first in Clovis, New Mexico, that were attached to spears thrown by *atlatls* (long rods) and used to hunt mammoth. Stone knives were also developed for butchering the prey. Following the Clovis culture, the Folsom culture developed shorter spears and stone tools for killing the bison that survived both climate change and paleo-Indian hunters. Throughout the West, mammoth and buffalo jumps have been found, where animals, herded by people on foot, were stampeded over the edges of cliffs and eroded gullies, where the kill was then butchered.

In the Southwest, following the age of large mammal hunting, a second type of culture emerged — the archaic culture, meaning old or early. The archaic culture, which existed from about 8000 B.P. to 4000 B.P., consisted of groups of hunting-gathering peoples who hunted the smaller animals that remained after mammoths and bison declined or disappeared from the region — deer, mountain lion, antelope, and rabbits. These groups moved continuously in small bands of related individuals as they foraged for plants and small animals, traveling with the seasons. They may have had sites to which they returned each year, or they may have found new locales where game was more abundant and where they knew that the plants they needed for survival could more easily be found. They developed snares and nets, stone and bowl chipping tools, and scrapping and cutting tools to process the fibers, tubers, and seeds found in the arid environment.

Around 2,500 B.P., possibly owing to population pressure or the advent of crops and technologies diffusing northward from Mexico, people began to live in settled communities, growing maize — a unique new world crop — together with squash and, later, beans. The three foods became known as the "three sisters," or the corn, beans, and squash complex. Melons and chili peppers were also added to the diet. People lived in settled communities, joining their labor together, women working with women and men with men.

In the area of the Southwest presently known as the Four Corners — where Utah, Colorado, Arizona, and New Mexico converge — a confluence of cultures existed. The Hohokam, who settled in southern Arizona, practiced irrigation-based horticulture, while the Anasazi dwelt in pueblos on the plateaus of the region and planted crops along rivers. These cultures reveal the development of sophisticated technologies for adapting to and manipulating the environment.

Knowledge of Hohokam irrigation systems is based on archaeological remains found in the area today, as well as on extrapolation from current cultures backward in time. Because the land was a desert, with rainfall amounts less than 10 inches a year, crops required catchment dams, irrigation canals, and water conservation. Environmental historian Donald Worster, in *Rivers of Empire* (1985), writes: "Water control relies on temporary structures and small-scale permanent works that interfere only minimally with the natural flow of streams.... [The users] are self-reliant, self-sufficient, and self-managing as individuals and as a community, though nature still sets in the main the terms on which their lives are lived."[1]

Between A.D. 300 and 900, the Hohokam people manipulated the arid environment by building large canals, some as wide as 30 feet across, 7 feet

deep, and 8 miles long, with as many as 8,000 acres at a time being fertilized
with water from the Salt and Gila river systems. Indians could grow crops
and store them throughout the year, so as to have a continual supply of food.
An environmental consequence of irrigated agriculture, however, is that water from the rivers leaches out salts, drawing them up from the subsoil and
leaving salinated topsoils. Just as the Mesopotamian cultures of the Near
East were abandoned owing to salinization, so — it is hypothesized — the
Hohokam were forced to abandon their villages and move elsewhere or to
remain, evolving and adapting to changed conditions. By the time the Spanish arrived, they met Piman-speaking, Akimel O'odham Indians, who were
still practicing a form of irrigation, but not on the enormous scale developed
by the Hohokam. Other explanations for the demise of the Hohokam include flooding, drought, earthquakes, and deforestation, some of these hy-
potheses influenced by environmental events occurring at the time the explanation itself was put forward.

A second group of southwestern peoples, the Anasazi (or "ancient ones"),
likewise developed sophisticated technologies for transforming the environment. Between A.D. 900 and 1150, in northwestern New Mexico, they constructed large multistoried communal houses in villages, or pueblos, in the
bottom of Chaco Canyon, along the edges of a river where they planted
crops. The communities consisted of many small villages with larger central
cities containing four-story apartments and ramrod-straight roads that linked
Chaco Canyon with other communities in the Southwest.

Not only did the Chaco people build roads and pueblos, they also developed an agricultural calendar that enabled them to utilize the southwestern
environment through an understanding of astronomy, a technology illustrated in films such as The Sun Dagger (1963). They attained the ability to
predict the advent of growing and harvesting seasons based on arrival of the
solstices and equinoxes — the solstices (or longest and shortest days) occurring around June 21 and December 21 and the equinoxes (equal days and
nights) around March 21 and September 23. They did so by building an
"observatory" on Fajada Butte in Chaco Canyon, a stone-slab structure as-
sembled so that a sliver of sunlight entered through two stones and fell on
an engraved spiral petroglyph. They were thus able to determine the best
times to plant and harvest their crops, gaining a measure of control over the
vicissitudes of an arid environment.

Mesa Verde (meaning "green table"), an Anasazi pueblo from the same
period, likewise illustrates the sophistication developed by southwestern peoples in manipulating their environment. Situated on a vast green plain with a

level top consisting of a juniper and piñon forest and located some 7,000 to 8,000 feet above sea level, Mesa Verde culture and agriculture developed gradually over time. The people of the earliest culture, from around A.D. 500 to 750, were basket makers who lived underground in pit houses. Around 750 to 1100, they added ceremonial circular kivas and began building houses above ground out of bricks. Between 1100 and 1300, they reached the peak of their complex culture, building ladders ascending to the cliff areas, as well as pathways descending to agricultural fields in the valleys. The vast agricultural network covered some eighty square miles of development, supporting numerous pueblos located about 1,800 to 2,000 feet above the river. Deforestation, social disruption, and the onset of an eleven-year drought around 1280 (as evidenced by the narrowness of tree-ring growth) were factors that may have caused abandonment of the pueblos.

A cooperative division of labor existed within these agricultural communities. Here, in contrast to the tribes along the eastern seaboard, men planted and tended the fields, while women and girls ground the corn into meal, an arduous and difficult task done on stone *metates* with handheld grinding stones, or *manos*, used for scraping across the corn and pulverizing it into meal. To keep the pace going and to inspire the women's work, elders too old to labor in the fields played flute music. The horticulturists also hunted deer, antelope, buffalo, and rabbit and traded corn for meat and hides with hunting bands from nearby mountains. But despite the level of culture and agriculture developed in the Southwest, a major ecological and social transformation began with the Spanish expeditions into the area in the sixteenth century. An ecological complex of introduced animals, plants, pathogens, and people radically changed the region.

The Pueblo Indians and Spanish Settlement of the Southwest

Into the Pueblo culture, which had evolved slowly over time to a very complex level, the Spaniards were suddenly injected. Their presence in the New World began in 1492 with the first of four voyages by Columbus. Earlier voyages by Vikings and Basques to Labrador, Newfoundland, and the Grand Banks along the eastern coast of Canada had taken place, but land-based explorations with lasting consequences stem from the post-1492 period. In the American Southwest, the first exploration was led by Marcos de Niza, who in 1539 traveled north from Mexico to the Zuni pueblos, where he reported on the riches he had viewed. He was followed by Francisco

Vasquez de Coronado in 1540, who spent two years in the region, traveling as far east as present-day Kansas.

After the initial explorations, two important additional expeditions occurred, the first by Francisco Sanchez Chamuscado in 1581 and the second by Don Antonio de Espejo in 1582, both of whom recorded their impressions in diaries. The Chamuscado-Rodriguez party explored the territory northward, to the Tiwa pueblos, where they left two friars, who were subsequently killed by Indians. Espejo followed two years later with fourteen soldiers and Franciscan priests. He explored many of the pueblos, including the Zuni, the Hopi, the Acoma, the Tiwa, and the Keres pueblos, and then returned by way of the Pecos River in New Mexico to avenge the death of the Franciscan friars.

The subsequent bloodshed was the beginning of a long series of problematic occupations by Spanish settlers. The first colonization of the area was by Don Juan de Oñate, who mounted an expedition northward from New Mexico to look for mines and to Christianize the Indians. He brought with him 400 colonists, 10 Franciscan missionaries, 7,000 head of cattle, sheep, and horses, and founded the first Spanish colony in New Mexico. Oñate treated the Indians brutally, setting up courts and cutting off the legs and arms of people who did not obey him. As a consequence, a Spanish court was convened to try him and to hear the complaints of his own soldiers.

The large animals, colonists, and missionaries changed the region forever. Environmental historian Alfred Crosby, in *Ecological Imperialism* (1986), has called the complex of European ecological introductions the "portmanteau biota, [a] collective name for the Europeans and all the organisms they brought with them."[2] Along with the portmanteau biota, the use of guns and swords and the written language, as evidenced by Spanish accounts and court documents, represented the organized power of Europeans to control and dominate the Indians and their environment.

In *When Jesus Came the Corn Mothers Went Away* (1991), historian Ramón Gutiérrez details the brutality of the expeditions that changed Pueblo land and life and argues that the Franciscan missions that followed actually provided a measure of security for the devastated Indians. The Franciscans established churches, transforming the religion of the native peoples from the worship of corn mothers to the worship of the male Christian God and the Virgin Mary. "Christianization," states Gutiérrez, "meant a reliable meat supply, iron implements of various sorts, and European foods."[3] The missionaries built churches and chapels, established workshops for weaving and smithing, and created irrigated gardens where they planted crops brought from the Old World. In introducing Christianity, the Franciscans sought to obliterate Indian animism and the fetishes of the Zuni and other peoples. They prohibited the masks and dances of the Pueblo tribes and substituted the Virgin Mary for the Corn Mother.

By the seventeenth century, the world of the southwestern Indians had been changed by a complex of ecological and social introductions: European diseases (which decimated Pueblo populations between 1638 and 1640), large domesticated animals, new techniques of warfare, the Christian religion, and alphanumeric literacy. But for the Spanish, as well as the Indians, the arid southwestern environment imposed constraints. Many colonists, along with their domesticated animals, died before irrigated fields with European grains and fruits were established and Indian rebellions contained. Throughout the process of colonization, however, Indian peoples retained much of their heritage, ritual practices, and sense of cultural identity.

Micmac Indians and French Settlement in the Northeast

A second case of ecological transformation in the New World is that of the hunting/gathering/fishing cultures of the Northeast, such as the Abenaki of northern Maine, the St. John's River Malecite, and the Micmac of southeastern Canada, all of whom shared a common culture. The Micmac lived in the area known as the Gaspé peninsula, south of Newfoundland, Labrador, and the St. Lawrence River. Micmac hunting, like that of other cultures in the Northeast, was based primarily on large animals, such as moose, deer, elk, and caribou, sources of both meat and hides. In addition, smaller animals, such as mink, muskrat, and beaver, were trapped for their furs.

Indian spiritual relations with animals derived from intimate everyday encounters. Indian peoples took their tribal clan names from animals, such as bear or deer, and often identified with or assumed the personalities of those animals. Animals were thought to live in separate societies with leaders, as did human hunter-gatherers, with a pale or white animal often thought of as the leader of the animal community. In addition, the animals had spirits, called *manitous* by the French Jesuits. These animal spirits were believed to communicate not only with the animals themselves, but with their dead ancestors and also with shamans — the spiritual leaders of the Micmac — and other hunting tribes.

The Micmac, like other hunting groups, went through ritual preparations before embarking on the hunt. An animal might appear to them in a dream, or the shaman might determine the time to set out. Or they might see a herd of deer and decide that this was the moment when the deer had come to them as a gift. Sometimes divination techniques were used to determine the direction of the hunt, such as burning the shoulder or a leg bone of an animal and then reading a pathway on the lines revealed on the bone. This act often randomized the hunt, giving animals time to reproduce.

In hunting, Indians believed the animal gave itself up to be killed so the hunter could survive, a process the Jesuits called ordained killing. Deer, for example, were autonomous subjects, in the sense that they were equal or superior to human beings. When a human hunter looked into the eye of a deer, an exchange of understanding occurred, an agreement or contract that this was the moment the deer had chosen to give itself up as a gift to the hunter. In addition to ritual preparations, there was ritual disposal of the remains. Land animals were to be disposed of on land, for example, and water animals were returned to their own watery abode. Mink were to be hung on trees and gutted to show other mink that the fur and meat had actually been used and not wasted. If Indians failed to follow a range of tabooed behaviors — from avoiding menstruating women, to not placing animal skulls properly in trees, to laxity in ritual preparations — drastic consequences might ensue. A species might make itself scarce, a hunter's luck could vanish, a sorcerer might inflict illness, or a family member might suffer an accident.

The fur trade that changed both Indian cultures and the ecology of much of Canada and the northern United States lasted from the 1580s to the late nineteenth century and was made possible by the fecundity of beaver. Naturalist Ernest Thompson Seton (1860–1946) estimated that, in 1600, there were upwards of 50 million beaver in North America. Beaver ponds varied in size from a small pond of several square feet to hundreds of acres surrounded by beaver dams. The resulting beaver ponds created habitats for many birds and animals, including muskrat, otter, and mink, also important in the fur trade. The peculiar fate of the beaver was to have a very soft coat underneath an outer coat of stiff guard hairs. The soft coat hairs had small barbs that would cling to a felt-hat base made from rabbit hair. Thus beaver fur was an ideal pelt from which to make the stylish beaver hats that were in demand in Europe and later in America.

The Micmac encouraged the trade, and as early as 1534 were reported as waving sticks with furs to attract Jacques Cartiér's voyagers. They were shrewd bargainers, desiring metal tools in exchange for pelts, realizing the value of metal fishhooks, traps, and guns, along with needles for sewing and kettles for cooking. But since they lacked the technologies to manufacture metal goods for themselves, Indians had to trap more beaver. Under pressure from the trade, beaver — along with lynx, otter, fox, mink, and other fur-bearing animals — rapidly disappeared from the northeastern region. Concurrently, the organically based economy of the Indians, rooted in disposable clothing, shelters, and utensils, began to be superseded by an inorganically based economy, tied to iron implements and other goods imported from Europe.

The advent of European diseases introduced by sailors, fishermen, missionaries, and settlers not only decimated tribal numbers, but also broke down cohesiveness and community. Some environmental historians, such as Calvin Martin, believe that disease initiated the breakdown of Indian spiritual relations with nonhuman nature. Indians broke faith with the animals and engaged in the fur trade. Martin writes: "No longer was [the Micmac] the sensitive fellow-member of a symbolic world; under pressure from disease, European trade, and Christianity, he had apostatized — he had repudiated his role within the ecosystem."[4] But Bruce Trigger asserts that it was the desire for European goods that provided the incentive to hunt beaver. "To date," he writes, "there is no hard evidence of major epidemics in the St. Lawrence Valley or southern Ontario during the sixteenth century."[5] Among the Micmac, smallpox epidemics began in 1639 and continued almost every decade throughout the seventeenth century. Measles occurred in 1633 and 1658. Historical and archaeological records show that these diseases appeared after the start of the fur trade. But when they did strike, they had an immediate and pronounced impact.

When disease decimated a hunting band, the relationships of trust, practice, and knowledge built up among its members over generations was broken down. As the survivors formed new groups, it was more difficult to hunt cohesively. The fur trade promoted individualism within the tribe and competition among tribes, as middleman tribes were created in the interior. As the trade expanded and dependencies on trade goods increased, some tribes became dominant and others declined in social status. Although Indians maintained a sense of integrity and retained tribal and oral traditions, the overall effect was to give Indian groups greater incentive to deplete animal populations.

The introduction of Christianity also transformed Indian life. The Jesuits were extremely shrewd in the methods they used to challenge native belief systems. Whenever a woman would pray with a priest for a child ill with smallpox or measles and that child survived, shamans would be ridiculed and their power undermined. If a man who was sick asked the priest for help, and was cured, the individual would then agree to convert to Christianity. When shamans themselves became ill, the priest would minister to them, asking them to join him in prayer. If a shaman survived, the priest made him agree to give up his drums, charms, and dances.

Baptism was substituted for Indian rituals, and those who took up the sacraments of baptism, confession, and the Mass were converted to Christianity and taught to read the Bible. In *After Columbus* (1988), historian James Axtell discusses "The Power of Print in the Eastern Woodlands": "The ability to read and write was awe-inspiring to the Indians largely because it duplicated a spiritual feat that only the greatest shamans could perform, namely, that of reading the mind of a person at a distance...." Indians often developed an extreme respect for the Bible, sent by a seemingly all-powerful God who prevented disease from falling on the priests, who in fact bore immunity.[6]

To Europeans, the power of writing over oral tradition reflected the supremacy of European culture. The Jesuits and colonists considered Indian stories as mere myths repeated over and over again. They believed alphanumeric literacy, used to record treaties and trading transactions, to be superior to the Indian oral traditions. Literacy allowed Europeans to engage in more powerful, centrally organized economic and political patterns with international ties than was possible among Indian cultures.

In comparing Indian and Western ways of relating to nature, Indians generally considered themselves to be just one among many entities in an animate world, living according to culturally defined canons of respect for other members, while nevertheless developing tools and technologies that allowed them to provide for their own subsistence. Seventeenth-century Europeans, on the other hand, emerging out of feudalism, were beginning to believe in the power of technology to control nature and advance an individual's status in life. The introduction of Christianity, new technologies, and market trading, reinforced by the power of literacy, therefore resulted in a transformed natural environment and Indian culture.

Plains Indians and the Westward Movement

A third case of Indian ecological modes of living and their transformation is that of the bison cultures of the Great Plains. Native Americans of the Great Plains centered their subsistence economy around the buffalo. But as Europeans moved onto the Plains, they developed policies toward Indians and buffalo that altered the region's ecology and native cultures.

The Great Plains presented unique ecological challenges to both native inhabitants and European Americans seeking to expand westward. They are divided into the tall grass prairies in the east and the short grass prairies in the west, with a division between tall and short occurring roughly around the 100th meridian. Westward of that meridian, the rainfall is less than 20 inches and tapers off further toward the western deserts. The 100th meridian is generally thought of as the 20-inch rainfall line. Agriculture is fairly reliable east of that line, but westward it becomes increasingly problematic, hence the Plains Indians were primarily nomadic.

The Great Plains produced severe weather conditions for all inhabitants, Indian and European alike. The region is subject to very rapid and sometimes violent weather changes. Chinook winds are warm winds coming off the mountains that can, within a half a day, dry up all crops growing in its path. Blizzards from the Canadian North sweep down across the plains, where there are few physical or vegetative barriers. "Northers" produce extremely low wind-chill conditions. Perhaps most importantly, the Plains are subject to periodic droughts, a condition that affects all life on the Plains.

The animal that dominated the Plains from earliest times was the bison. A single herd of bison could contain as many as 12 million head and cover up to 50 square miles, with the average herd being about 4 million. Recent estimates put the bison population on the short-grass Plains at 30 million head. Herd sizes varied with drought, disease, and predation by wolves, coyotes, and Indians. The bison played a significant role in shaping the ecology of the soil, water, and grasslands and provided Native Americans with food, clothing, and tools, with hides also being used for pictographs that recorded tribal histories.

Indians achieved great sophistication in obtaining and using the buffalo as a primary source of protein. Folsom men, who inhabited the Plains 9,000 years ago, herded the buffalo, walking behind the often erratic and unpredictable herds. Buffalo, unlike cattle and sheep, could not be domesticated, but early Indians developed an understanding of herd psychology. Moving with them along level ground, they could herd the animals into an eroded cul-de-sac, where they could readily kill them with bows and arrows. Or they could cover themselves with mud in a wallow while they waited for the herd to come close enough to throw spears. Indians on foot also used the surround technique for capturing bison: hunters from several villages encircled a herd, set fire to the grass, and shot the trapped animals with spears or arrows.

The cliff drive, however, was the primary method Indians used to kill buffalo before the advent of the horse. They placed lines of stone along the ground, making a V-shaped path toward a cliff. These funnel-shaped pathways often extended outward a mile and a half from the cliff, over which the buffalo would be driven on foot or by setting fires. The fall over the cliff would either stun or kill the animals, after which they could be butchered and the meat dried. A herd of buffalo might provide a group of Indians with fresh meat for two to three weeks, with the rest being dried for future use. At one ancient site in Colorado (Olson-Chubbock), 193 carcasses were excavated, which could have supplied a band with 75,000 pounds of meat. A group could, therefore, live off one stampede for a month or more before moving on.

Hunters evolved ritual methods of killing the buffalo, and sacrificed the first buffalo of a hunt. They took the bone from the humps of bull buffaloes and used the fat around the hump for celebrations. They used buffalo skulls for rituals, filling the skulls with grass and setting them in places of worship. Buffalo bones were often used in Indians' sacred bundles — complex and varied bundles of stones, bones, and other objects that held particular meaning for individual Indians.

The arrival of the horse on the Plains dramatically changed life for Indians and altered ecological conditions. Horses present in the New World had become extinct on the Plains after they migrated eastward over the Bering land bridge. They were domesticated in Asia and then adopted by Europeans thousands of years ago. They were reintroduced into Mexico by Hernando Cortes in 1519, and also came north with the explorations of Hernando DeSoto in the Southeast (1539–42) and Coronado (1540–42) and Oñate (1598) in the Southwest. Many escaped the expeditions and became feral. In 1680, the Pueblo Indians revolted against the Spanish, captured their horses, cattle, and sheep and, along with the Navajo, established a livestock trade with other tribes. By the 1690s the plains of Texas had horses, and by the 1780s horses had dispersed northward to the Columbia plateau. While there is debate over whether Indians obtained horses by taming feral horses or trading domesticated horses, or both, the introduction of the horse changed the ecology and economy of the Plains.

On the high, short-grass Plains, the horse gave rise to tribes of nomadic Indians. Before the advent of the horse, most tribes lived along the edges of the short-grass Plains, where they could hunt bison as well as game, gather plants, grow crops, or trade with the horticultural tribes of the Mississippi Valley. The horse-mounted nomads — Arapaho, Assiniboine, Atsina, Blackfeet, Cheyenne, Comanche, Crow, Kiowa, and Sioux — now became skilled riders and efficient buffalo hunters, but, at the same time, more dependent on both the horse and fluctuations in the populations of bison. Mounted Indians could now efficiently drive buffalo into the same rock pounds that they had previously used on foot and could pursue animals while mounted. They could also construct bison pounds from lodgepoles — logs from mountain-grown lodgepole pines — dragged behind their horses. But with greater reliance on the bison, they were locked into the cycles of drought and disease that affected bison numbers, and became more dependent on trade.

The horse also changed Indian lifestyles. With the horse as transportation, they could use poles for teepees and for travois to pull their goods. They built temporary pounds and camps, to which they regularly returned. A family of two Indians needed five horses in order to travel — one horse for each of the two people to ride, a third to haul the teepee, a fourth to haul the travois loaded with gear, and often a fifth loaded with meat or other provisions. A family of seven or eight, therefore, might need ten or eleven horses. A village of 300 people could potentially have more than a thousand horses. This meant that Indians had to move continually in order to find grass for their horses. In the winter, when the Plains were covered with snow, they had to find winter forage. They, therefore, congregated in areas such as the valleys of the Wind River Mountains in Wyoming. The horse could, therefore, become a liability, limiting movement and campsite locations.

Horses, moreover, became signs of wealth and prestige. The more horses and other forms of property an individual had, the greater that person's social status. If a chief had 300 horses, his tribe was admired more than that of a chief who had only 100 horses. Thus, greater social differentiation emerged as a result of the arrival of the horse.

If the horse gave the Plains Indians an advantage over the bison by the eighteenth century, Indian susceptibility to introduced diseases undercut that advantage by the nineteenth century. An epidemic of smallpox in the pueblos of New Mexico in 1780–81 was probably diffused by horse-mounted Indians to the Plains, where its consequences among the Arikara were noted in 1795 by French traders, an outbreak reported to have also killed many Mandan. Horse-mounted nomadic Indians, however, suffered fewer losses than those in sedentary villages. But a devastating smallpox epidemic swept the Plains in 1837, spread from an infected deckhand on a trading vessel to three Arikara women and thence to other Arikara, Sioux, Mandan, Gros Ventres, Pawnee, Blackfeet, Piegan, and Assiniboine Indians. As many as 17,000 people may have died. Then in the 1840s and 1850s, Indians equipped with horses, devastated by disease, and already engaged in the beaver and hide trades with Europeans, encountered waves of miners, ranchers, and farmers moving west. The encounter permanently changed the Plains environment and Indian ways of life.

The European Transformation of the Plains

A number of theories exist as to the historical process by which Europeans transformed the Plains and the Indian cultures that had evolved there. In *Guns, Germs, and Steel: The Fates of Human Societies* (1997), Jared Diamond uses the "guns, germs, and steel" complex as a shorthand for the immediate causes responsible for Indians succumbing to Europeans. "History followed different courses for different peoples because of differences in peoples' environments, not because of any biological differences among peoples themselves."[7] Despite the apparent natural bounty of the American continent at the time of European contact, the Americas were severely disadvantaged in terms of candidate species for domestication. The Americas had only the llama and alpaca, and lacked horses and cattle until European explorers introduced them. Isolation from Old World diseases, especially diseases associated with domestic animals and plants (such as smallpox, measles, and bubonic plague), combined with the European American capacity for settled agriculture and ranching, spelled the downfall of Indians. Europeans had highly developed forms of political organization and superior military capabilities, as well as writing based on letters and numbers, all set in motion by the crop agriculture and animal domestication that evolved 10,000 years ago in the ancient Near East. The underlying, or ultimate, cause, therefore, of the different levels of development among European and American Indian cultures, according to Diamond, lay in environmental differences that gave rise to European domestication, settled agriculture, and complex sociopolitical organization.

Environmental historians Daniel Flores and Andrew Isenberg offer additional environmental and market-oriented explanations for the decline of the Plains Indians and the buffalo on which they depended. Flores argues that Plains buffalo populations were weakened by climate changes that diminished the grasses on which buffalo fed, by bison diseases such as brucellosis, and by Indian participation in the early hide trade, in which buffalo were killed in order to obtain food, clothing, blankets, gunpowder, pots, pans, and other staples. Flores writes: "a core [bison] population, significantly reduced by competition with horses and by drought [was] quite susceptible to human hunting pressure."[8] Fewer buffalo meant fewer resources on which Indians could depend for subsistence. Isenberg states: "[By] embracing the emerging Euroamerican market, the plains nomads bound their fate to the Euroamerican economic and ecological complex. In the nineteenth century, the dynamic grassland environment, commercial exploitation of the bison, and epidemic disease would bring an end to the nomads' dominance of the western plains."[9]

Walter Prescott Webb, in *The Great Plains* (1931), provides a technological explanation for the European impact on Plains Indian cultures and buffalo populations. The Europeans needed "a weapon more rapid than the Indian's arrows, of longer reach than his spear, and, above all, one adapted to use on horseback."[10] The Colt six-shooter allowed them to pursue Indians while mounted and to discharge six shots in sequence. The rifle allowed buffalo hunters to kill bison by the hundreds.

Webb discussed the great encounter between two horse cultures — Indians and Europeans — on the Plains. Webb's assessment was that the Indians were better horsemen than the Europeans. Indians shooting arrows on horseback, or firing spears from beneath a horse's neck, had an initial advantage over Europeans. Early explorers noted that Indians, who carried a hundred arrows in a quiver case and were able to keep ten arrows in the air at once, were superior to a man with a rifle. The Indian, moreover, was able to drop down along the side of the horse, using the horse's body for protection. Many Indians preferred bows and arrows, which they could make themselves, thus controlling their own technology, whereas they lacked the ability to manufacture guns or ammunition. Nevertheless, guns began to have an impact on Plains cultures. The eastern Plains Indians acquired rifles from whites, while the southwestern Indians and those living in the mountains lacked them because the Spanish refused to supply ammunition. Differentiation therefore occurred between those tribes who had guns and those who did not. Ultimately, Webb argues, Indians were subdued by introduced technologies, including not only guns but barbed wire, windmills, and railroads, which forever changed the Plains.

Environmental and technological factors were supplemented in the 1860s through 1870s by a U.S. government policy of ridding the Plains of both Indians and buffalo, dealing the *coup de grâce* to Indian cultures. In 1867, one member of the U.S. Army is said to have given orders to his troops to "kill every buffalo you can. Every buffalo dead is an Indian gone."[11] In 1875, General Phil Sheridan, the military commander in the Southwest, urged that medals — with a dead buffalo on one side and a discouraged Indian on the other side — be created for anyone who killed buffalo. He allegedly said, "Let them kill, skin, and sell, until the buffalo is exterminated.

It is the only way to bring a lasting peace and allow civilization to advance."[12] But Flores questions the statement, noting that, in 1879, Sheridan sent a telegram to Washington, D.C. stating, "I consider it important that this wholesale slaughter of the buffalo should be stopped." Isenberg concurs: "The army was happy to see hide hunters, but they were not commanding them to kill bison.[13]

Nevertheless a massive slaughter of bison by white hunters heralded the demise of the Plains Indian cultures. Buffalo war parties went west by train, shooting buffalo from train windows; throughout the West, people killed buffalo. Records showed 120 buffalo killed in 40 minutes and 2,000 in a month. The average buffalo hunter killed one hundred a day. One hundred thousand buffalo were killed each year, until they were on the verge of extinction, removing the subsistence base from Indian cultures. Buffalo remains were used in the hide and bone trade. In 1876, some 3 to 4 million buffalo killed on the Plains supplied hides and bones for robes and fertilizers. Three thousand hides were loaded onto each boxcar and 350 boxcars went east. In a space of 10 to 15 years, buffalo were removed from the Plains and the remaining Plains tribes relocated to reservations. In conclusion, therefore, a combination of environmental and climatological factors, human and animal diseases, domesticated animals, technologies ranging from six-shooters to railroads, the buffalo trade, and government policies all contributed to the rapid reduction of the Plains Indian tribes by the 1870s.

The Ecological Indian

Environmental historians have disagreed on the extent to which American Indians were ecologists and conservationists both before and after European contact. Some have called them "native conservationists," the "first ecologists," and peoples who "lived in harmony with the land," contrasting them with European colonists who "exploited" the land, "wasted" resources, and degraded the environment. Native Americans — such as Lakota lawyer and author Vine Deloria, Jr., Anishinabe conservationist Winona LaDuke, and O'odham environmentalist Dennis Martinez — have defended the idea that Indians lived in accordance with ecological limits and limited the number of animals they hunted. LaDuke writes, "We have a code of ethics and a way of living on this land which is based on being accountable to [natural] law. That is the understanding of most indigenous peoples."[14]

In assessing whether Indians can be called ecologists or conservationists, in the sense of "showing concern for the environment and acting out of that concern," anthropologist Shepard Krech III writes:

On one hand, Native people understood full well that certain actions would have certain results; for example, if they set fire to grasslands at certain times, they would produce excellent habitat for buffaloes one season or one year later. Acting on their knowledge, they knowingly promoted the perpetuation of plant and animal species favored in the diet. Inasmuch as they left available, through these actions, species of plants and animals, habitats, or ecosystems for others who came after them, Indians were "conservationists." On the other hand, at the buffalo jump, in the many uses of fire, in the commodity hunt for beaver pelts and deerskins, and in other ways, indigenous people were not conservationists. Yet their actions probably made little difference for the perpetuation of species … until Europeans, with their far greater numbers, commodified skins, pelts, and other animal and plant products.[15]

The relationship of Indians to the land and the ways in which that relationship was altered by European colonization thus remains a topic of discussion and debate.

Several additional case studies would be required for a more comprehensive discussion of Native and European encounters and their ecological effects. These might include: the Eastern woodland horticultural tribes, the salmon-centered potlatch cultures of the Pacific Northwest; Tlingit encounters with Russian seal and otter hunters in the Alaskan Panhandle; Hawaiian Island cultures and the advent of New England whalers; the Seminole Indians of Florida and Spanish explorers; and the Ojibwa of the Upper Midwest. Nevertheless, many of the same environmental and cultural factors and causal connections discussed above would be pertinent. Also important are questions of Indian resilience and adaptation to altered circumstances and transformed environments and the ways in which their own cultural traditions were retained and reinforced.

Conclusion

Indians developed sophisticated cultures in different environments throughout North America, using hunting/gathering/fishing and horticultural techniques by the time of European exploration and settlement. These environments were radically transformed by introduced animals, plants, pathogens, and European peoples who migrated to and altered native cultures and ecosystems. In the Southwest, the Spanish overcame the Pueblo Indians, introducing settled agriculture around the missions and converting Indians to Catholicism. In the Northeast, the French engaged Indian hunters in the fur trade and used the devastation of introduced diseases to challenge their animistic religions and Christianize them. On the Great Plains, where Indians had domesticated feral European horses and used them to hunt buffalo — and where environmental conditions were less favorable to European agricultural methods — climatic changes, technological innovations, and government policies aided Euro-American settlement and hastened Indian decline. Between the mid-sixteenth and mid-nineteenth centuries, the North American continent had been changed from a environment inhabited by American Indians engaged in hunting, gathering, fishing, and horticulture to one occupied by vastly reduced numbers of Indians, along with European immigrants and their introduced animals, plants, diseases, and more complex agricultural, social, political, and cultural systems.

Notes

1. Donald Worster, *Rivers of Empire: Water, Aridity, and the Growth of the American West* (New York : Pantheon, 1985), 31.

2. Alfred Crosby, *Ecological Imperialism: The Biological Expansion of Europe, 900–1900* (New York: Cambridge University Press, 1986), 270.

3. Ramón Gutiérrez, *When Jesus Came the Corn Mothers Went Away: Marriage, Sexuality, and Power in New Mexico* (Stanford: Stanford University Press, 1991), 94.

4. Calvin Martin, "The European Impact on the Culture of a Northeastern Algonquin Tribe: An Ecological Interpretation," *William and Mary Quarterly* 31 (January 1974): 23.

5. Bruce G. Trigger, "The Ontario Epidemics of 1634–1640," in *Indians, Animals, and the Fur Trade: A Critique of Keepers of the Game,* ed. Shepard Krech III (Athens: University of Georgia Press, 1981), 22.

6. James Axtell, *After Columbus: Essays in the Ethnohistory of Colonial North America* (New York: Oxford University Press, 1988), 93.

7. Jared Diamond, *Guns , Germs, and Steel: The Fates of Human Societies* (New York: Norton, 1997), 25.

8. Daniel Flores, "Bison Ecology and Bison Diplomacy: The Southern Plains from 1800–1850," *Journal of American History* (September 1991): 482.

9. Andrew Isenberg, *The Destruction of the Bison: An Environmental History, 1750–1920* (New York: Cambridge University Press, 2000), 62.

10. Walter Prescott Webb, *The Great Plains* (Boston: Ginn, 1931), 170.

11. Colonel Richard Irving Dodge (1867), quoted in *Heads, Hides, and Horns: The Compleat Buffalo Book*, by Larry Barsness (Fort Worth: Texas Christian University Press, 1985), 126.

12. General Phil Sheridan (1875), quoted in Barsness, *Heads, Hides, and Horns*, 126.

13. Flores and Isenberg, quoted in "Historians Revisit Slaughter on the Plains," by Jim Robbins, *New York Times*, 16 November 1999, D3.

14. Winona LaDuke, "From Resistance to Regeneration," *The Nonviolent Activist* (September–October 1992): 3–4.

15. Shepard Krech III, *The Ecological Indian: Myth and History* (New York: Norton, 1999), 24, 212–13.

2 The New England Wilderness Transformed, 1600–1850

The environmental history of the New England forests focuses on
three stages of use: Indian subsistence; the colonial forest
economy; and wilderness appreciation. It also explores two core
themes — the human labor needed to extract useful commodities,
and the transformation of the idea of wilderness. Indians used the
forest for hunting and cleared openings for horticulture. Colonists
introduced European livestock and crops and established
permanent settlements, while extracting forest products for
overseas trade. Human settlement and resource depletion brought
about ecological changes in the forest, fostering a transformation
in the perception of wilderness from savage to sublime. This
chapter investigates how Native Americans and European
immigrants both used and viewed the forest environment.

The New England Forest and Indian Land Use

The New England forest provided rich, although different,
resources for Native Americans and European colonists. It is made up of
three primary ecological regions. The northern forest is composed of coni-
fers, such as balsam, fir, and spruce, and hardwoods, such as aspen and
birch — food for the beavers prized by Indians and colonists for their furs. In
the middle band, where Indians established horticulture and hunted deer,
immigrants found white pine for ship masts, red and white oak for barrel
staves, and hickory for farm tools. The southernmost band of the forest is the
oak and pitch pine region, suitable for agriculture and for naval stores —
products such as pitch, tar, and turpentine, needed by the colonial shipping
industry. The bands vary with topography and blend together in transition
zones, but foster different patterns of settlement and economic use.

The New England forest at the time of colonial settlement was an open,
park-like terrain created by Indians accustomed to burning the land and
clearing the underbrush in the spring and fall. Indians set fires to ease their

passage through the woods and to attract deer to browse areas. Burning was a sophisticated technology developed by Indians to manage their forests.

The New England Indians were horticulturists, growing corn, beans, squash, and pumpkins. This agricultural complex had diffused northward from Mexico, reaching New England about A.D. 1000. The corn, beans, and squash complex had a number of ecological advantages that benefited both Indians and colonists. Corn and beans were planted together in the fields, so that beans twined up the corn stalks. When the leaves began to come out, the corn and beans shaded the ground, where the pumpkins and squash were planted. This ecological system of polycultures kept down insect damage and enhanced yields. Indian groups traditionally remained in one place for seven or eight years while they planted their fields, and then, when the fields were exhausted and crop yields declined, moved the entire settlement. They may have prolonged the move by adding fish as fertilizer on the fields toward the end of the cycle. It is unclear, however, how often this procedure was used, because large loads of fish had to be carried from the brook in the springtime, a period known as the "starving time," when Indians were dependent on fish for food.

In 1616, disease struck the Indians of the New England area, decimating their numbers to a fraction of their former population. The historical records are somewhat unclear as to whether that disease was bubonic plague or smallpox, since descriptions of the Indians' symptoms make diagnosis difficult. Bubonic plague, however, is transmitted by rodents, particularly rats, which were prevalent on the ships of the explorers who had traded along the coast, with fleas serving as an intermediary and transmitter. It is known from the historical records that the Indian villages did have many fleas during the summer months. Thus bubonic plague could have had a mechanism of transmission. Whether smallpox or bubonic plague was the cause, however, the forest at the time of European settlement was depopulated of many of its former inhabitants, their fields lying abandoned for appropriation by the new colonists. A smallpox epidemic in 1636 likewise took immense tolls on New England Indians.

The Settlement of New England

The settlement of New England by English Pilgrims and Puritans was motivated by religious desires and economic needs. The two groups — Pilgrims and Puritans — who came to New England differed slightly from each other. The Pilgrims (who were religious separatists) arrived onboard the *Mayflower* and settled Plymouth Colony in 1620, whereas the Puritans (who had hoped to purify the church from within) left England in a vast migration in 1630 and settled the Massachusetts Bay Colony. These two groups were not the first to attempt to found a colony in New England; there were several abortive attempts in the two decades prior to the arrival of the Pilgrims. They were, however, the first to settle successful, lasting colonies in the New England area.

The Pilgrims came for two reasons. One was to seek religious freedom; the other was economic. They wanted natural resources and products from the New World to trade with England, including timber, fish, furs, and medicines. The Pilgrims landed first on Cape Cod, near present-day Provincetown on the northeastern tip of the Cape. Once there, they made three expeditions out onto the land. On one of those trips, they found and confiscated a cache of seed corn that the Indians were preserving for the following year. They took that corn back to the ship to plant in their ultimate settling place. The landscape on Cape Cod, as they described it, was "open and without underwood," consisting mainly of sandy plains. As such, it differed from the environment they found on the mainland coast at the colony they established near Plymouth Rock.

William Bradford, author of *The History of Plymouth Plantation, 1620–1647*, is the source of much of the knowledge about Pilgrim perceptions of the New England forest. Bradford writes that the Pilgrims encountered "a hideous and desolate wilderness full of wild beasts and wild men." This was the perception of a people terrified by their new environment and the challenges of survival. To them the country had a "wild and savage hue."[1] Such ideas of wilderness and wildness were prevalent in the literature of New England settlers from the beginning. A distinction between nature and culture, wild and civilized, was thus implicit in the ways in which Pilgrims recreated themselves and their community.

Wood from the vast forests was the primary resource used by the colonists for shelter and fuel. They built and lived within a wooden stockade at the edge of the forest, against the ocean shore, symbolizing the demarcation between the wilderness outside and their own identity and central community inside. Each house was fenced, emblematic of the individual family's own home and sense of separation from other colony members, even as they engaged in a cooperative venture together. The colony was also fixed in space, which was different from the shifting hunting-gathering and horticultural economy found among Indian survivors in the local forest.

The Puritans followed the Pilgrims to New England, settling in Massachusetts Bay, to the north of Plymouth colony, in a large migration in 1630. John Winthrop led the Puritan migration at a time when tensions between the British king and the Puritans had greatly intensified in England.

Before Winthrop's ship, the *Arabella*, left England, Winthrop noted the reasons for migrating: "The whole earth is the Lord's garden," he said, and quoted God's instruction in Genesis 1:28 "to increase, multiply, replenish the earth and subdue it."[2] This verse gave the colonists permission to subdue nature. They believed that they brought with them a mandate from God giving them dominion over the earth. They looked to the New World to supply economic items to trade, as well as necessities for subsistence. Prior to settlement, the natural resources needed to subsist in the New World had been catalogued — fish, deer, furs for trade and clothing, salt, trees for lumber, masts, pitch, and tar for shipbuilding. They knew where iron for future mills was located and that there would be sufficient grass for cattle and goats, as well as acorn mast for pigs released into the woods.

Colonial Land Use

Several features differentiated European land use patterns from those of the Indians. Colonial land tenure was based on the concept of private property. Colonists entered into treaties with the Indians in order to exchange trade goods for land, obtaining title in perpetuity. While Indians often believed they were only giving hunting rights and planting rights, colonists wrote up agreements that took title "forever." In addition, Europeans claimed vast tracts of land for their kings, who in turn granted title to colonies for settlement. There was thus a vast difference in understanding as to what was actually being negotiated at the time.

Colonial agriculture combined Indian and European patterns. The settlers first planted the corn/beans/squash complex (including pumpkins) that had been domesticated by the Indians. They added to the Indians' triad their own tetrad of European cultivated grains — wheat, rye, oats, and barley. Wheat did not do well after the first few decades because it was struck with a "wheat blast." God, they said, had seen fit to "smite the wheat," killing it with what was later found to be a fungus hosted by the surrounding barberry bushes. Rye and corn, therefore, became the basic ingredients of the bread used in the colonies.

The colonists also added livestock to the ecosystem — components that environmental historian Alfred Crosby called the Europeans' "portmanteau biota" and biologist Jared Diamond called "the major five" — i.e., the five domesticated animals that spread worldwide with European colonization. These five large animals brought by the colonists — cattle, pigs, goats, oxen, and eventually horses — had been domesticated in Eurasia over a period of several thousand years and were essential to the Europeans' settled agricultural complex. Oxen and horses — used in plowing — combined with broadcast grains and the three-field rotation system, promoted permanent colonial settlements. Pigs and cattle roamed in the woods and fields, disrupting Indian's shifting horticultural and hunting patterns, and consumed large quantities of meadow grasses, altering New England ecology.

During the seventeenth century, nature in New England was commodified in terms of specific resources. Animals such as beaver, fox, and lynx were trapped for their pelts and traded. White pine masts and other forest products, such as turpentine, pitch, and tar, were extracted from the New England forest. These resources became part of an emerging system of world trade.

Marketing the Forest

European colonists came from countries that were undergoing transformations from organically based to inorganically based economies. In the Middle Ages, wind, water, animal muscle, and human labor were integral parts of an organic economy, used to supply human needs. But by the seventeenth century, when New World colonization was taking place, inorganic resources were increasingly beginning to be exploited, such as iron ore for guns, spades, and kettles and silicon for glass-making. Charcoal from timber (an organic product) was necessary to fuel the forges and furnaces at the base of the inorganic economy of the iron industry. Forest products were also essential to building ships for trade and for the barges, locks, and warehouses used to transport and store commodities.

During the seventeenth century, capital from Old England was infused into New England. The colonists were financed by companies in England that provided money for ships and for the transport of settlers to the New World in order to establish colonies. In return, the colonists were expected to extract natural resources and return them to the mother country. Mercantile capitalism, or long-distance trade financed by merchant capital, was the earliest form of capitalism that developed. It began to link the continents together, as the explorations surveyed and cataloged the natural resources of the world.

In order to obtain products that were part of the rising mercantile economy, the settlers began to identify resources for trade from their new environment — furs, fish, and forest products. Colonists trying to survive in the struggle against nature developed the ability to extract commodities from it. In exchange, they not only needed iron products, they also wanted particular foods obtained through long-distance trade, such as coffee and tea, spices from the Orient and Africa, and sugar and molasses — commodities not produced in New England.

Beginning in the 1640s, wood products from the New England forest were traded with Barbados, the Canary Islands, Madeira, and England. Ships following triangular trade routes started by loading manufactured items in England, then sailed to the Canary Islands for wine, and finally to Boston for barrel staves. Or, ships might go directly to Barbados to pick up sugar and molasses and then take these products to New England to exchange for red and white oak barrel staves, made from timber extracted from the New England forest.

Perhaps the most important forest product was the white pine used in the mast trade. The forest trade was an extension of the fur trade northward into New Hampshire from the original Massachusetts Bay Colony. On the bays that led into the ocean were mast landings for logs brought down the rivers and floated into coastal bays for loading onto ocean-going ships. By 1700, along the New Hampshire and Maine coasts, there were ninety sawmills and thirty teams of oxen. Three to four ships a year sailed for England, loaded with masts for the British navy, which was vying with France, the Netherlands, and Spain for supremacy of the seas in an emerging era of nationalism. Ships commanded the seas, not only for commercial purposes, but for warfare. New England's forests provided essential naval stores, such as white pines for tall straight masts; oak roots for "knees" to build ship hulls; tar and pitch for sealing the hulls; and spars and yards on which to hang the sails from the masts.

In extracting products from the forest, an economic system developed, based on the forest as a natural resource, on logging labor, and on merchant capital. Forest laborers included cutters, who chopped down the trees; barkers, who stripped the bark from the trees; swampers, who prepared the way for logs to be transported downstream; and teamsters, who hauled the logs with oxen. Other skilled laborers included drivers, raftsmen, scalers, and sawyers. At the sawmill, logs were cut into planks to be sold on the market. The mill owner, who supplied the capital to construct the mill, was often one of the wealthiest people in town.

The economic system depended on a structural split between labor and capital. By the early eighteenth century, class distinctions had formed in the American colonies between people at the bottom of the economy (such as the lumberers, whose skilled labor often took years to develop), the mill owners who transformed logs into boards, and the Boston and Portland timber merchants who marketed the extracted products. At the upper levels of the economic hierarchy were the mast agents, sent by London contractors to the New England colonies to supervise the mills and lumberers. George Tate, mast agent in Falmouth (now Portland) Maine, for example, had an expensive three-story house built in 1755 on the edge of the Fore River, and was quite wealthy for an eighteenth-century New Englander.

At the very top of the hierarchy was the Surveyor of Pines and Timber, the king's agent, who upheld British laws pertaining to the forest. He was the enforcer of the Broad Arrow Policy, instituted in the 1690s in New England, which reserved all white pine trees 24 inches in diameter, measured at 12 inches off the ground, for the British Crown. The King's Broad Arrow was cut into the tree by mast agents and sub-agents who went through the forest, blazing the arrow into tall pristine white pines. The policy was an early conservation measure inasmuch as it preserved a natural resource for later economic use.

The Broad Arrow Policy was modified over the next forty years through various proclamations and extensions. It was first expanded to include not only white pines, but also pitch pines necessary for naval stores. The area of reserved trees was then extended into New Jersey and New York. By 1721, the proclamation was further modified to include all young trees, even the smallest trees bordering a township. The objective was to preserve all pines more than two feet in diameter at one foot above the ground as potentially useful for ship masts for the British navy.

The Broad Arrow Policy was a source of contention between the colonists and the British. It was among the numerous complaints that ultimately led to the American Revolution, and was also the source of skirmishes in towns such as Weare, New Hampshire, where the mast agents who rode into town were pulled from their horses and their horses' tails and manes sheared. The agents were then tarred, put back on their horses, and run out of town.

By the early eighteenth century, a major transformation had taken place in the ecology of the New England forest as a result of European colonization and the marketing of forest products. Hundreds of acres of timber had been cut, creating climatic change. The air was less moist, the hillsides drier, and there was the potential for more wind, especially for hurricanes that came up the coast. The colonists had not only created possibilities for more runoff and erosion, but they had depleted many of the animals and birds that were integral parts of evolved ecosystems. They radically reduced the numbers of beavers, otters, bears, foxes, and deer. They killed black-birds, woodpeckers, and crows as pests (which the Indians had refused to kill because they were the sacred bearers of corn). The native wildflowers were sharply diminished, and English flowers and weeds had been introduced. Many of the marshes and beaver ponds had been drained, and fish stocks in the rivers depleted. The forest was a vastly different place than that occupied one hundred years earlier by Indians.

The Forest Economy

The economic system that developed in New England was part of an emerging capitalist system that linked nature, labor, and capital, turning natural resources into commodities to be traded on the market. The commodification of nature occurred as part of the so-called triangular trade, involving Europe as a source of manufacturing and management, Africa as a source of slaves, and the New World as a source of natural resources. Natural resources were used for subsistence, providing people with food, clothing, shelter, and energy. Alternatively, nature could be used for exchange value — that is, as a means of trading and profit-making. There were, therefore, two forms to the exploitation of nature, one for use and one for profit.

The Marxian approach clarifies the relationship between nature and capital that emerged in the New World. Marx linked land, labor, and capital together in a framework centered on modes of production (society's infrastructure) and ideas (society's superstructure). His idea of land can be broadened to include the environment, and more generally nature itself, providing a framework for environmental historians to use in analyzing the human relationship to nature.

In his foundational work, *Das Kapital*, written in the 1860s, Karl Marx analyzed the structural split between labor and capital, such as that emerging in the New World's timber industry. The lumberer became part of the process of appropriating nature for profit by adding surplus economic value to the log. Although nature gives humanity free resources in the form of free timber, the capitalist appropriates that timber for himself. The capitalist takes the timber, to which the laborer has already added value, and adds further value by cutting the log into planks at the sawmill. The worth of the timber therefore increases. By adding additional value to the timber, it can be sold at a profit. But that profit is created at the expense of the wage laborers, who are paid less than the profit obtained by the capitalist.

The social effects of the split between labor and capital in the New England forest economy were discussed with concern by the New England elite. Jeremy Belknap of Yale College, in his 1812 *History of New Hampshire*, noted that the contractors and agents were the ones who made huge fortunes, but the laborers who actually spent their time in the woods spent their money on credit. They bought liquor, or engaged in other forms of indulgence, and lost most of their earnings. They were paid at the end of the season, but anticipated their earnings on credit and were henceforth kept in a state of debt and poverty. This helped to create a class of exploited wage workers.

Belknap's ideas were reinforced by Timothy Dwight, President of Yale College, in his book *Travels in New England and New York* (1823). Dwight said: "Those who are lumbermen are almost necessarily poor. Their course of life seduces them to prodigality [and] thoughtlessness of future wants."[3] Dwight echoed a concern, prevalent at the time, that people on the frontier had degenerated into living like animals, and because they subsisted primarily on animal meat, they themselves would became wilder, losing their civilized characteristics. Belknap and Dwight thus expressed concern about what was happening to laborers in growing industries such as the timber trade.

A third commentator on the forest industry was Edward Kendall, who traveled in the U.S. in 1807–8. Kendall observed that many settlers had "degenerated ... into lumberers." They wandered through the forest "making spoil ... of the wealth of nature." They gave nothing back to nature, there being no sense of reciprocity between them and the forest. In contrast to seventeenth century ideas that timber was a gift of the earth, the lumberer simply took for himself what nature gave. "What nature plants he enjoys, but he plants nothing for himself."[4] Despite an emerging consciousness about conserving and replanting timber by the time of the American Revolution, Kendall nonetheless noted and deplored the enormous waste. But not only was the structural split between labor and capital apparent by the end of the colonial period, a split between mind and nature was also evident.

Mind, Labor, and Nature

In his 1995 article, "Are You an Environmentalist, or Do You Work for a Living?" environmental historian Richard White observed that most people (such as the New England colonists and timber workers), through most of human history, have known nature through their bodies, by working in it directly. In the New England forest economy, cutters, barkers, teamsters, and raftsmen cut down trees, moved logs across rough terrain, and risked their lives to float them downstream, working daily from sunrise to sunset. They put out enormous energy, interacting with nature through bone, brain, and muscle to produce physical work. Ordinary people, throughout the settlement of the country, labored directly in nature, whether in forests or on farms. Only the elite (such as Belknap, Dwight, and Kendall) were exempt from the bodily toll taken by work. As the country was industrialized and more people worked in urban industries, an increasing physical and intellectual distance between people and nature opened up. Artists, writers, and intellectuals who came to appreciate nature aesthetically and to advocate its preservation were often removed from the necessities of daily labor in forests and on farms. This distinction between physical work in nature and intellectual work about nature is still apparent today.

As an intellectual in the university, White points out that much of his own connection to nature comes from recreation. Whereas today's lumberers (like their counterparts in colonial New England) work directly in the environment and know nature through their laboring bodies, today's environmentalists (like the colonial elite and early preservationists) know nature through the mind, through aesthetic appreciation, and through recreation. White offers not only an historical analysis of the split between mind and nature, but suggests a means of healing it as well. He concludes that if we "could focus on our work rather than on our leisure, then a whole series of fruitful new angles on the world might be possible.... We may ultimately find a way to break the borders that imprison nature as much as ourselves. Work, then, is where we should begin."[5] White's view about the split between mind and nature is reinforced by the transformation of the image of wilderness in the minds of the Pilgrims and Puritans who lived directly in the New England forest, to that of the elite and urban dwellers who appreciated and preserved it for its aesthetic and recreational values. What did wilderness mean at the time of New England colonization, and how did it change?

The Idea of Wilderness

The concept of wilderness is one of the most complex ideas in environmental and human history. As environmental historian Roderick Nash pointed out in *Wilderness and the American Mind* (1967), the word wilderness comes from old English and Germanic words, *wildern* and *wildeor*. Nash notes, "The root seems to have been 'will' with a descriptive meaning of self-willed, willful, or uncontrollable. From 'willed' came the adjective 'wild' used to convey the idea of being lost, unruly, disordered, or confused."[6] To many people living in medieval Europe, the ancient forests contained dense trees and underbrush, deep shadows, and frightening beasts. The word was likewise associated with barrenness and the desert. In the Bible, the book of Deuteronomy is called "In the Wilderness," and many of the connotations of wilderness come from the desert lands in which the Israelites sought enlightenment and in which they wandered for forty years. Wilderness, therefore, means not only a dense forest, but also a barren land. In addition, there is the sense of fear and bewilderment. Such connotations associated with the concept of wilderness came with the Puritans to the New World as cultural baggage. The wilderness is the antithesis of the garden, or of paradise. The biblical story says that Adam and Eve were driven out of the Garden of Eden into the wilderness, a desert. They had to labor in the earth to change it into a garden, just as the colonists had to change the New England wilderness into a cultivated garden.

The idea of wilderness appeared in the sermons of many of the ministers of New England churches during the 1640s and 1650s. Hartford, Connecticut minister Thomas Hooker (1586–1647) preached that Puritans "must come into and go through a vast and roaring wilderness" before "they could possess that good land which abounded with all prosperity [and] flowed with milk and honey."[7] Peter Bulkeley preached in 1646 that God had dealt with the Puritans as he had "dealt with his people Israel" for "we are brought out of a fat land into the wilderness."[8] Rhode Island founder Roger Williams (1603–83) likewise spoke of a "wild and howling land" as a reminder that his people had fallen from grace and that their souls were spiritual wildernesses.[9] Thus purification of the wilderness in the soul of the Puritans was an important part of the religious experience. The wilderness provided symbolic justification for Puritan land conversion as well. They could take over the wilderness from the Indians and transform it into a garden through their own ecological additions, even as they transformed the spiritual wilderness in their own souls.

The idea of wilderness changed in this country toward the end of the eighteenth century as a result of ideas developed in England, Germany, and France. In England, Edmund Burke's *Origins of Our Ideas of the Sublime and the Beautiful* (1757) began to look at forests, mountains, and waterfalls as beautiful places. The German philosopher Immanuel Kant wrote *Observations on the Feeling of the Beautiful and the Sublime* in 1761, in which nature and wilderness began to take on meanings of reverence and awe. God's action in the land through thunderstorms and lightning was now looked upon not as the work of the devil, but as a manifestation of God's goodness. The sublime was manifested in waterfalls, mountains and canyons, and in sunsets, rainbows, and oceans. The idea of the sublime as a religious experience became important in the romantic period of the late eighteenth and nineteenth centuries in America. Nature was now cathedral, temple, and Bible.

New Englanders began to treasure their mountains, forests, and waterfalls and to look upon them as exemplars of the sublime rather than as obstacles to be surmounted, as the Puritans had viewed them. In her poem "An Hymn to Evening" (1773) African-American poet Phillis Wheatley of Boston eulogized the sunset as God's glory in the world below: "Through all the heav'ns what beauteous dyes are spread! But the west glories in the deepest red: So may our breasts with every virtue glow, The living temples of our God below!" In "Thanatopsis" (1817), Massachusetts poet William Cullen Bryant (1794–1878) wrote of "the venerable woods — rivers that move in majesty," while his poem "A Forest Hymn" began with the lines "The groves were God's first temples." Massachusetts teacher Caroline Barrett White wrote in 1850: "When I contemplate nature my heart expands with an intensity and feeling of love, of admiration, of reverence for that Being who has spread out before us the sublime works of creation...."[10]

The American wilderness began to be appreciated in the nineteenth century as the nation's forests began to disappear. What remained of nature that had not been used for economic purposes or settled was eulogized as wilderness. Wilderness was also connected with a sense of racism against Indians. Indians had lived on the North American continent for at least 10,000 years. They had managed the land, made their presence known, and transformed it through hunting, gathering, and fire. But by the late nineteenth century, Indians were being moved to reservations, and national parks were being designated as people-free reserves. Although William Bradford had found the New England forest "full of wild beasts and wild men," his "wilderness" nevertheless contained Indians. In contrast, the Wilderness Act, passed in 1964, stated that in a wilderness, the earth and its life are untrammeled by "man," and that "man" himself "is a visitor who does not remain." Wilderness was thus defined as devoid of human presence.

The concept of wilderness, therefore, has changed over time. In 1977, Roderick Nash summed up the changes, noting that "historians believe that one of the most distinguishing characteristics of American culture is the fact that it emerged from a wilderness in less than four centuries."[11] In 1995, however, environmental historian William Cronon wrote a controversial article entitled "The Trouble with Wilderness" in his edited book *Uncommon Ground* (1995), arguing that wilderness is a social construction. The idea of "wilderness," Cronon noted, is "a profoundly human creation."[12] Like "nature," "wilderness" is a cultural construct. As an idea, it is a product of civilization, created in contrast to civilization. It is only through our own civilization that we know something that is not civilized to be wilderness. Cronon observed that "There is nothing natural about the concept of wilderness. It is entirely a creation of the culture that holds it dear, a product of the very history it seeks to deny."[13] There is a sense of an opposite, or otherness, to the idea of "wilderness." In the article, Cronon argued that "we mistake ourselves when we suppose that wilderness can be the solution to our culture's problematic relationships with the non-human world, for wilderness is itself no small part of the problem."[14] Numerous ecologists, environmental advocates, philosophers, and historians, however, differed with his interpretation on the grounds that nature was a real, evolved, ecological system rather than a historical construct. The Pilgrims and Puritans, therefore, have left not only the legacy of a transformed New England, but also a transformed idea — that of wilderness.

Conclusion

Between the seventeenth and nineteenth centuries, the New England forest was transformed from a land perceived as a wilderness to a land of useful commodities, to be extracted from nature by human labor. For Pilgrim and Puritan settlers, sawmill entrepreneurs, and the English king, the forest played a central role in organizing ways of life and perceptions of nature. It provided timber for masts, naval stores, barrel staves, and construction; habitat for beaver, deer, and ducks; and a source of fertile soils for agriculture. It was perceived variously as a threat to settlement, the home of the devil, and a place of religious purification; as a land of sublime goodness and picturesque beauty; as a vanishing national asset, and later as a place where "man is a visitor who does not remain." The forest as wilderness is also connected to human labor as a means of transforming natural resources into commodities, as well as a place where people do not labor at all, but simply observe the beauties of nature. Environmental historians, in probing the meanings of changing concepts such as wilderness and labor, cast light on their historical complexity.

Notes

1. William Bradford, *Of Plimoth Plantation* (Boston: Wright and Potter, 1901), 94–95.

2. John Winthrop, "Conclusions for the Plantation in New England," in *Old South Leaflets*, no. 50 (1629), 4–5.

3. Timothy Dwight, *Travels in New England and New York*, ed. Barbara M. Solomon and Patricia M. King (1823; reprint, Cambridge: Harvard University Press, 1969), 2:160–61.

4. Edward Kendall, *Travels Through the Northern Parts of the United States in the Year 1807 and 1808* (New York: I. Riley, 1809), 3:75–76.

5. Richard White, "Are You an Environmentalist, or Do You Work for a Living?" in *Uncommon Ground: Rethinking the Human Place in Nature*, ed. William Cronon (New York: Norton, 1995), 185.

6. Roderick Nash, *Wilderness and the American Mind*, rev. ed. (New Haven: Yale University Press, 1973), 1.

7. Thomas Hooker, *Application of Redemption — The Ninth and Tenth Books*, 2d ed. (Cornhil, England: Peter Cole, 1659), book 9, 5.

8. Peter Bulkeley, *The Gospel-Covenant, or the Covenant of Grace Opened* (London: Benjamin Allen, 1646), 143.

9. Roger Williams, quoted in "The Wilderness and the Garden: Metaphors for the American Landscape," by Peter Fritzell, *Forest History* 12, no. 1 (April 1968): 16–32, at 22.

10. Phillis Wheatley, "An Hymn to the Evening," in *Poems on Various Subjects, Religious and Moral* (London: E. Johnson and A. Bell, 1773), 59; William Cullen Bryant, *The Poetical Works of William Cullen Bryant* (New York: D. Appleton,

1906), 22, 79; Caroline Barrett White, *Diary*, Worcester, Mass.: American Antiquarian Society, 1849–1915, manuscript, September 9, 1850.

11. Roderick Nash, "The Value of Wilderness," *Environmental Review* 3 (1977): 14.

12. William Cronon, "The Trouble with Wilderness, or Getting Back to the Wrong Nature," in *Uncommon Ground: Rethinking the Human Place in Nature*, ed. William Cronon (New York: Norton, 1995), 69.

13. Ibid., 79.

14. Ibid., 70.

3 The Tobacco and Cotton South, 1600–1900

The study of the Tobacco and Cotton South from the perspective of environmental history considers the relationships among soil, slavery, and the plantation and sharecropping systems. In the southern United States, human activity profoundly affected an environment of long growing seasons, fertile soils, and abundant rainfall. These advantages resulted in massive outputs of staple crops, primarily tobacco and cotton, as well as rice and sugar. This chapter examines changes in the land that occurred in the Chesapeake area under tobacco cultivation in the seventeenth and eighteenth centuries, and in the Deep South as cotton production spread inland following the 1793 invention of the cotton gin. On large plantations, slavery, and later sharecropping, fostered class distinctions and racism, generating social and political tensions. The system was also threatened by the vulnerability of soils to depletion and erosion, and the susceptibility of tobacco and cotton crops to insect infestation.

The Chesapeake Environment and Indian-European Relations

Nature's fecundity greatly impressed the English who first settled on the southeastern coast of North America — first on Roanoke Island (present-day North Carolina) in 1585 and then permanently, in 1607, further north at Jamestown (present-day Virginia) near the mouth of Chesapeake Bay. Off the coast of the future state of North Carolina, Arthur Barlowe reported in 1584 that his expedition encountered "so sweet and so strong a smel, as if we had bene in the midst of some delicate garden." And on Roanoke Island, the soil was "the most plentifull, sweete, fruitful and wholesome of all the world." The following year, Governor Ralph Lane of Roanoke Colony told of a land that abounded in "sweete trees," "pleasant gummes," and "grapes of such greatness" as not found in all of Europe.[1]

The first colonists in North Carolina and Virginia encountered not only a fertile environment, but flourishing populations of Native Americans living in densely settled, communal villages. Artist John White's drawings at Roanoke (1585 and 1587) and Captain John Smith's descriptions at Jamestown (1607) vividly portrayed the eastern woodland Indians' hunting, fishing, and horticulture centered on the "three sister" triad of corns, beans, and squash that had migrated north from Mexico.

Through a variety of rituals and practices, Indians extracted a livelihood from the natural dower. When the first edible ears of green corn appeared in August, a festival of all-night dancing celebrated and propitiated the corn mother. In the forests, they practiced a long-fallow strategy of preserving fertility by moving their villages and clearing new fields while the old fields recovered. Although hunters periodically burned the forest undergrowth to increase the deer population, they prepared for the hunt with rituals that imitated their prey. In 1564, for example, French artist Jacques Le Moyne portrayed the Timuca hunters of Florida, disguised with deer heads and skins, engaged in a face-to-face encounter with actual deer across a stream.

The first intimations of irrevocable change for Native Americans came from introduced diseases. In 1588, Thomas Hariot noted in his descriptions of Roanoke: [S]trangly it happened that within a few days of our departure the people began to die very fast. In some towns twenty people died, in some forty, in some sixty, and in one sixscore.... And the strange thing was that this occurred only in towns where we had been ... and it happened always after we had left."[2] The reasons, unknown at the time, were the first epidemics of contagious diseases (smallpox, whooping cough, measles, scarlet fever, chickenpox, mumps, and influenza) associated with Europeans and their livestock that would sweep through indigenous communities. A century later in 1705, Robert Beverley cataloged the diseases and devastation that had "wasted" the Indians of Virginia, reducing their number to less than "five hundred fighting men." He noted further that "They have on several accounts reason to lament the arrival of the Europeans.... The English have taken away great part of their Country and consequently made every thing less plenty amongst them. They have ... multiplied their Wants, and put them upon desiring a thousand things, they never dreamed of before."[3]

The Anglo-American transformation of southern soils and forests began around Chesapeake Bay, where colonial plantations and docks sprang up along a labyrinth of navigable rivers, which made easy both local travel by boat and direct trade with Europe. The first permanent settlement, Jamestown, was established in 1607 on the James, southernmost of the great rivers that flowed from the Appalachians eastward into the bay. Paralleling it to the north on Virginia's western shore were the York, the Rappahannock, and the Potomac. Still farther north, ringing the head of the bay in Maryland, were the eastward flowing Patuxent, the southward flowing Susquehanna, and the westward flowing Choptank, Nanticoke, and Pocomoke.

The original settlers — many too genteel to work and dazzled by dreams of finding gold — might have been starved out but for a fortuitous discovery. A craze was sweeping Europe for smoking the dried leaves of the Native American tobacco plant, which Spanish conquerors were bringing back from the West Indies. In 1614, an experimental planting by John Rolfe (also known for marrying the Powhatan Indian princess Pocahontas) proved that the sandy loams of the Chesapeake tidewater produced a sweeter smoke, and within a few years a fabulously profitable tobacco boom was transforming Virginia society. By 1617 Governor Thomas Dale had to stave off famine by decreeing that all who planted tobacco must also plant at least ten acres of grain.

The Powhatan Indians, whose crops and seed reserves had been raided by the English, turned against the colonists in the Virginia Massacre of 1622, destroying many of the James River settlements and killing a third of the English population. Revenge was only temporary, however, and the Indians, decimated by disease, were soon drawn into the European trading system, supplying the colonists with deer hides in exchange for kettles, knives, blankets, and clothing. According to environmental historian Timothy Silver, "Indians showed an uncanny ability to hang onto traditional methods of exchange, incorporating Europeans and their goods into the long-established native culture and economy.... Although both groups participated willingly, disease and liquor — the effect of which white traders might not have foreseen but exploited nonetheless — helped tip the balance in favor of the European system."[4]

Tobacco Cultivation

Tobacco culture — though it saved the colonists — was hard on both soils and the forests that replenished them. Sandy loams enriched by silt and clay made the tidewater floodplains ideal for growing tobacco, corn, and wheat. But these soils could be cultivated only after cutting away the richly diverse coniferous/deciduous forests on which their fertility depended. Most serious in the long run was the loss of the deciduous oaks and flowering understory trees — poplar, tulip, dogwood, sweet gum — that for eons had brightened the forest with blossom in spring, kept it shaded and moist through summer heat, and enriched its thick humus of leaf litter each fall. Plantations used pines (loblolly, longleaf, and pitch) for building and to supply the tar, pitch, and turpentine required by the sailing ships that carried their products to market. The Atlantic white cedar and bald cypress of wet lowlands were converted into roofing shingles. Diverse hardwoods — black, red, and white oak, red maple, and hickory — became barrel staves, farm implements, and fuel.

Tobacco plants used up, in three or four years of successful harvests, the fertility so slowly banked by forests. Then fresh land had to be cleared and planted. For several generations, however, forests and fresh soil were so abundant that their depletion did not seem a problem. With tobacco markets insatiable and land readily available, the only barrier to wealth was finding workers for the laborious processes of planting, cultivating, pruning, harvesting, and curing.

The process of producing the finest tobacco demanded meticulous attention. First, flat trays of a rich mould sown with seeds produced seedlings to be set out in early spring, when the danger of frost had passed. After the soil was loosened and pulverized by plow or hoe, it was heaped into hillocks, into which the seedlings were individually set. When the plants were four to five inches high, they were thinned, transplanted, and allowed to grow to about a foot in height. After a month, they were pruned, topped, and replanted about four feet apart. Painstaking labor was required to hoe the soil around them and to strip away the suckers that sprouted on their stems and the hornworms that could destroy their large leaves. As the ripening leaves turned brown, starting with those lowest on the plant, workers harvested the leaves and heaped them in piles to sweat. They then hung the leaves in bundles from the rafters of a tobacco barn and dried them for several weeks over a fire, which circulated warm air around them. Two more weeks of sweating in heaps brought the tobacco to its peak of quality, after which it could be packed tightly into large barrels called hogsheads for shipment to Europe.

Talk of tobacco dominated the lives of planters. Everyone knew, shared, and conversed in common cultural terms, employing a language and set of assumptions about the characteristics of the tobacco crop. A shared vocabulary pertained to techniques, methods, and times that certain tasks were to be performed, such as removing suckers and tobacco hornworms, a major pest for the crop. Most planters had several fields at some distance from each other, often accessible only by boat. The planters or their overseers made regular circuits around their holdings to inspect the crop. Since tobacco plants were set out in sequence at different times, planters kept careful records of planting and production schedules for each field. A planter's entire life was caught up in the routine of planting, harvesting, and marketing tobacco.

The labor for these important tasks was supplied, at first, by young English immigrants who wanted to get in on the tobacco boom but were too poor to pay for their passage to the New World. These "indentured servants" signed an indenture or contract promising to work, for four to seven years, for the planter who paid the ship captain for bringing them over. For a time this system enabled servants, after working out their indentures, to graduate into an increasingly comfortable society of small to middling-scale tobacco planters. Gradually these planters learned to build more comfortable houses and to insure their subsistence and broaden their diets by raising corn and wheat, and by importing from Europe historian Alfred Crosby's "portmanteau biota" of plants for vegetable gardens and orchards, as well as cattle and pigs that could graze in the forest.

But opportunities for indentured servants to graduate into planterhood soon dried up as population increased and the more successful planters began locking up vast tracts of the remaining fresh lands to insure perpetuation of their use-up and move-on mode of tobacco production. Increasingly, former servants had to content themselves with subsistence farming on small tracts of less fertile land farther inland, often beyond the "fall line," where getting tobacco to market became difficult as rivers became unnavigable, plunging over the last shelf of upcountry rock to the flat tidewater plain. More fatefully, the rising class of tobacco nabobs were discovering a new source of cheaper, more exploitable labor, supplied by black Africans.

Slavery and Southern Agriculture

The change in the labor system from indentured servants to slaves had been foreshadowed in 1619, when twenty black Africans were sold as indentured servants from a Dutch ship anchoring at Jamestown. It did not take planters long to discover that black servants were more exploitable than white servants. Unfamiliar with the dominant language and culture, they were less able to know and defend their legal rights under the indenture system. If abused by masters, they were less able to run away and blend in with

the general population. More serious was the white majority's racist prejudice against blacks. As historian Winthrop Jordan argues in *White Over
Black* (1968), the English of this period associated blackness with dirt and
foulness, putrefaction of meat and garbage, black magic, and the Black
Death (bubonic plague) that had decimated European populations in the
fourteenth century.

How slavery emerged from indenture is indicated by scattered legal documents and court decisions. When three servants, one of them black, ran
away in 1640, a Virginia court sentenced the two whites to three additional
years of indenture and the black to servitude for life. In 1682 another court
ratified widespread practice by declaring that all blacks not Christian when
purchased were enslaved for life; by this time their offspring were consigned
to slavery also. Slavery had emerged as an institutionalization of racism, a
system in which skin color and physical characteristics legitimated bodily
ownership. Planters promoted this development because it enabled them to
buy, for not too much more than a seven-year indenture, a lifetime of de-
fenseless labor, along with that of the laborer's descendants. With these developments, blacks multiplied from 1.9 percent of Virginia's population in
1620 to 22 percent by the end of the century

The wealth produced by this system of labor is suggested by Virginia
planter William Fitzhugh's letter describing his tobacco plantation in 1686.
He had a thousand acres, most of it in marshes and thickets not yet cleared,
but 300 acres of which were plantable. Beyond this plantation, he had another 22,000 acres on which the soil had not yet been broken and another
1,500 acres in several different places. He had thirty young, vigorous slaves
capable of mating and reproducing and therefore of adding to his capital.
His thirteen-room plantation house was surrounded by dairies, stables,
barns, and hen houses. He had a large orchard with apple and other fruit
trees, which could be harvested to feed plantation workers, and a large vegetable garden, as well as cattle, hogs, horses, and sheep. He also had a grist
mill for grinding corn and wheat for the plantation's grain supply. He had
250,000 pounds of tobacco on hand, and his plantation could be expected
to produce 60,000 pounds per year.

The eighteenth century brought the opulence of this plantation system
to a peak. By the time of his death in 1732, planter Robert "King" Carter
(1663–1732) was the richest man in colonial Virginia and a member of the
House of Burgesses. He owned 300,000 acres of land in numerous parcels
along the Rappahannock and Potomac Rivers and as far west as the Blue
Ridge Mountains. As a middleman between traders and planters, he pos-

sessed seven hundred slaves, a twenty-five-fold increase over those held by Fitzhugh a few decades earlier. He had 2,000 head of cattle and swine and several hundred sheep scattered over dozens of holdings. Carter's barns also accommodated a hundred horses, used as draft animals, for overseeing his holdings, and for travel. His numerous offspring intermarried with others in the planter gentry, cementing the family's power and influence. Carter's grandson, Robert Carter III (1730–1804) maintained a townhouse in Williamsburg and a plantation at Nomini Hall on the Potomac River. The plantation annually consumed 27,000 pounds of pork, twenty beef cattle, 550 bushels each of wheat and corn, four hogsheads of rum, and 150 gallons of brandy. In the winter three pairs of oxen daily hauled enough wood to keep twenty-eight fires burning continuously in the main house and outbuildings. Carter had inherited 100 slaves from his grandfather, and by 1785, through their natural reproduction, he possessed 466 slaves.

The southern slave system suggested a dichotomy between mind and body, culture and nature. Historian Ronald Takaki, in *Iron Cages* (1979), argues that slavery exhibited a major contradiction between culture and nature, exemplified in William Shakespeare's play, "The Tempest" (1611). Caliban — a dark, deformed individual — lives on an island taken over by Prospero, who exiles Caliban and forces him to work. Caliban, whose mother is African, is portrayed as resembling the devil — he is dark and vile in nature, eliciting fears of violence. Such fears could be quelled, it is implied, by developing the intellect's control over the base nature of the body, the savage animal, and indeed nature itself. "America," argues Takaki, "became a larger theater for the *The Tempest*.... Far from English civilization, [colonists] had to remind themselves constantly what it meant to be civilized — Christian, rational, sexually controlled, and white."[5] A dichotomy between civilization and the natural world thus began to emerge, in which the white civilization of England represented the highest level of the intellect and purity, while the African slave stood for the unruliness of nature and the body. Such cultural constructions lay at the root of a plantation system that exploited both slave and soil.

Southerners depended on slave bodies and slave knowledge for cultivation not only of tobacco, but also of rice, which became a mainstay of agriculture in tidewater South Carolina and Georgia in the eighteenth century. Varieties of rice from West Africa and Madagascar were introduced into South Carolina during the 1690s via the Chesapeake and the West Indies. Planters employed African cultivation methods, including clearing the land with fire, threshing with flails, and husking the grain with mortar and pestle.

Historian Daniel Littlefield states: "[B]efore Carolina was settled, English-
men were aware that Africans possessed the technical knowledge to produce
this crop and ... from the earliest period of successful rice production in
South Carolina a relationship developed between this region and rice-
growing regions in Africa."[6]

After early prohibitions against slavery were lifted in Georgia in 1751,
rice planters from South Carolina and the West Indies moved into the tidewater low country. Georgia planters grew rice in swamps, employing methods that South Carolinians had learned from their slaves, including diking
rivers to create impoundment ponds and building floodgates to regulate water flow. Planters also expanded into sugarcane, indigo, and sea island cotton production, creating several integrated landscapes of production. "As
they were molded out of the low-country environment by planters and their
slaves," writes environmental historian Mart Stewart, "plantations constituted agroecological systems that restructured biological processes for agricultural purposes.... Those who created these systems had to manage them
carefully to maintain the balance of energy inputs and outputs necessary for
continued productivity."[7]

Southerners defended slavery on both biological grounds — that blacks
showed more resistance to diseases such as malaria and yellow fever — and
on environmental grounds — that they were more suited than whites to
working in hot humid climates. Both arguments were problematic. Of the
former, Silver points out the cost paid by southerners: "Although planters
could use African biological defenses to good advantage and sometimes
cited these characteristics as justification for using slave labor, ... newly arriving blacks served as carriers for new strains [of disease]."[8] And Mart Stewart, in "Let Us Begin with the Weather?" (1997) challenges arguments that
"hitched together the cultivation of certain plants, the institution of slavery,
and a climate [southerners] also deemed 'peculiar.'" He asserts instead that
southerners invented a regional weather at odds with local weather observations in order to justify the use of slaves in fields and swamps. "Those farmers and planters who kept records ... have left rich documentation of the extraordinary diversity of climates in the region." Stewart concludes that "the
regional weather they made was more distinctive than the weather they got.
Indeed, when Southerners used climate to legitimize a social order, they did
not *begin* with the weather, but *ended* with it, and ended ... with an argument of such force and conviction that it long survived the storm of the
Civil War."[9]

Despite their degradation as a race, African Americans maintained a cultural identity, making significant contributions to southern agriculture and hence to environmental history. They introduced important food crops into southern society. African foods were stowed on slave ships and grown in provision gardens. Slave traders, as well as slaves, introduced crops from other parts of the world. Yams were brought by slaves from Africa. Eggplant came from Africa to South America, from whence it was brought by Portuguese slave traders to the United States. Peanuts from South America were introduced into Virginia by African cooks who arrived onboard slave ships.

Slaves grew foods for consumption by their own families in dooryard gardens that, within the larger system of oppression, afforded some sense of self-worth and autonomy. Most slave owners allowed blacks to have garden patches outside their cabins, but slaves doubled their workloads to maintain them. By working on weekends or late at night after returning from the fields, they often produced a surplus to sell to their owners and to trade in town markets. Slave subsistence consisted of cornmeal and bacon, with garden foods being supplemented by hunting, gathering, and fishing. Crops included black-eyed peas, cabbages, yams, sweet potatoes, squash, and collards, augmented by barnyard chickens and fresh eggs, all grown and harvested by slave families. Gardens planted with such complementary crops often resulted in relatively higher yields per acre, and were less depleting of nutrients than fields of single crops grown alone. As such, slave subsistence was less destructive of the soil than was large-scale plantation agriculture.

Soil Exhaustion in the Tobacco South

Historian Avery O. Craven, in *Soil Exhaustion in the Agricultural History of Virginia and Maryland, 1606–1860* (1926), maintained that "soil exhaustion and tobacco cultivation went hand in hand."[10] Tobacco rapidly depleted the soil, hence luxuriant crops could be grown for only three or four years. Soon after planting, soil nutrients — especially nitrogen and potassium — began to decline and soil microorganisms created toxins that poisoned tobacco plants. Soil fungi and root rot resulted from continual planting in the same soils. Manure, which could have supplied nitrogen, was in short supply, as cattle were left to roam in the woods; in any case, planters believed that it spoiled the taste of the tobacco. With the loss of potassium, the soil became acidic, the land was abandoned, and pine, sedge, and sorrel — indicators of acid soils — took over. Because of the use of the hoe and the continuous scratching of the surface of the soil, erosion became common, resulting in the vast, deep gullies observed on abandoned tobacco lands by eighteenth-century travelers.

Timothy Silver points out that colonists adapted to the problems of soil depletion by changing their farming methods: "As the landscape around them changed, colonists frequently had to adjust and readjust their goals and methods to correspond to ecological reality. When Virginia and Maryland colonists planted tobacco, the demanding weed exhausted their fields in only a few years. Planters adjusted first by planting corn on the worn tracts and then by allowing them to lie fallow. That worked until the population and labor force grew too large to allow depleted fields adequate time to recover. Colonists then shifted tactics again, growing more wheat and seeking to fill eroded ditches or replenish their fields with manure."[11]

By the late eighteenth century, an agricultural improvement movement began to take shape. Planters, such as George Washington and Thomas Jefferson, experimented with rotation systems, using wheat, corn, and clover fallows to refurbish worn-out lands. Nitrogen-restoring crops, such as beans, cowpeas, and lucerne, were also planted on exhausted plots. Barnyard manure and green fertilizers were spread on fields, and contour plowing was initiated. To counter acid soils, plaster of Paris, guano, and marl were purchased and applied to fields. Virginia planter Edmund Ruffin (1704–1865) advocated using marl or "calcareous manure," a form of lime made from oyster shells, to add alkalis that would reverse soil acidity. Using products such as marl, however, was both expensive and labor-intensive, and failed to find favor with most planters.

The extent of southern soil exhaustion has been questioned by geographer Carville Earle in "The Myth of the Southern Soil Miner" (1988). Earle argued that some eighteenth-century tobacco farmers did rotate crops, following tobacco with rotations of corn, wheat, peas, and beans. They also used a long fallow system in which fields were left in fallow to recover for some twenty years. Earle maintains that small farmers had first-hand knowledge of local conditions and used the method of trial and error to combat soil exhaustion, rather than relying on agricultural improvers. Those who worked the land, rather than scientists, were better able to devise solutions. "A variety of evidence documenting southern sensitivity to agronomic practice and soil conservation," Earle asserts, "contradicts the image of an historically invariant soil miner."[12]

By the nineteenth century, southern tobacco production had moved into the Piedmont and Appalachians and included tobacco for cigar and cigarette manufacture. In 1839, a slave named Steven on Alisha Slade's Caswell County plantation put charcoal embers on a dying fire in a tobacco barn in the North Carolina Piedmont, creating a light golden tobacco called bright leaf. The use of Carolina Bright became highly popular, especially when beaten into a powder and rolled into cigarettes. The American Tobacco Company, founded by the Duke family of North Carolina, held a monopoly on cigarettes from the Civil War through the early twentieth century, when it was broken up by the application of antitrust laws.

The Tobacco South thus produced a unique culture and agriculture, rooted in a slave system that degraded both African-Americans and southern soils. It left a legacy of class and racial differences intimately tied to growing tobacco for profit. It resulted in an opulent lifestyle enjoyed by a few planters at the top of a social hierarchy that mimicked the European aristocracy, while producing an underclass of laborers whose bodies bore the scars of servitude, even as they made lasting contributions to southern economy and culture.

The Cotton South

By the end of the eighteenth century, slavery was becoming moribund. The soils of the Tobacco South were moving toward depletion, and other crops such as wheat and corn were being grown in their place. The African slave trade was banned in 1807. Some southern slave owners wanted to free their slaves, but after the invention of the cotton gin in 1793, the cotton boom renewed the spread of slavery. The slave population increased through natural reproduction, augmented by smuggling and the illegal slave trade.

An important period in southern agriculture was the one-hundred-year interval between the advent of the cotton gin in 1793 and 1893, when the boll weevil crossed the Mexican border and moved into Texas. The cotton gin had enormous influence on the development of the South, helping to prolong slavery and allowing the South to develop a strong regional economy. The boll weevil helped to undercut that dominance, forcing post–Civil War farmers to diversify and pushing many sharecroppers off the land and into the North and West.

The cotton gin was the idea of the widow of Revolutionary General Nathaniel Greene and of Eli Whitney, a young Yankee inventor who visited Greene's Georgia plantation. Mrs. Greene proposed that he construct a machine to separate the sticky seeds from short-staple cotton. Sea island cotton, grown on islands along the coast of Georgia, had long fibers (1.625 to 2 inches long), with easily removed, smooth seeds. But inland short-staple cotton (with 0.625- to 1-inch fibers) and long-staple cotton (1.125 to 1.75 inches) had sticky seeds that were difficult to separate from the cotton fibers. Eli Whitney, perhaps following the suggestion of Mrs. Greene, constructed a machine to do just that.

The cotton gin greatly accelerated the process of removing the seeds from the cotton bolls. Prior to the invention of the gin, slaves — under the direction of an overseer — sat in groups, picking the seeds out of the cotton by hand. The gin allowed cotton to be deposited by slaves in a container on its top, where wire brushes separated the sticky seeds from the lint. It emerged free of seeds and ready to be bailed and shipped. Early gins were operated by horses or mules, which turned the cranks, but were replaced during the nineteenth century by steam engines.

Cotton production spread rapidly across the lower South. In 1791, just two years prior to the invention of the cotton gin, southern cotton production was 5 million pounds per year, 33 percent of which planters exported. By 1860, just prior to the Civil War, it had skyrocketed to 1.75 billion pounds per year, 80 percent of which was sold abroad. Until 1812, inland cotton production was focused primarily in Georgia and South Carolina, but by 1860 the entire region of the South, as far west as eastern Texas, contained acres of cotton. The United States was the largest cotton-producing area in the world, and cotton was the country's leading export.

Cotton played an important role in the emergence of an internal market economy in the United States. The market revolution and transportation revolutions of the 1820s made sectional economic development possible. New England began to focus on textile manufacturing, using the South's cotton. In 1790 Samuel Slater brought the idea for a spinning jenny from England to New England, and in 1815 Francis Cabot Lowell introduced the power loom for weaving. The two inventions provided an efficient method of producing cotton thread and weaving it into cloth. The Northeast became the first industrial sector to specialize in manufacturing textiles. Concurrently, the Midwest and the Old Northwest, north of the Ohio Valley, began to develop wheat, while Pennsylvania and the Great Lakes areas focused on coal, iron, and copper production. Sectional development made it possible for the South to concentrate on cotton, as well as tobacco, indigo, rice, and sugar.

Environment and Society in the Cotton South

Cotton required a 200-day frost-free growing season and rainfall amounts of 50–60 inches per year. The area of commercial cotton production extended south and west of the northeastern boundary of North Carolina, bypassing the Appalachians and skirting the northwestern border of Tennessee, from where it crossed the Mississippi River to the Arkansas Ozarks and north-central Oklahoma, and then southward into eastern Texas. North of this line (for example, in Virginia and the Nashville region of Tennessee), it could be grown in mild years or for household use. The cotton belt encompassed four major soil regions: the eastern coastal plain, the gulf coastal plain, the central alluvial valleys, and the western prairies.

Despite the transformation initiated by the cotton boom, throughout the nineteenth century the South remained primarily rural. Landscape architect Frederick Law Olmstead (designer of Central Park in New York City), who traveled through the South in 1853 and in 1856 published *A Journey in the Seaboard Slave States,* observed that the South was in such a poor state of development that even in Virginia one could travel for hours without seeing evidence of habitation. "One has to ride through the unlimited, continual, all shadowing, all-embracing forest, following roads in the making of which no more labor has been given, than was necessary to remove the timber."[13] Throughout the nineteenth century, that situation continued to be the state of nature in much of the rural South.

The cotton plantation system required level or gently rolling hilly land. The big house was situated on a hill to catch the breezes in the hot summer months and to survey the owner's domain. Slave cabins were located below, near the fields and often near the marshes and meandering rivers that drained the land. The system itself depended on a supervisor, overseer, or manager, working under the direction of the plantation owner, and a supply of cheap, controllable labor. Production centered around single crops easily cultivated in fairly routine ways, in what environmental historian Albert Cowdry has called the "row-crop empire." In the American South, these crops were primarily tobacco, cotton, rice, indigo, and sugar.

The word plantation originally meant a colony, such as Plimoth Plantation in Massachusetts. The term, however, evolved to mean a privately owned estate. Southern plantations were often hundreds of acres in extent. Mules were the preferred beasts of burden, and small landowners were often called two-mule farmers. Mules as well as horses pulled the heavy wagons loaded with bales of cotton, each of which weighed 250 pounds. Larger plantations had gin houses, with geared wheels operated by mules. Other important technologies for manipulating the environment included bull-tongue plows, shovels, hoes, cultivators, and harrows.

As the Cotton Kingdom boomed and planters expanded their holdings, slaves were marched across the South in gangs under the supervision of overseers, along trails on either side of the Appalachians or crowded onto boats that sailed around Florida to Louisiana and Mississippi. The journeys were sometimes depicted as happy travels, with people playing violins and banjos. They were not. Conditions were gruesome and families were separated.

By 1860, slaves were housed on plantations all across the South. Twenty-five percent of whites were slaveholders — approximately 10,000 families. Of those, 88 percent owned fewer than 20 slaves, the rough cut-off point for employment of an overseer. Most families who did own slaves had fewer than 20, and some had only two or three. Yeoman farmers, who comprised two-thirds of southern families, for the most part owned no slaves at all. The rest of the southern population was made up of mulattos (often the sons or daughters of white plantation owners) and free blacks, many of whom were the descendants of slaves manumitted in the eighteenth century.

Cotton Production

Southern life was centered on cotton planting and harvesting. Whether by slaves, yeoman farmers, or sharecroppers, cotton was planted in April, with the seed scattered at about 100 pounds per acre. Seedlings came up about 10 days later and, after the third leaf appeared, workers hoed the ridges to remove the weeds and loosen the soil closest to the plant. Short, light bull-tongue plows were used to turn up the earth between the rows. Plowing continued to alternate with hoeing to remove the weeds around the young plants. The cotton began to appear on the lowest parts of the plant in July, with the main cotton-harvesting season beginning in September and continuing into December. On larger plantations, black slaves picked most of the cotton. Yields on good soil were about one and a half bales per acre, a bale equaling 250 pounds of cotton.

The main cotton pest, before the advent of the boll weevil in 1893, was the cotton bollworm, a moth larva of the order Lepidoptera. The worm penetrates the outer leaves and destroys the cotton boll before it can develop. The bollworm goes through several stages of larval development, after which it pupates and metamorphoses into a moth. Control of the bollworm was managed by gangs of slaves moving through the fields and removing and killing the larvae, a labor-intensive form of pest control similar to methods used by Indians. Prior to the early twentieth century, there were no insect control poisons available.

The production system in the Cotton South differed from that of the Tobacco South. In the Tobacco South, slave labor was task-oriented and individualized, while in the Cotton South it was gang-driven. Tobacco was planted in hills and squares on fresh lands that were then abandoned after 3 to 4 years owing to worn-out soil, whereas cotton was planted as a row crop, often in rotations alternating with corn. The major pests for tobacco were tobacco hornworms, fungi, and soil toxins, while those for cotton included the cotton bollworm and later the boll weevil. The Tobacco South was fueled by the African slave trade, whereas slavery in the Cotton South was the result of natural reproduction and the illegal slave trade. The Tobacco South was a part of early mercantile capitalism, or long-distance coastal trade with England and the Netherlands, while the Cotton South rose with industrialization, supplying raw materials for factories in England and New England. Tobacco was a luxury item and narcotic, while cotton was a fiber that became an everyday necessity for clothing.

As in the Tobacco South, soil exhaustion was a problem for cotton farmers. Environmental historian Albert Cowdrey attributes the degradation of southern soils to single crop agriculture: "Row crops bared the soil.... [A]ny system which covers too many fields with the same plant falls afoul of the ecological principle which states that the simplest systems are apt to be the most unstable."[14] Soil toxins, parasites, and erosion worked to deplete the land of its nutrients. Historian Eugene Genovese instead places the blame on slavery, arguing that "slavery and the plantation system led to agricultural methods that depleted the soil."[15] Slaves were worked to exhaustion in gangs to produce profits for their owners at the expense of the soil. With declining soil fertility, profits also declined, leaving little labor or funds to invest in agricultural improvement. Despite knowledge of the benefits of marl, gypsum, guano, and lime, the slave system was too inefficient and the slaves too overworked for planters to use fertilizers. With soil degradation, Genovese concludes, planters reaped their just desserts for enslaving and degrading black people. Despite their failure to use more expensive solutions, however, Carville Earle argues that cotton planters used cotton and corn rotations, spread manure on soils, and turned cattle into the fields to restore nitrogen. Soil exhaustion, therefore, was not as widespread as the mythology suggests. Moreover, soil exhaustion did not cease with the end of slavery in the post–Civil War era.

Post–Civil War Sharecropping

After the Civil War (1861–65), black people were freed, the plantation system was reorganized, and many blacks became small farmers or sharecroppers. Because the plantation owner no longer had slaves to work the land, many owners divided their holdings into smaller plots and leased them to poor black or white farmers in exchange for a percentage of the crop. Between 1880 and 1920, the total population engaged in sharecropping expanded dramatically. According to historian Pete Daniel in *Breaking the Land: The Transformation of Cotton, Tobacco, and Rice Cultures since 1880* (1985), "the new labor system was a varied but unpatterned blend of illiteracy, law, contracts, and violence.... The sharecropping arrangements varied — from state to state, crop to crop, county to county, and farm to farm — and changed over time with the passage, enforcement, and understanding of laws."[16]

Sharecropping was a loose term for several farming methods in which southern farmers engaged in some method of borrowing money, tools, or land from an owner. In sharecropping, someone else, such as the planter or absentee landlord, owned the land. Often the owner also supplied the tools, seed, farming equipment, mules, and even the food, in exchange for a percentage of the crop. In tenant farming, which was related to sharecropping, the tenant owned some of the equipment — for example, the mules and the tools — and rented the land in exchange for a portion of the crop. Tenant farming was a step up from pure sharecropping, in the sense that the tenant owned some equipment and tools and could take his capital with him. A third method was the crop lien system, in which the farmer owned the land but borrowed the seed, fertilizers, and perhaps the equipment. In each case, about a third or a quarter of the crop went back to the merchant or landowner who also appropriated clients' assets for failure to repay loans.

During the Civil War, the South had lost much of its market share in the world export system when Great Britain began to import its cotton from countries such as India. But by 1878, the South began to recover its share of the market, and did well for the next two decades. In 1895, however, the almost exponential growth in demand for cotton was beginning to slacken. The total world demand was proportionately less, with the annual increase dropping to about 1 to 3 percent. With the drop in cotton prices, southerners were not as well off as they had been in the prior two decades. On top of this decline, the arrival of the boll weevil created an additional factor of uncertainty, reducing cotton yields by about 50 percent.

The Impact of the Boll Weevil

The boll weevil crossed the border from Mexico into Texas in 1893; ten years later, in 1903, it was poised on the Louisiana border. The boll weevil, a beetle of the order Coleoptera, differed from the cotton bollworm, a moth larva. The weevil went through four stages of development, taking about 25 days to mature. The insects reproduced very rapidly and were extremely resistant to all kinds of weather conditions. If a single pair mated in the spring, it could produce as many as 250,000 offspring by fall. Up to 50 percent survived the winter.

The boll weevil attacked the boll itself. Cotton has three outer leaves, or bracts, surrounding a four-sided husk, or square, in the center of which the boll grows. The weevil made a hole in the square, sucking out the fluids. It then laid its eggs in the square's center, where the new larvae developed. The larvae ate the interior, the cotton boll died, and the bracts began to curl. The interior, instead of producing a cotton boll, began to rot and fell to the ground, releasing the weevils, which then moved on to a new plant. One of the methods of controlling the weevil, therefore, was for farmers to pick up and destroy the squares containing the larvae and mature weevils.

The weevil represented an enormous threat to the economy of the South. Louisiana opened its Boll Weevil Convention in 1903 with a manifesto against the weevil: "The state of Louisiana is threatened on the west by an insect known as the cotton-boll weevil.... If we consider the amount of money that is in circulation, we realize the immense importance of the crop."[17] In short the entire economy of the South was at risk.

In 1887, the U.S. Department of Agriculture established Agricultural Experiment Stations in the land grant colleges and hired Cooperative Extension Service agents to work directly with farmers. With the appearance of the weevil, the Service began to distribute information on how to control the insect and to educate farmers on techniques that would increase cotton yields and farm income. At the time the weevil first arrived in the South, farmers had no chemical controls, hence management focused on other methods. Scientists worked with farmers to develop labor-intensive approaches. Farmers plowed up fields after the fall harvest and burned the stalks and litter that harbored the weevils, thereby preventing the weevil from over-wintering. They released cattle into the fields to eat the leaves, stalks, and litter, at the same time fertilizing the plots with manure. Farmers also initiated new methods of planting. They planted earlier in the season so that the cotton would set its bolls before the weevils multiplied and became too numerous. They also made use of early-blooming and early-setting cotton varieties. And finally, in 1910, planters began to use chemical pesticides, the first being Paris Green (copper acetoarsenite). The results were limited in scope, however, owing to variations in commitment and practice among farmers and the resilience of the weevil.

Historian Pete Daniel points out that "the boll weevil did not discriminate by the race of the farmer," forcing both white and black farmers to seek aid in combating the new agricultural threat.[18] Black extension agents worked with black farmers to bring to their attention improved methods of controlling the weevil and raising crop yields. Black scientist Booker T. Washington, who headed up Alabama's Tuskegee Institute, developed it into one of the primary institutions in the South to benefit African-American farmers. Through the Tuskegee Negro Conference, county fairs, short courses, and leaflets, he helped to disseminate improved techniques. Another educational method was the Jesup Agricultural Wagon, backed by New York banker Morris Jesup, which traveled throughout the countryside with information for black farmers. Although the new methods were helpful, they were also expensive, and the combination of declining yields and higher costs drove many farmers out of business.

The effects of the boll weevil on southern agriculture were not entirely negative, inasmuch as it forced the diversification of crops and the improvement of farming methods. In Enterprise, Alabama, for example, farmers began to raise peanuts, which, by 1917, had become the area's major crop. Farmers prospered, harvesting more than a million bushels a year, and marked their success by erecting a statue of the boll weevil, the world's only monument celebrating a pest.

The boll weevil remains a pest today. In 1993, the Texas legislature, along with other southern states such as Florida and North Carolina, began a boll weevil eradication program. They received funds from the Department of Agriculture for a $3.9 million program in Texas. As a demonstration, 500,000 acres were sprayed with malathion. Boll weevils, along with other insects, were indeed killed. But removal of insect pests led to an outbreak of armyworms that destroyed 90 percent of the cotton crop. As a result, farmers went into debt. Rachel Carson's 1962 book, *Silent Spring*, and the research that followed it made clear the ecological problems caused by pesticides. Broad-spectrum pesticides kill many insects in addition to those targeted and, when chemicals are concentrated in the food chain, they can negatively affect the numbers of and relationships among other organisms in the environment and create health problems for humans who work the land and consume the products.

Conclusion

The study of the environmental history of the Tobacco and Cotton South from the seventeenth through the nineteenth centuries provides a window on the complex relationships among people, crops, labor, and soil. The plantation system depended on marketable crops such as tobacco, cotton, rice, sugar, and indigo; a source of cheap labor found in slavery and later in sharecropping; soils that were exceptionally fertile and whose fertility could be maintained over time; and the control of insect pests to maintain crop yields. The extraordinary yields produced by the tobacco and cotton systems were reinforced by a market system that relied on the South to produce staple agricultural commodities; by a cultural system that justified racism on the ground of biological differences; by power instilled over slaves through violence and the threat of violence; by the planting of monocultures and row crops; and by labor-intensive and later chemical pest controls. In exploring the history of the South from the perspectives of planters and slaves and the roles played by soils and insects, environmental history offers new interpretations of traditional history as well as lessons for the future of nature and humanity.

Notes

1. Hugh Talmage Lefler and Albert Ray Newsome, *North Carolina: The History of a Southern State* (Chapel Hill: University of North Carolina Press, 1954), quotations at 4, 5.

2. Thomas Hariot, *A Brief and True Report of the New Found Land of Virginia* (1588), in *The New World: The First Pictures of America*, ed. Stefan Lorant (New York: Duell, Sloan and Pearce, 1965), 272.

3. Robert Beverly, *The History and Present State of Virginia* (London: R. Parker, 1705), 232–33.

4. Timothy Silver, *A New Face on the Countryside: Indians, Colonists, and Slaves in the South Atlantic Forests, 1500–1800* (New York: Cambridge University Press, 1990), 192-93.

5. Ronald Takaki, *Iron Cages* (New York: Knopf, 1979), 12.

6. Daniel C. Littlefield, *Rice and Slaves: Ethnicity and the Slave Trade in Colonial South Carolina* (Baton Rouge: Louisiana State University Press, 1981), 113–14.

7. Mart Stewart, *"What Nature Suffers to Groe": Life, Labor, and Landscape on the Georgia Coast, 1680–1920* (Athens: University of Georgia Press, 1996), 90.

8. Silver, *New Face on the Countryside*, 161.

9. Mart A. Stewart, "'Let Us Begin with the Weather?': Climate, Race, and Cultural Distinctiveness in the American South," in *Nature and Society in Historical Context*, ed. Mikuláš Teich, Roy Porter, and Bo Gustafsson (New York: Cambridge University Press, 1977), 241, 243, 251.

10. Avery Odelle Craven, *Soil Exhaustion as a Factor in the Agricultural History of Virginia and Maryland, 1606–1860* (1926; reprint, Gloucester, Mass.: Peter Smith, 1965), 24.

11. Silver, *New Face on the Countryside*, 195.

12. Carville Earle, "The Myth of the Southern Soil Miner," in *The Ends of the Earth: Perspectives on Modern Environmental History*, ed. Donald Worster (New York: Cambridge University Press, 1988), 177.

13. Frederick Law Olmsted, *A Journey in the Seaboard Slave States* (New York: Dix and Edwards, 1856), 79.

14. Albert Cowdrey, *This Land, This South: An Environmental History* (Lexington: University of Kentucky Press, 1983), 79.

15. Eugene Genovese, *The Political Economy of Slavery* (Middletown, Conn.: Wesleyan University Press, 1989), 97.

16. Pete Daniel, *Breaking the Land: The Transformation of Cotton, Tobacco, and Rice Cultures since 1880* (Urbana: University of Illinois Press), 4–5.

17. *Proceedings of the Boll Weevil Convention* (Baton Rouge, La.: Bureau of Agriculture and Immigration, 1903), 5.

18. Daniel, *Breaking the Land*, 9.

4 Nature and the Market Economy, 1750–1850

By the eighteenth century in America, two types of economies
existed in interaction but also independently of each other — a
coastal exporting economy along the eastern seaboard and an
inland subsistence-oriented economy, where access to
transportation and export markets was limited and costly. During
the nineteenth century, a dynamic market-oriented economy arose
throughout the United States westward to the Mississippi River
that integrated the two sectors. This chapter explores the
transition from the coastal exporting and inland subsistence-
oriented economies of the eighteenth century to the market
economy of the nineteenth century. It investigates the ways
writers, poets, philosophers, and artists reacted to the economic
development of the country and the ways they perceived nature,
wilderness, and civilization.

The Inland Economy and the Environment

The environmental costs of commercial production did not
reach most of America until the nineteenth century. Above the fall line and
beyond the reach of coastal markets, retreating Indians were supplanted by
Euro-American subsistence farmers attracted by cheap land. Their small
farms spread over the hills of upland New England, the woodlands of western Pennsylvania, the southern Piedmont, and the valleys of the Blue Ridge
and Appalachian Mountains. In these areas, limited production supplied
the rude comforts of subsistence, and transportation costs prohibited open-
ended production for the market. Economic and social relationships were
based largely on bartering and cooperation, as opposed to the commercial
exchange found along the coast. By the early nineteenth century, this subsistence culture of small farmers comprised the majority of free Americans.

The virtues of this independent and land-owning citizenry were soon
being celebrated as an "agrarian ideal" by French immigrant J. Hector St.

John de Crèvecoeur and American statesman Thomas Jefferson. Crèvecoeur, a member of the French lesser nobility, came to the American
colonies in the mid-eighteenth century. In 1759, after traveling in western
Pennsylvania, he settled beside New York's Hudson River, where he wrote
Letters from an American Farmer (1782). In his letter "What is an American: This New Man," he made a fundamental distinction between hierar-
chical, aristocratic European society and egalitarian American society. The
people of America, he wrote, were simple farmers, "tillers of the soil," and
as such typified a new American ideal based on ownership of property and
the work ethic. Those willing to live a frugal but comfortable life could obtain title to land held in "fee simple." This agrarian-minded society was
supported politically by what Crèvecoeur called the "silken bonds of mild
government," a *laissez-faire* economic system untrammeled by government
regulations.

The same ideal for America was held up in Query 19 of Thomas Jefferson's *Notes on the State of Virginia* (1787). "Those who labor in the earth
are the chosen people of God," he asserted. Independent yeoman farmers —
those who owned a small piece of land in their own name and controlled
their own labor — formed the very foundation of American democracy. In
reality, this status applied primarily to white male property-owners, and did
not include slaves or women (although some women did own property or
were conduits for the passage of inherited property to male heirs of the next
generation). Jefferson warned that if manufacturing became established in
America, the country's workshops would soon resemble the sweatshops of
Europe, destroying the ideals of democracy and farm ownership. He therefore recommended that manufacturing stay in Europe and that the United
States remain the home of independent farmers who produced goods and
used resources for domestic subsistence.

Environmental historians have debated the extent to which eighteenth-
century farmers were involved in the market economy. William Cronon in
Changes in the Land (1983) states that "land in New England became for
the colonists a form of capital, a thing consumed for the express purpose of
creating augmented wealth."[1] Carolyn Merchant, on the other hand, sees
land as a source of family and community subsistence. In *Ecological Revolutions* (1989), she argues that "between 1700, when the inland towns were being settled, and 1790, when ecological crisis and European markets stimulated agricultural intensification, an economy oriented to subsistence and
family preservation flourished in inland-upland New England."[2] Both agree,

however, that the market economy did ultimately transform the environment of New England and other eastern states by the nineteenth century.

Land Use in the Inland Economy

The inland economy was based on an ecological system derived from both Europe and Native Americans. After clearing a small patch in the woods, perhaps two to five acres, farmers typically used it two or three years for crops and then five to eight years for pasture before allowing it to revert to woodland. When the soils in the first plot were exhausted of nutrients, they cleared and planted another two-to-five-acre plot. The plot itself was often put through a rotation system (developed in medieval Europe) of three sub-fields, in which Indian corn was grown on one or two acres the first year and a European grain, such as rye or barley, the second year. Each sub-field in succession was then allowed to lie fallow for the third year, to recover its nutrients. This short-term three-field rotation system allowed restoration of the soil, while the long-fallow "swidden" system allowed the regrowth of forests.

Once fields were carved out of the eastern pine and hardwood forests, subsistence ecology prescribed using them, spatially and sequentially, for both crops and animals. While crops were growing, rail fences kept out the pigs and cattle grazing in the woods. Then, after the harvest, cattle were let in to clean up the refuse and, in the process, manure the land for the coming year. Thus soil, water, and light combined with crops and animals to recycle nutrients and maintain crop yields. Meanwhile, the surrounding forest moderated the climate, reduced winds, and provided habitat for beneficial birds and insects.

Farmers also cleared one or two acres of forest a year to obtain fuel for the family, reserving a 40-acre woodlot that would reforest itself over a period of 20 to 30 years. People bartered crops, tools, and labor with neighbors. They used produce as if it were cash, exchanging food for shoes, bricks, or help in building fences and butchering hogs. Women exchanged cheese, eggs, and vegetables with other housewives, sometimes keeping accounts on the pantry door. In these ways, inland farmers formed cooperative communities. Some farmers kept written account books in which money appeared in the debit and credit columns, but no cash actually changed hands. Money, which was scarce in rural areas, was simply a notation for recording the exchange of products and labor between individuals.

Despite the transportation barrier, subsistence farmers found various ways to get enough cash for paying nominal taxes and buying high-utility store goods such as guns, crockery, and metal utensils or farm implements. One farmer might drive cattle to market. Another might burn wood from cleared fields to produce an easily transportable kettle of potash, which fetched a good price as an essential ingredient of glass and fertilizer. Others who lived close enough might haul a cart of firewood or surplus produce to the villages that grew up around country stores. Linking farm to market was the country storekeeper, who, like the miller, was one of the more well-to-do people in the rural community. Coopers, broommakers, and shoemakers (who were also farmers) bartered their wares directly with neighbors or traded them to the storekeeper in exchange for imported wares.

White subsistence farmers pressured the environment differently from Native Americans. Euro-American livestock and three-field short fallows made permanent settlement possible, whereas Indians' long fallows entailed moving their villages to fresh ground every seven or eight years. White farmers' free-ranging cattle trampled the forest floor and their pigs rooted in the soil, creating erosion but also manuring the woods. Whereas Indians' greater reliance on hunting and gathering entailed firing the woods for easier passage and more browse for deer, whites burned only to clear fields. Whites' firearms were more destructive than Indian arrows and spears, and their hunting decimated "pest" species such as bears, wolves, foxes, and hawks; reduced such subsistence prey as squirrels, possums, doves, quail, and grouse; and helped to exterminate passenger pigeons, Carolina parakeets, and heath hens. Their overall effect on the environment eventually exceeded that of Indians, as their populations grew faster and occupied the land more densely. Nevertheless, production for subsistence, whether by Indians or whites, made far lighter demands on the natural dower than would the impending market economy.

The Inland Economy and the Worldview of Its People

American farmers inherited an organic worldview from their forebears in Renaissance Europe, who likewise drew livelihood directly from the land. Through this prism they saw themselves as interacting with a nature that was alive in all its interconnected parts. Historian Herbert Leventhal describes this perspective in his book entitled *In the Shadow of the Enlightenment: Occultism and Renaissance Science in Eighteenth Century America*

(1976). A chain of being, as he explains the organic worldview, linked together all the parts of a living cosmos, from fixed stars and planets and moon down to earth, animals, plants, and even the lowliest stone. "By using the chain of being," Leventhal writes, "man was able to place any entity in its proper place, and determine its relationship to all other beings."[3] God as ultimate creator acted through an animate entity, often characterized as "Mother Nature," who carried out his dictates in the mundane world. People lived out an interactive I-Thou, rather than an instrumental relation with an animate natural order in all its manifestations. Therefore failed harvests, storms, or droughts were interpreted as punishments for improper actions, bountiful harvests as rewards for good behavior. The organic worldview thus contained within it an ethic of reinforcement or retaliation for human actions.

The organic worldview's hold on American subsistence farmers was most evident in their devotion to the almanac. "Except for the Bible," writes historian Charles M. Andrews, "probably no book was held in greater esteem or more widely read in the colonies in the eighteenth century than the almanac."[4] In *Early American Almanacs: The Colonial Weekday Bible* (1977), historian Marion Barber Stowell explains that the almanac comprised mainly "astrological predictions, advice on husbandry and health, and humor," a format dating back to medieval Europe.[5] Based on the idea that the macrocosm, or larger cosmos, influenced the microcosm, or human body, each almanac contained a "man of the signs," a diagram depicting the part of the body influenced by a particular sign of the zodiac. The moon's daily location in a given zodiacal sign rendered the associated part of the body especially vulnerable. Within the organic framework, therefore, possessing an almanac for the farmer's particular location was critical to predicting human health and determining herbal remedies.

Almanacs were also essential for timing agricultural activities. "The waning or declining moon," Leventhal writes, "helped a plant set its roots down in the earth just as the rising moon helped plants grow upward toward the sky."[6] When the moon was full or waxing, it was believed to exert a pull on water and hence on the fluids in plants, an extrapolation from observations of the moon's tidal action on oceans. When the moon was full, farmers planted crops that grew upward, such as corn, rye, and wheat, and when the moon was new, and presumably had the least influence, they planted root crops, such as carrots, turnips, and beets. They bred pigs in the full of the moon, believing that its influence would produce more offspring. They grafted fruit trees or cleared bushes in the "new of the moon" because the sap would bleed less and the graft would heal faster. Such practices were consistent with the microcosm-macrocosm theory that the heavens above influenced the earth below. An organic worldview thus guided the behaviors of farmers and explained their place in the cosmos.

Market Farming

The subsistence culture was ultimately doomed by the growing difficulty of providing viable farms for the sons of traditionally large families. Although many farmers adapted to shrinking land by selling out and moving to larger tracts of cheaper lands farther west, it was becoming clear by the early nineteenth century that cheap land was not as inexhaustible as it had once seemed. In long settled eastern areas — especially New England — where farms had been undergoing subdivision longest and western lands were less accessible, landless sons and daughters resorted to infertile lands, or to putting-out systems of household manufacture, such as spinning and weaving, shoemaking, and broommaking, or to wage labor for port cities and the swarm of small manufactories spawned along the fall line by abundant water power and cheap labor.

While this agrarian crisis crept inland at the turn of the century, a bonanza galvanized the coastal market economy. Western Europe, disorganized by the wars of the French Revolution and Napoleon, turned to the United States for shipping and foodstuffs. As a commercial boom pushed wheat prices high enough to bear the cost of wagoning from the interior, production of wheat for the market caught up a broad swath of subsistence farmers, stretching from northern Virginia to New York.

Leading the shift to market farming were the Pennsylvania "Dutch" (from "Deutsch" or German). Schooled to a painstaking husbandry in Germany, they had migrated to the rich soils of Chester and Lancaster counties outside Philadelphia. As described by Philadelphia physician Benjamin Rush (1745–1813) in his 1789 article, "The German Farmers of Pennsylvania," they constructed substantial barns to protect their animals, built high fences to keep cattle and pigs out of their gardens, and transported wheat and vegetables in large wagons to the city's markets.

Inspired by this example, agricultural improvements based on the use of fertilizers spread elsewhere to wealthier farmers with access to an urban market. Only they could afford the additional hired hands required to restore their soils and take up agriculture designed for market profit. To do so, they began raising cattle in order to obtain manure for fertilizing fields and planting higher-yielding crops. They seeded legumes in fallow fields and enriched their soils with fish fertilizers and expensive guano, the dung of seabirds from islands off the coast of Peru. Educated market farmers kept account books, numbered fields, weighed products, and calculated costs and profits for each field, crop, and agricultural practice. Produce was taken to county fairs, where prizes were given for the largest vegetables and heaviest cattle.

As pioneers of market farming, the Pennsylvania Dutch had capitalized on the country's first major transportation project — a turnpike, or improved wagon road, financed by tolls, from Philadelphia to Lancaster. Its success inspired a craze for turnpike building, as capital generated by the commercial boom reached out from the great ports and smaller commercial centers to stimulate and engross trade with the countryside. The advent of market farming, based on the dynamic interplay of profit motive, wage labor, and turnpike construction, heralded the interlinked transportation and market revolutions of the early nineteenth century.

The Transportation and Market Revolutions

The commercial boom inspired dreams of a comprehensive national transportation system, and in 1808, Thomas Jefferson's Secretary of the Treasury Albert Gallatin proposed a federally financed network of roads and canals to "shorten the distances into the remote corners of the United States."[7] The War of 1812 and fears of a powerful central government stymied this grandiose plan. But when the war ended in 1815, an outburst of entrepreneurial zeal, technological ingenuity, and financing by state and local governments produced a series of transportation developments that transformed American life.

Leading this transportation revolution was the steamboat, which Robert Fulton had perfected on the Hudson in 1807. After the war, steamboats spread rapidly along the vast network of rivers beyond the Appalachians, where there was no fall line and therefore almost no rapids to impede navigation. Adapted for this purpose with a multi-deck superstructure built on a flat raft and paddlewheels skimming the surface, the western steamboat could navigate up to the shallow headwaters of the Mississippi-Ohio river system and its multitude of tributaries, as well as lesser rivers flowing to the Gulf of Mexico. Transporting bulky commodities upstream as well as downstream far more rapidly and at a fourth of the former cost, steamboats spread market production across the West, from the bustling river ports of Pittsburgh, Cincinnati, St. Louis, Memphis, and the great western entrepôt at New Orleans, to new domains of the Cotton Kingdom on the fertile black loams of Alabama and Mississippi.

New forms of transportation affected the environment not only directly, but also by fostering population growth and production for the market. To supply steamboats with fuel, squatters along the rivers felled great tracts of forests. "Consumption of firewood by steamboats," writes environmental historian Andrew Isenberg, "was probably the primary cause of riparian deforestation in the United States in the first half of the nineteenth century."[8] Traveling the great western rivers in the 1820s, the inspired painter of American birds, John James Audubon (1785–1851), expressed admiration for the industrious squatters along the Mississippi, but amazement at the rapidity of the land's transformation farther upstream on the Ohio. "This grand portion of our Union," he lamented, "instead of being in a state of nature, is now more or less covered with villages, farms, and towns, where the din of hammers and machinery is constantly heard; ... the woods are fast disappearing under the axe by day, and the fires by night; ... [and] hundreds of steamboats are gliding to and fro over the whole length of the majestic river, forcing commerce to take root and to prosper at every spot."[9] His *Birds of America* (1827) brilliantly captured the beauty and abundance of American birdlife at a time when much of it was vanishing before axe and rifle.

The rising commodity production fostered by steamboats in the West sharpened competition among East Coast ports to engage in western trade by surmounting the Appalachian transport barrier. With state financing in 1817, New York City exploited the unique water-level gap between the Adirondack and Catskill Mountains for the western world's most ambitious engineering project, a canal linking the Hudson and Mohawk Rivers with Lake Erie at Buffalo. Generating an amazing flow of commodities upon its completion in 1825, the Erie Canal set off a canal-building mania. Most important were the ambitious systems of Ohio, Indiana, and Illinois, connecting the Great Lakes with the Ohio and Mississippi Rivers. By linking the western rivers to the Erie Canal in cheap water transport, they guaranteed the Empire City's dominance as the great American entrepôt.

Rounding out the transportation revolution was a British technological innovation — the railroad. After 1830, railroads reached out to tap interior trade from Charleston, Baltimore, Boston, Philadelphia, and New York. By the 1860s, trunk-line railroads connected the East Coast with St. Louis on the Mississippi and Chicago on Lake Michigan, and 30,000 miles of track brought cheap transport for bulky market commodities within a few days' wagoning of most settled areas. In 1869, the Union Pacific met the Central Pacific at Promontory Point in Utah to complete the first transcontinental railroad.

Railroads cut deep scars through the landscape and made heavy demands on forests and mines for firewood, timber, coal, and iron. They contaminated their routes with noise, smoke, ashes, and threats of fire. William Cronon brings their larger impact to life in *Nature's Metropolis: Chicago and the Great West* (1991), by demonstrating how capital reached out hundreds of miles from the railroad hub of the Midwest to extract lumber, cattle, and wheat from the countryside for the Chicago market.

The railroad fired the mid-century American imagination as nothing else. For Concord's sage, Ralph Waldo Emerson (1803–1882), it was a "work of art which agitates and drives mad the whole people, as music, sculpture, and picture have done in their great days," and it "introduced a multitude of picturesque traits into our pastoral scenery." Emerson waxed most eloquent, however, about the railroad's practical benefits. He exulted over "railroad iron" as a "magician's rod" with "power to awake the sleeping energies of land and water."[10] "A clever fellow," he wrote, "was acquainted with the expansive force of steam; he also saw the wealth of wheat and grass rotting in Michigan. Then he cunningly screws on the steam-pipe to the wheat-crop. Puff now, O Steam! The steam puffs and expands as before, but this time it is dragging all Michigan at its back to hungry New York and hungry England."[11]

Emerson's Concord neighbor and protégé, Henry David Thoreau (1817–1862), was characteristically otherwise-minded. "The whistle of the locomotive penetrates my woods summer and winter," Thoreau complained, " … informing me that many restless city merchants are arriving…. Here come your groceries, country; your rations, countrymen! Nor is there any man so independent on his farm that he can say them nay…. All the Indian huckleberry hills are stripped, all the country meadows are raked into the city."[12]

Both men sensed the historic shift the railroad announced. The transportation revolution was part cause and part effect of a broader market revolution that brought with it a profound transformation of American economy, society, values, and environment. The entire nation was becoming a unified, dynamic market of interdependent sections, each specializing in commodities as advantaged by its natural resources. The South continued to produce tobacco and rice and expanded its cotton and sugar production. The Middle Atlantic states and the Northwest focused on wheat, livestock, iron, and coal. The Northeast, while retaining its dominance of international trade, finance, and commercial services, moved increasingly into the manufacture of textiles, clothing, shoes, and hats.

Tradition-bound farmers did not convert easily to the competitive, get-ahead culture demanded by the market. According to historian Charles Sellers, "Radically new imperatives confronted people when they were lured or pushed from modest subsistence into open-ended market production. By the 1820s rapidly spreading channels of trade were replacing an unpressured security of rude comfort with an insecurity goaded by hope of opulence and fear of failure. Within a generation in every new area the market invaded, competition undermined neighborly cooperation and family equality."[13] Everywhere American society was wrenched in new directions to reshape people for careers of calculation and competitive striving.

The market revolution triumphed politically only after a final struggle that produced the democratic subsistence culture's most lasting legacy. "When market stresses climaxed in the Panic of 1819," Sellers writes, "... subsistence farmers and urban workers rose in political rebellion against banks, conventional politicians, and 'aristocrats.'" Seizing on a popular hero scorned by market elites, "the people" elected Andrew Jackson president in 1828 by "the largest popular majority in the nineteenth century." Jacksonian democracy was characterized by its exclusion of women and its racist animus against banished Indians and enslaved blacks, and it eventually bowed to the market's inexorable imperatives. Nevertheless, by attacking banks as engines of capitalist transformation, Jackson mobilized for the first time a mass national electorate and ushered in modern democratic politics. Every American president since has had to run the gauntlet of his popular democracy. "Democracy," Sellers concludes, "arose in resistance to capitalism, not as its natural political expression."[14]

The market revolution threatened the American environment more than any other development in modern history. It threw open land, water, air, and all the life they contained to unrestrained development in the pursuit of wealth and status. It made profit-and-loss the sole criterion for dealing with nature, conceived as inert matter. A mechanistic worldview based on the quantification of matter and energy, interchangeable parts, mathematical prediction, and the control of nature replaced the animate cosmos of the colonial farmer. In its wake lay cut-over forests, smoky air, polluted streams, and endangered wildlife. But the market's triumph evoked the first stirrings of an environmental consciousness that would eventually challenge its excesses through the democratic process.

Nature and Ambivalence About the Market Economy

As the market economy moved across the land, the comments of Audubon, Emerson, and Thoreau evinced a growing preoccupation with nature and its relation to humanity. In reaction to loss of the wild, poets, philosophers, novelists, and artists lamented the rapidity with which the country was being developed and eulogized sublime and picturesque vistas found in mountains, oceans, sunsets, and rivers.

Taken collectively, however, this body of thought betrayed an ambivalence between dismay at devastated nature and fear of nature's untamed wildness, both nonhuman and human. It moved toward preference for a middle ground, the cultivated garden, as prefigured in Crèvecoeur's *Letters from an American Farmer*. Crèvecoeur maintained that people took their character from the environments in which they lived. Along the Atlantic coast, fishermen were energetic and independent, taking their "nature" from the boisterous ocean. Farther inland, people were simple farmers, nurtured, like plants, by the soil. On the frontiers, however, they donned buckskin clothing and lived by consuming the meat of wild beasts, taking on animal-like characteristics. Crèvecoeur's ideal environment, therefore, was the middle ground inhabited by the simple farmer.

Explorers and scientists in the eighteenth century were among the first to record their appreciation of nature. Virginia planter William Byrd (1674–1744) described the delights of wilderness camping on a survey of the North Carolina boundary in his 1728 *History of the Dividing Line*, and depicted the Appalachians as ranges of blue clouds rising one above the other in vistas of increasing perfection. He lamented, however, that wilderness valleys lacked "nothing but cattle grazing in the meadow, and sheep and goats feeding on the hill, to make ... a complete rural landscape."[15] Botanist William Bartram (1739–1823), after a trip through the southern Appalachians in 1773, reported that he was "seduced by these sublime, enchanting scenes of primitive nature." Similarly, in 1791, British publicist William Gilpin (1724–1804) found in American forests "the pleasing quality of nature's roughness and irregularity and intricacy."

An urban appreciation of nature also began to appear. The poetry of Boston's Phillis Wheatley (1753–1784) delighted in the natural world as a rational, harmonious whole rather than a tumultuous wilderness. Arriving on a slave ship at age eighteen and studying ancient literature and Latin at the behest of her impressed mistress, she mastered the fashionable classical allusions and tropes that were conventions of the eighteenth-century Enlightenment — the nine muses including Calliope, muse of poetry; Aurora, goddess of the dawn; the sun as "illustrious king of the day"; and the "gentle zephyr," or wind. In "An Hymn to the Morning" and "An Hymn to the Evening" (1773), she evoked the beauty of nature, the transitory quality of life, and the power of God's action in the natural world. Presaging the more romantic nature poetry of Philip Freneau and William Cullen Bryant, Wheatley became the country's first published black poet.

The middle-ground ideal of the cultivated garden was most definitively articulated in the Leatherstocking novels of the upstate New York squire James Fenimore Cooper (1789–1851), the most popular American writer of the early nineteenth century. He pitted the nature-nurtured nobility of backwoods scout Leatherstocking, or Natty Bumppo, against the artificiality of advancing civilization on the one hand, and against the barbarism and "wasty ways" of frontiersmen on the other. Although Cooper, in *The Pioneers* (1823), described with anguish the massive slaughter, for sport and market, that drove vast flocks of passenger pigeons into extinction, he was forced to admit, concludes literary historian Annette Kolodny, "that a non-exploitative white community, living harmoniously within the embrace of nature, was no longer a possibility."[16]

Nature's tensions and affinities with American market society were most deeply explored by the "transcendentalists." This coterie of clerics, poets, and philosophical essayists, centered in Boston and Concord, Massachusetts and led by Emerson, had absorbed in Harvard classrooms the abstruse philosophy developed in Germany by Immanuel Kant, as made intelligible for them by the English poet Samuel Taylor Coleridge. It offered these youthful rebels an alternative to the soul-withering rationality of Boston's Unitarian commercial elite, which had leached away the passion and commitment that their forebears derived from the Puritan God. In place of Jehovah as ultimate reality, it posited transcendent ideas — those above the material world — that sustained and were mirrored in natural reality and implanted in every soul. Nature became the source of spiritual insight, as symbol or emblem of these ideal truths. Revering wilderness as bespeaking God rather than denigrating it as evil, like their Puritan ancestors, many transcendentalists came to feel that it should be cherished and preserved.

Thoreau saw this new faith as reinforcing the organic cosmology of rural America against threatening market forces. He was influenced, as well, by the romantic poets of Europe — Johann Wolfgang Goethe in Germany and William Wordsworth, along with Coleridge, in England — who viewed nature as a source of spiritual insights. In 1845 he "retired" for two and a half years to Walden Pond, just outside Concord, to think through, in this new way, the values of the natural landscape. Drawing also on folk and traditional cultures, on the sages of India, and on the Native American animist view that deities and spirits exist in all things, Thoreau published his masterpiece *Walden* in 1854. Here, luminously homely metaphor expressed an I-Thou relationship with nature that made plants and animals equal, animate beings.

Thoreau spoke of looking into the depths of Walden Pond — the earth's eye — as a means of absorbing higher truths. Acutely aware of how rapidly nature was disappearing under the aegis of the market, he wished to live lightly on the land. His beanfield exemplified his desire to save and restore nature by infusing subsistence farming with an ethic of preservation. If the land were left alone, the forest would return through the process of plant succession. Native plants and animals would reappear, and perhaps even the Indian would once again walk along the paths of a restored wilderness. The axiom "in wildness is the preservation of the world" distilled his personal wilderness ideal.

Walden Pond, for Thoreau, represented the middle ground between civilization and wilderness. Concord and Boston symbolized the market, Mount Katahdin (or Ktaadin) in Maine the distant wilderness. The railroad, which traveled along the edge of his pond, brought commodities and excitement from Boston and beyond, disrupting the serenity of his woodland retreat. But the pond was also a safe haven against the wild. On a trip to Mount Katahdin, Thoreau experienced the terror and loneliness of the wilderness and paused to ask, "Who are we? What are we?" Here nature, in contrast to the pastoral setting of Walden, was savage and titanic, inhuman and lonely. He relished the community, not just of people, but of other animals and plants in the middle landscape of Walden Pond.

Emerson took a different tack. Bred in the Boston bosom of New England's commercial elite and finding an audience for his lectures among the businessmen of northern cities, he strove to fuse enthusiasm for nature with enthusiasm for the market. "The Over-Soul" (1841) linked the human soul with "the eternal One," the transcendental "soul of the whole ... the universal beauty, to which every part and particle is equally related." As offshoots of an "everlasting nature," we "see the world piece by piece, as the sun, the moon, the animals, the tree; but the whole, of which these are the shining parts, is the soul." The worship of nature as manifestation and emblem of the Over-Soul instilled in every human soul an appreciation for the earth as part of the "Unity" of all creation. But simultaneously, Emerson's great manifesto, *Nature* (1836), lauded "commodity" or usefulness to humans as one of nature's preeminent qualities, and his most popular lecture (and essay), "Wealth" (1844), rested on the proposition that man "is born to be rich" and will not "content himself with a hut and a handful of dried pease." The most moral individuals, it seemed to follow, were those "men of the mine, telegraph, mill, map, and survey" who "esteem wealth to be the assimilation of nature to themselves."[17]

As national sage, Emerson elaborated his vision of national destiny in "The Young American" (1844). "Trade planted America," he proclaimed, and trade was its high calling. "This great savage country should be furrowed by the plough, and combed by the harrow; these rough Alleghenies should know their master; these foaming torrents should be bestridden by proud arches of stone; these wild prairies should be loaded with wheat; the swamps with rice; the hill-tops should pasture innumerable sheep and cattle."[18] Emerson's popularity at mid-century suggested that his commodification of nature was drowning out concerns about its degradation in a flood of exhilaration about progress.

The Hudson River School of Painters

The dialogue between Thoreau and Emerson became richly visual in the landscape paintings of a group of artists called the Hudson River School. They were so called because many of them lived in New York City and found their favorite locale for painting in the nearby Catskill Mountains region of the Hudson River Valley. Here, and later in New England, the West, Europe, and even South America, they sought out landscapes through which to capture on canvas their preoccupation with nature and its invasion by civilization.

Environmental historian William Cronon interprets the paintings of the Hudson River School, along with later representations of the westward movement, as capturing narrative moments within a larger history. "Reading the history of environmental change," he writes, "requires us to place each painting in a dynamic continuum that encompasses not just the painting's present moment but the past from which the landscape emerged and the future toward which its artist believed it was heading."[19] These paintings are set in a narrative of progression through three stages. The first stage, the wild, representing untamed nature, gives way to the second stage, the pastoral, representing the middle ground of the cultivated garden, which gives way to the third stage, the urban, representing the railroads, factories, and cities of the triumphant market. The narrative can be a progressive saga of material progress or a declensionist tale of nature's degradation.

Hudson River painting, like transcendentalism, was part of a shift in European and American culture from eighteenth-century rationalism to nineteenth-century "romanticism." Spiritual truth, according to romantic doctrine, was implanted in every soul, evoked by nature, and experienced intuitively and emotionally rather than intellectually and analytically. This tendency in art and thought, along with the simultaneous upsurge of an emotional Protestant revivalism, expressed yearning for a spirituality and feeling denied by the market's cash calculus. Thomas Cole (1801–1848), the tutelary genius of the Hudson River School, preached that scenes of pristine nature "affect the mind with a more deep toned emotion than aught which the hand of man has touched. Amid them ... the mind is cast into the contemplation of eternal things."[20]

The Hudson River artists stood on the cusp between the first two stages of Cronon's narrative, the wild and the pastoral. Wild nature was the special theater of "the sublime," meaning that its untamed grandeur, beauty, and power evoked the strongest emotions of reverence, awe, and dread. Cole and his closest associates, Thomas Doughty (1792–1856) and Asher B. Durand (1796–1886) were tenaciously attached to the wild and sublime. "Landscape painting," said Durand, "will be great in proportion as it declares the glory of God, by representation of his works, and not of the works of man." Cole's "scenes of wild grandeur ... never touched by the axe, ... never deformed by culture," brought delight to his bosom friend William Cullen Bryant, the nature poet and Jacksonian Democratic editor of the New York *Evening Post*.[21]

The Hudson River painters captured the sublimity of the wild, as in Cole's "The Clove" (1827), by constructing chaotic landscapes of dark forests, menacing clouds, jagged or looming peaks, twisted tree trunks, and foaming cataracts. They were especially accomplished in the dramatic use of light, and their dark landscapes of the wild were backlit by brilliant skies of divine luminescence. They feared, as Cole put it, "that with the improvements of cultivation the sublimity of the wilderness should pass away"; and they were trying, said the *Literary World*, to rescue from the grasp of "Yankee enterprise . . . the little that is left, before it is too late."[22] The sublime reached its apogee of extravagance when western exploration carried later members of the Hudson River School to the irresistibly suggestive spectacles of the Rocky and Sierra Nevada Mountains. Albert Bierstadt (1830–1902) painted enormous canvases of Estes Park, Colorado; Yosemite Valley; and other majestic showcases; and Thomas Moran (1837–1926), who accompanied the 1871 Hayden expedition to Yellowstone, produced overwhelming paintings of spectacular canyons and peaks in Yellowstone and the Grand Canyon.

Many feared, however, that the untamed sublimity of the wild might release the untamed passions of the human heart. As wild nature retreated, the Hudson River painters gave equal time to evoking a more tranquil sublimity from warmly sunlit representations of the pastoral's cultivated garden. Doughty's "Autumn on the Hudson" (1850) captures the pastoral mood in a tranquil scene of a modest home on the edge of the river, with sailboats drifting along the water, and people in the foreground preparing a field for crops.

More characteristically these "priests of the natural church,"[23] as art historian Barbara Novak names them, juxtaposed the wild and pastoral stages, often with disquieting intimations of the third or urban stage to come. Cole's "The Oxbow" (1836), portraying a nearly circular bend in the Connecticut River near the town of Holyoke, reveals stark contrasts. Wild nature, depicted by forests and dark clouds on the left, moves toward a pastoral landscape on the right, where sheep graze in a meadow and corn husks are stacked along the river's edge. The land has been cleared, the setting is tranquil, and the river wends its way through a peaceful scene. But there are intimations of what is to come. A haze on the hillside indicates the burning of the forest by people settling the land, and scars among the trees represent the cut-over forest. Cronon suggests that the river, painted in the shape of a question mark, is Cole's way of asking, "What will the future hold?"

Cole framed the problematic of nature most explicitly. Through painting he sought "to walk with nature as a poet."[24] Durand portrayed him and Bryant together as "Kindred Spirits" (1848) communing with nature in the sublimity of a Catskills glen. When Cole departed for Italy to hone his skills amid paintings by the old masters, friend Bryant distilled their vision of America as nature's nation in a farewell sonnet. The painter would carry with him "a living image of thy native land," as he viewed Europe's "fair, but different" scenes, bearing "everywhere the trace of man." "Gaze on them, till the tears shall dim thy sight," the poet admonished, "But keep that earlier, wilder image bright.[25]

Cole's most ambitious project for keeping the wilder image bright was "The Course of Empire" (1836), a series of five paintings that turned the three-stage progressive narrative of material progress on its head by adding a two-stage declensionist conclusion. Drawing on his Italian experience, and using ancient Rome as a metaphor for the United States, he painted the same setting as changing through five stages: "The Savage State" (hunters in the wilderness), "The Pastoral State" (shepherds and their sheep in an Arcadian landscape), "Consummation" (bustling commerce amid an imposing array of classical buildings), "Destruction" (the same scene swept by fire and pillage), and "Desolation" (deserted ruins engulfed by wild nature).

Cole's fantasy of recovering the wild became increasingly hard to sustain when the urban menace claimed industrial sites and raw materials in the Catskills. As geographer Michael Heiman points out, the Hudson River painters "often overlooked or screened out with vegetation the burned-over fields, stinking tanneries, polluted streams, clamorous sawmills, and other production intrusions" on their temple of the sublime.[26] Around that temple the country's particularly noxious tanning industry first concentrated. This "environmental malignancy," writes historian Isenberg, fouled streams with organic contaminants that overwhelmed the capacity of aerobic bacteria to break them down, and chewed up for tannin dense forests of ancient hemlocks that had taken three centuries to mature. "Deforestation," he adds, "had widespread environmental as well as economic consequences, ranging from the destruction of local wildlife habitat to increased erosion and runoff of soil into streams."[27]

Bowing to reality, Hudson River artists began to include in their paintings occasional renderings of towns (Durand's "View of Troy"), as well as railroads (George Inness's "Delaware Water Gap"), but these were subsumed in the distant backgrounds of pastoral landscapes. "So it is with a shock," Novak writes of Inness's "Lackawanna Valley" (1855), "that we find a train forcing its way toward us out of the middle distance to become the main protagonist." No other picture in American art, she concludes, so "aptly embodies the moment of juncture between nature and civilization." The "busily smoking" locomotive charges into a pastoral landscape prominently and incongruously scarred by stumps.[28] This painting was doubtless seen as a celebration of progressive technology by directors of the Delaware, Lackawanna, and Western Railroad, which commissioned it, but perhaps by the artist as an elegy for vanishing pastoral nature. From here it was only a

small step to unequivocal celebration of the onrushing urban and industrial
order by later artists and illustrators.

Artists and the Vanishing Indian

As the Hudson River School lamented a vanishing nature, other artists
were recording the lifeways of a vanishing Indian, demonstrating in the
process that American Indians were far more complex and interesting than
contemporary stereotypes allowed. Swiss painter Carl Bodmer (1809–1893),
who accompanied the German scientist Prince Maximillian to the western
prairies in 1832–34, portrayed the varied Indian uses of nature with greater
detail and precision than any other artist. His "Interior of a Mandan Earth
Lodge" (1833–34) depicts the strong timbers dragged by Mandan and their
horses from the mountains to the plains to frame their lodges. The lodges
themselves had roofs of earth and prairie grasses. Indians, along with their
dogs, are shown sitting in the center of the lodge at a campfire, their tethered
horses visible outside. The painting includes a basket intricately woven by a
woman of the tribe, spears used by men in the buffalo hunt, and shields for
warfare made from buffalo skins and decorated with pictographs.

A chance encounter with a group of Indians visiting Philadelphia was an
epiphany for George Catlin (1796–1872). Entranced by these "lords of the
forest," and alarmed that they were "rapidly passing away from the face of
the earth," he resolved to spend his life painting them in their native wildness before they were gone. "Wild" was not a pejorative word to Catlin. "A
wild man may have been endowed by his Maker with all the humane and
noble traits of a tame man," he insisted, and he delighted in "the vast and
pathless wilds" he first experienced in 1832 on the Trans-Mississipi prairies.
He advocated a national park, in which the lifestyle and lands of "a truly
lofty and noble race" might be preserved. His numerous paintings of varied
Indian interactions with nature included a group extracting red-brown shale
(named Catlinite) for pipes ("Pipestone Quarry," 1848), and women scraping and dressing buffalo hides to make robes and hanging them to dry along
with buffalo meat ("Comanche Village," 1834).[29]

The growing perception that Indians were doomed brought an alternative into prominent view in 1853, when a sculpture by Horatio Greenough
(1805–1852) was placed in the United States Capitol. "The Rescue" depicts a white woman and her child shrinking in terror from an Indian with
tomahawk raised. The central figure, however, is a giant frontiersman firmly

restraining, but not harming, the Indian. Both the woman and the Indian are being rescued. White men, the sculpture seems to say, could save the vanishing Indian by quelling his imputed "savagery" and transforming him into a "civilized" being.

By mid-century, differences over vanishing western Indians and vanishing western wilderness were meshing with differences over the wild, the pastoral, and the urban that derived from the East. Increasingly, contending narratives of nature and national destiny framed, and were framed by, dramatic change in the Trans-Mississipi West.

Conclusion

One narrative suggested by pre–Civil War environmental history may be summarized as follows. The advancing market not only devastated nature, but also stirred a widespread hunger for an emotionally compelling spirituality and connectedness that the competitive ethic of profit and loss could not satisfy. For much of the market world's new middle classes, this hunger focused on nature, as indicated by the popularity of Cooper's Leatherstocking tales, transcendentalism, and the Hudson River painters. This yearning could not be satisfied for long by Emerson's arranged marriage between commodified nature and the elusive Over-Soul. When, at century's end, peaking assaults on nature finally aroused a conservationist majority, it would draw inspiration from the tradition of Thoreau and Cole, and the means for checking market excesses from Jacksonian democracy.

Notes

1. William Cronon, *Changes in the Land* (New York: Hill and Wang, 1983), 169.

2. Carolyn Merchant, *Ecological Revolutions: Nature, Gender, and Science in New England* (Chapel Hill: University of North Carolina Press, 1989), 152.

3. Herbert Leventhal, *In the Shadow of the Enlightenment: Occultism and Renaissance Science in Eighteenth Century America* (New York: New York University Press, 1976), 223.

4. Charles M. Andrews, *Colonial Folkways* (New Haven: Yale University Press, 1919), 151.

5. Marion Barber Stowell, *Early American Almanacs: The Colonial Weekday Bible* (New York: Burt Franklin, 1977), 36.

6. Leventhal, *In the Shadow of the Enlightenment*, 40.

7. Albert Gallatin, "A Report on Roads and Canals, April 6, 1808," in *Government Promotion of American Canals and Railroads, 1800–1890*, by Carter Goodrich (New York: Columbia University Press, 1960), 31.

8. Andrew C. Isenberg, *The Destruction of the Bison: An Environmental History, 1750–1920* (New York: Cambridge University Press, 2000), 93.

9. John James Audubon, "The Ohio," in *Delineations of American Scenery and Character* (London: Simpkin, Marshall, Hamilton, Kent, 1926), 4.

10. William H. Gilman et al., eds., *The Journals and Miscellaneous Notebooks of Ralph Waldo Emerson* (Cambridge: Harvard University Press, 1960–82), 10:353; Ralph Waldo Emerson, *The Complete Works of Ralph Waldo Emerson*, with a biographical introduction and notes by Edward Waldo Emerson (Boston: Houghton Mifflin, 1903–1904), 1:370, 451–55.

11. Ralph Waldo Emerson, "Wealth," in *The Complete Essays and Other Writings of Ralph Waldo Emerson*, ed. Brooks Atkinson (New York: Modern Library, 1940), 694–95.

12. Henry David Thoreau, *Walden* (Boston: Ticknor and Fields, 1854), 126.

13. Charles Sellers, *The Market Revolution: Jacksonian America, 1815–1846* (New York: Oxford University Press, 1991), 152–53.

14. Charles Sellers, "Capitalism and Democracy in American Historical Mythology," in *The Market Revolution in America*, ed. Melvyn Stokes and Stephen Conway (Charlottesville: University Press of Virginia, 1996), 322.

15. William Byrd, *The Writings of 'Colonel William Byrd of Westover in Virginia, Esqr,'* ed. John Spencer Bassett (New York: Doubleday, 1901), 242; William Bartram, *The Travels of William Bartram: Naturalist's Edition*, ed. Francis Harper (New Haven: Yale University Press, 1958), 69; William Gilpin, "Remarks on Forest Scenery and Other Woodland Views" (1791), quoted in *Wilderness and the American Mind*, rev. ed., by Roderick Nash (New Haven: Yale University Press, 1973), 46.

16. Annette Kolodny, *The Lay of the Land: Metaphor as Experience and History in American Life and Letters* (Chapel Hill: University of North Carolina Press, 1975), 90.

17. Atkinson, ed., *Complete Essays and Other Writings of Emerson*, 261, 695, 698.

18. Ralph Waldo Emerson, "The Young American," *Dial* 4 (April 1844): 484–507.

19. William Cronon, "Telling Tales on Canvas," in *Discovered Lands, Invented Pasts*, by Jules David Prown et al. (New Haven: Yale University Press, 1992), 45. Many of the paintings discussed below are reproduced in Cronon's article and in *Nature and Culture: American Landscape and Painting, 1825–1875*, rev. ed., by Barbara Novak (New York: Oxford University Press, 1995).

20. Novak, *Nature and Culture*, 4–5.

21. Ibid., 5, 9.

22. Ibid., 4–5.

23. Ibid., 9.

24. W. L. Nathan, "Thomas Cole and the Romantic Landscape," in *Romanticism in America*, ed. George Boas (Baltimore: Johns Hopkins University Press, 1940), 34.

25. William Cullen Bryant, *The Poetical Works of William Cullen Bryant* (New York: D. Appleton, 1906), 127.

26. Michael Heiman, "Production Confronts Consumption: Landscape Perception and Social Conflicts in the Hudson Valley," *Environment and Planning* D: *Society and Space* 7 (1989), 170.

27. Isenberg, *Destruction of the Bison*, 132.

28. Novak, *Nature and Culture*, 171–74.

29. Peter Matthiessen, ed., *North American Indians* (New York: Viking, 1989), 2–8.

5 Western Frontiers: The Settlement of California and the Great Plains, 1820–1930

This chapter focuses on the westward movement of Americans, including the settlement of California and the Great Plains, as examples of the interactive relationships between the environment and humanity in the western United States. Core ideas — such as the frontier and Manifest Destiny — helped to shape the ways in which land was acquired and developed. This chapter explores the multicultural relationships, opportunities, and oppression experienced by the many ethnic groups that developed California during the Gold Rush, and examines the environmental effects of gold mining. It looks at the subsequent settlement of the Great Plains and the impact of various technologies on the transformation of the Plains environment. Finally, it offers some theories put forward by environmental historians as to why the West developed as it did.

Westward Expansion and the Settlement of California

The western United States was acquired in several stages during the first half of the nineteenth century. The Louisiana Purchase in 1803, which extended westward from the Mississippi River to the Continental Divide, had inspired President Thomas Jefferson to send Merriwether Lewis and William Clark to explore the Mississippi and the Missouri Rivers. They crossed the mountains in Idaho and followed the Columbia River to the Pacific. Following their successful return, numerous expeditions were mounted by fur traders and explorers into the Rocky, Sierra Nevada, and Cascade Mountains. By the 1840s, most of the land west of the Mississippi had been explored, and wagon trains of settlers were following the Oregon Trail westward to Oregon's Willamette Valley, just south of the Columbia River. President James K. Polk envisioned the United States as extending across the entire breadth of the continent, and his election to the presidency in 1844 triggered the acquisition of additional lands west of the Mississippi. In a controversial move in 1845, Congress annexed Texas, which had been asserting its independence from Mexico for nine years. Following a border skirmish, President Polk declared war on Mexico the next year. Then, in 1846, Oregon territory up to the 49th parallel was acquired by treaty with Great Britain, and, in 1848, in the Treaty of Guadalupe Hidalgo, a defeated Mexico ceded to the United States the present states of California, Nevada, Utah, and most of Colorado, Arizona, and New Mexico. The Gadsden Purchase, completed in 1853, acquired additional land from Mexico for a level train route to the west through southern New Mexico and Arizona. Gold was discovered in California on January 19, 1848, and in 1849, miners began the rush to occupy California's gold country, a movement that would irrevocably alter its native peoples and natural environment.

California Native Peoples and the Advent of Europeans

Prior to the Gold Rush, the native peoples of California lived within an organic economy based on relations among Indians and fish, deer, bear, acorns, and other living things. Although California's Mediterranean climate was highly suited to agriculture, California natives west of the Sierras remained hunters, gatherers, and fishers, many living in the densely settled Central Valley. East of the Sierras, the Paiute practiced irrigated horticulture, but the corn, beans, and squash complex of the American Southwest was not established west of the mountains.

The staple food of the California Indians, especially those in the Central Valley, the Coast ranges, and the Sierra foothills, was the acorn. California contained many species of oak, and Indians ground the acorns into meal to make a variety of breads. They also harvested bulbs and seed-bearing plants that grew in the coastal chaparral, the Central Valley, and the Sierras. They set fire to the hillsides to clear the underbrush for better passage, to create browse areas for deer, to encourage the growth of edible plants, and to force out jackrabbits, a major source of food.

Indian life underwent a major transformation with the advent of European colonizers. Historian George Phillips states, "After 1769 and the establishment of permanent Spanish settlements, Indian culture came under enormous pressure. But the first Californians did not readily submit to for-

eign domination as is sometimes thought. Rather, they implemented strategies of evasion, selective acculturation, and military resistance."[1]

As Spanish, Mexicans, and Russians explored and settled California,
ranching, trade, and missionary work ensued. Cattle released into the
coastal valleys by the Spanish gave rise to a vigorous hide and tallow trade,
colorfully described by Richard Henry Dana in *Two Years Before the Mast*
(1840). Workers rounded up cattle, stripped their hides, and waded through
water with the heavy layers on their heads to load them on boats waiting offshore. The hides were used for leather goods, while tallow made from cattle
fat was needed for candles in an era before electricity. Trade also began with
the Russians, in collaboration with fishermen from Boston, for otter pelts
and whales.

A number of rationales existed for the settlement of California and
the American West. America's "Manifest Destiny," proclaimed journalist
John L. O'Sullivan, was for Anglo-Saxon America "to overspread and to possess the whole of the continent which Providence had given us for the development of the great experiment of liberty."[2] Senator Thomas Hart Benton argued that the white race should expand its civilization to the West
Coast and take over land occupied by the red race. Other rationales included the concept of "American Progress," depicted in an 1872 painting of
that name by John Gast. Imbedded within the painting of settlers moving
west is the ideal of transforming "wild" nature and "wild" Indians into "civilization." The painting represented an Enlightenment narrative of progress,
the idea of bringing light to a dark land.

The Multicultural Character of the Gold Rush

Most of the development initially associated with gold mining occurred
in various towns in California's Sierra Nevada Mountains during the second
half of the nineteenth century. Molten gold, formed in the Sierras, Rockies,
and Cascades during the ancient uplifting of the mountains, poured into fis-
sures, designated as veins. This organic term, coming from the idea of veins
within the ancient "earth mother," was rooted in an older concept that both
gold and the earth were alive. Veins that were close together formed a lode,
or a set of veins that could be worked together. The idea of a mother lode
was based on the belief that several veins within the earth rose to the surface, where they could be tapped. Placers were places where gold had
eroded into small particles through the action of water. It could be recov-

ered by panning, or washing the heavier gold nuggets out of the gravel by ro-
tating a flat-bottomed pan. The area known as the Mother Lode in the western Sierra foothills was noted not only for the early process of gold panning,
in which prospectors employed pans, cradles, and sluice boxes to wash and
separate out the heavy gold particles from the sand and gravel, but also for
river mining, hydraulic mining, hard rock mining, and dredging. Silver was
found in conjunction with gold in areas east of the Sierras, such as Nevada's
Comstock Lode.

During the Gold Rush, people came to California by three major routes.
One was around Cape Horn by boat from the East Coast. Another was to
sail to Panama and cross the hot humid rain forest, an area that harbored
tropical diseases, and then travel by boat up the West Coast to California.
Most gold seekers, however, came by overland trail through the west. The
main California trail left Council Bluffs on the western edge of Iowa, traversed Nebraska to Fort Laramie in Wyoming, and continued over the Continental Divide at South Pass, whence it followed the Humboldt River
across Nevada and the Sierras to the gold fields. Those who started late in
the season turned southward to Salt Lake City and crossed the Salt Lake
Desert to the Humboldt River Valley, in order to ascend and cross the Sierras before the snow fell.

People of numerous ethnic and racial origins came to the gold country.
American miners, along with Chinese and Mexicans, mingled with European and South American immigrants. Walter Pigman, a gold miner in
1859, commented on the ethnic diversity of the gold country: "Had the opportunity of seeing the mixed multitude of human being that are in this
country. The Americans take the lead, the Spanish next, then comes the
poor degraded native Indian, the Chinese, the Chilean, the Mexican, and,
in fact, some from every nation of the earth are to be found here. All in
search of gold!"[3]

Indians initially participated in the process of panning for gold. Samuel
Ward, who worked on the Merced River in 1849, wrote: "[T]he members of
the wigwam kept together in their own watery pew; the father scooping into
his batea the invisible mud and sand of the river bed, and the mother bearing it to the shore to perform those skillful gyratory manipulations by which
the water is made to carry away ... the earthly matter, until ... there remains
in the bottom of the pan the yellow spangels" of glittering gold.[4] But soon
the Indians were pushed out of the gold country, their people massacred,
and their ways of life devastated. Miners blasted out rocks, hunted animals
for meat, and chopped down oak trees, depriving Indians of their staple

foods. Mining debris spoiled the rivers on which they depended for fish, and depleted salmon runs.

In 1849, California declared itself a slavery-free state, and its constitutional convention that year outlawed slavery. In the Compromise of 1850, the United States agreed that California was to be admitted to the Union as a free state. But as part of that compromise, the U.S. also passed the Fugitive Slave Law, requiring people to return slaves determined to be fugitives by federal magistrates. Northerners in the East were mandated to comply by returning slaves to the South; many not only disobeyed the law, but were outraged by its passage.

Many blacks believed that their best chance to secure freedom was to travel west through Canada and then proceed south into California. There they formed mutual aid societies to assist each other economically and socially and to avoid the Fugitive Slave Law. Many found opportunities to work in the gold mines, and those who arrived as slaves were often able to purchase their freedom and that of their families. A white forty-niner from Ohio wrote in his journal: "I saw a colored man going to the land of gold, prompted by the hope of redeeming his wife and seven children.... His name is James Taylor."[5] Reuben Rudy, friend of abolitionist William Lloyd Garrison, came to California by way of Panama and made $600 after digging for four weeks on the Stanislaus River. Other blacks headed for places such as Auburn Ravine, where an African-American miner working alongside whites was photographed in 1852. By then there were 2,000 blacks in the state of California. In some cases, they returned to the East, were reunited with their families, and remained in the free North.

Chinese immigrants also came to California in great numbers, many from Guangdong Province. They arrived on credit tickets organized by Chinese steamer lines and set out for the gold country on ferries through the San Joaquin Delta to Stockton and Marysville. In Auburn Ravine, white miners worked alongside Chinese as well as black miners. By 1852, there were 5,000 Chinese along the Yuba River, and by 1855 their number had swelled to 20,000. They lived in China camps, where they retained their own customs, foods, and religious practices. But they fell victim to the 1850 Foreign Miner's Tax, which charged them $20 a month, a hardship for most Chinese. By 1870, their numbers in the mines had dropped by half, and many moved into the Central Valley, where they began to establish agriculture. Those who stayed behind worked the river bottoms abandoned by other miners. Others worked for railroad companies, helping to lay track through the Sierras and across the continent. In 1882, the Chinese Exclusion Act was passed, and vigilante violence began to discourage many Chinese from further activities. From California, they moved into other western states, where gold and other mining activities were prevalent. Others went north to Oregon and Washington or through the Rockies into Montana, where they established Chinese communities.

Types of Gold Mining

Gold panning was the source of most of the gold for the first two or three years of the Gold Rush. Gold nuggets were found in placer deposits in riverbeds. All one needed was a pan or a cradle with riffle bars on the bottom and a box on the top, into which sand, rock, and silt from the riverbeds were scooped. As mining advanced, a chain of sluice boxes with riffle bars was set on an incline so the water would run off, leaving the gold, which was heavier than water or sand, in the bottom of the trough. As the placer mines containing the nuggets were depleted, other forms of mining followed. In river mining, cooperatives built dams to divert the water away so the streambed could be mined. Hard-rock mining used shafts and underground tunnels with rails and ore carts to extract gold-bearing rock, while in the devastations of "hydraulicking," entire hillsides were washed away.

Hydraulic mining began in 1852 through the efforts of Anthony Chabot, creator of the enormous "monitor" nozzles that blasted water onto hillsides and captured nuggets in chains of sluice boxes. Miners, forced out of panning for gold or victimized by failed river-mining cooperatives, became wage laborers for the owners of the new mines. In 1879, a newspaper reporter described the scene: "We stand at the brink of the mine.... Around us are naked rocks and well-scraped furrows, piles of pine-wood blocks for use in the flume, rusting joints of condemned water pipe, and shops where soot-covered men are lifting joints of new pipe." One experienced a cacophony of sounds. "As we turn to descend, a measured succession of sounds begins.... [T]he next thing is to get rid of the large boulders often weighing tons. They must be blasted into fragments so small that when the water is turned on here again they will be swept down and out through the tunnel.... [H]ere are thirty or forty men, busy with drills, in a great hammering company. It is at this instant, wild music."[6]

Hydraulic mining depended on vast systems of canals and ditches that diverted water from lakes and rivers, which then fell by gravity through pipes and hoses, producing powerful streams of water that could be directed against hillsides. By 1882, there were 6,000 miles of flumes, ditches, and sluices in the Sierras. Reservoirs constructed along the Yuba, Bear, Feather, and American Rivers held 50 to 500 acres of water on their surface. Sluice gates across streams and reservoirs regulated the force and quantity of water sprayed on hillsides, releasing dirt and rock that flowed through sluice boxes to collect the gold.

Photographer Carleton Watkins (1829–1916), who traveled throughout California taking pictures of mining and agricultural activities, took a fa- mous photograph of hydraulic nozzles blasting away entire hillsides. The enormous effect was described by the 1879 newspaper article: "There's real pleasure … about this gigantic force. [The water] is beaten into foam un- til … it comes out with … a wicked vicious unutterable indignation.… [I]t requires much experience and judgment to know how to use this stream to best advantage, and with greatest safety.… [R]ocks two feet in diameter fly like chaff when struck by the stream."[7] Anyone on whom the hydraulic blasts might turn did not survive.

"Hydraulicking" came to an abrupt end in 1884 when Judge Lorenzo Sawyer of the Ninth Circuit Court issued the famous "Sawyer decision," halting mining using hydraulic nozzles. The golden era of hydraulic mining lasted only thirty-two years, but, during that time, it had major human and environmental consequences.

Environmental Effects of Hydraulic Mining

Hydraulic mining daily displaced 50 to 100 tons of debris, consisting of muddy soil and rock, an enormous amount compared to that from hard- rock mining, which followed the hydraulic era. The force of the water washed out entire hillsides. The debris then washed downhill, carrying the topsoil with it, and creating huge fans of tailings at the base of the mines. Cones of debris dotted the hillsides, and colorful, vertical cliffs were formed from water and wind erosion. Debris filled the mountain streams and rivers, creating sandbars and islands and producing banks of yellow mud.

As the debris flowed down to the valleys below, it covered acres of farm- land, filled the beds of streams, created shoals in the rivers and bays, and halted tidal action in the cities of Marysville and Sacramento. Historian Robert Kelley, in *Gold vs. Grain: The Hydraulic Mining Controversy in Cali- fornia's Sacramento Valley* (1959), described the effects on farmlands: "Farm- ers talked of how the rivers were filling. How each small rise in the rivers produced floods, how hundreds of acres of fruit orchard along the rivers were dying as debris spread slowly, imperceptibly, out into the valley."[8]

The Yuba riverbed rose at an annual rate of a third of a foot per year and the Sacramento at the rate of a quarter of a foot per year. The debris in the Yuba spread out along the shores of the river in a three-mile-long fan, containing 600 million cubic feet of debris from the mines. The streets of Marysville were originally 25 feet above the Yuba River, but, by 1879, so much debris filled the riverbed that the streets had to be protected by levees. Debris began to fill the Suisun Bay and Carquinez Straits below Sacramento, with its effects felt as far west as San Francisco Bay.

In 1917, engineer Grove Karl Gilbert (1843–1918) made a study for the United States Geological Survey of the amount of debris flowing from the Sierras through the San Joaquin and Sacramento Rivers and its effect on the Suisun, San Pablo, and San Francisco Bays. His report was replete with photographs showing debris-filled creek beds and buried forests. The debris spread out on either side of streams and rivers, confining the water to a narrow central channel and changing its flow. After the Sawyer decision halted hydraulic mining, tree trunks formerly under water began to reappear as the accumulated debris began to wash downstream.

The effects of the debris-clogged rivers on the tides at Sacramento remained until the early twentieth century. Gilbert's data showed that, in the 1880s, the Sacramento River was filled with debris from hydraulic mining, but that by 1902, the tides had returned as far upriver as Sacramento itself and, by 1920, had reached 20 inches, just below the 24 inches they had been originally. But at Marysville, upstream from Sacramento, the lack of tides remained. Today, Marysville — originally above water — is below water level and protected by levees. Nevertheless, trees began to regrow along riverbanks, and restoration of the river ecosystem began to take place. As the effects of hydraulic mining diminished, much of the vegetation grew back, and some of the conditions that existed prior to the hydraulic era were reestablished. Nonetheless, permanent effects remained, and hillsides still bear the visual marks.

The period of hydraulic mining that effectively ended with the 1884 Sawyer decision was followed by hard-rock mining. New technologies were developed to remove the gold and silver imbedded in rock, rather than in the more accessible riverbeds and gravel-filled hillsides. The rock had to be pulverized and mercury added to bond with the gold in the form of an amalgam. Mercury was mined in the Salinas Valley south of San Francisco and in the Napa Valley to the north, whence it was transported to the Sierras. The gold-mercury amalgam was heated in bullion furnaces at high temperatures to form liquid gold metal and to vaporize the mercury. The gold was then poured into bullion bars and transported to the mining exchange in San Francisco, while globules of mercury washed down the rivers.

Major environmental effects resulted from the discharge of mercury. The waste mercury washed down the streams, poisoning the rivers for fish and creating unhealthy water for human use. Today that mercury is still embedded in mountain streams and in tributaries of the Sacramento River. Although covered by deep silt, it nevertheless represents a hazard for humans and fish, especially when the rivers are dredged or the silt is disturbed.

Environmental Change in the Sierras

Environmental changes from hydraulic and hard-rock mining occurred in the Sierra Nevada mountains and foothills. The original cover of oaks in the foothills was extensive. Miners cut the woodlands for fuel, for building materials for cabins and sluice boxes, and for range and pasture lands. After cutting and grazing, oak seedlings were eaten by cattle and deer, hence the trees did not reestablish themselves, and chaparral — which is fire adaptive — moved higher into the foothills than in the pre–Gold Rush era.

Poet and commentator Joaquin Miller (1837–1913) described the changes in the Sierra that resulted from mining activities: "I stood in ... the placer mines. Smoke from the low chimneys and log cabins began to rise and curled through the cool, clear air on every hand, and the miners began to come out."[9] He sketched a scene of gloom and darkness in the valleys where the miners were working as they toiled to enrich themselves for private profit. The land around them became pockmarked with holes dug for mining shafts, and entire rivers were diverted out of their streambeds.

Sierra wildlife was also transformed as a result of Gold Rush activities. The grizzly bear, which had been native to the western slope, became extinct from the area in 1924. Mountain sheep, which had been common in the high Sierras, were reduced by hunting and human presence. The wolverine, fisher, and marten were trapped out, while coyotes and wildcats were killed to protect sheep. Mule deer occurred throughout the Sierra, migrating to the high meadows in the summer and returning to the foothills in autumn, creating trails that humans also used for traveling. Cougar, the major predator on mule deer, was drastically reduced, hence deer populations increased, causing ecological damage by browsing on shrubs.

The populations of small mammals and birds were affected by mining and logging, as well as by fire and grazing. Logging reduced the number of tree squirrels and native birds, which lost ground to other species that tolerated the changed conditions. The tree squirrels were replaced by ground squirrels and chipmunks; Sierra birds, such as woodpeckers, chickadees, nuthatches, creepers, and Steller's jays, by fox sparrows and green-tailed towhees, which adapted to the chaparral that took over the habitat.

Many fish, including most of the salmon runs, were sharply reduced by mining debris, streamside tree removal, and habitat destruction by cattle grazing. They were replaced by native trout, propagated by the Fish and Game Department, and by brown and eastern brook trout introduced to supplement native species.

After the Gold Rush, the Sierra foothills and mountain valleys were further transformed by grazing. By the turn of the century, Swiss dairy farmers were taking their cows into the mountain valleys in the spring and bringing them back to the lowlands in the fall. Large herds of beef cattle and sheep overgrazed the meadows after the snow melted. Cattle numbers are now regulated by the U.S. Forest Service.

Although restoration of landforms, trees, and wildlife is possible to a certain extent, the area will never revert to the pre-mining era or to the ecosystem known by native Americans. Population effects on the Sierras further decrease that possibility. Mining thus left an indelible impact on both the landscape and on the multiethnic populations of California, only partially mitigated by ecological restoration and the enforcement of civil rights for minorities.

European Settlement of the Great Plains

The main settlement of the Great Plains occurred after the 1840 migrations to Oregon and the 1849 Gold Rush to California. Environmental historian William Cronon has interpreted the history of the Great Plains in terms of narrative. The grand narrative of America, Cronon argues, is a story of progress. The frontier narrative depicts that formative story and, as such, is the master narrative of American culture. A hostile environment, initially conceptualized as a Great American desert, was gradually brought under control and transformed into a garden, making the Great Plains a Garden of the World. That transition in perception occurred as people increasingly settled the Plains and gained control over nature. Two formative accounts reveal the environmental history of the Great Plains as a progressive narrative: Frederick Jackson Turner's "Significance of the Frontier in American History" (1893); and Walter Prescott Webb's *The Great Plains* (1931).

The opposite, or declensionist narrative, according to Cronon, relates history as environmental decline. A pristine grassland, at first uninhabited, was then occupied by nomadic bands of Indians. White settlers who came into this natural Garden of Eden, or nearly pristine nature, transformed it over a period of 150 years into a desert, exemplified by the Dust Bowl of the 1930s. Donald Worster's *The Dust Bowl* (1979) illustrates the ecological decline of the Plains that came about through capitalist agriculture and ranching and resulted in the ecological disaster of the Dust Bowl.

These two story lines, however, are both linear — the first uphill, the second downhill, masking nuances and irregularities. "When we choose a plot to order our environmental histories," Cronon notes, "we give them a unity that neither nature nor the past possesses so clearly. In so doing, we move well beyond nature into the intensely human realm of value."[10] Stories that focus on white settlers often leave out other people's perspectives — for example, those of black cowboys and homesteaders and women settlers, as well as the stories of native Americans and the bison (told in chapter 1). How are these stories illustrated in the environmental history of the Great Plains?

Frederick Jackson Turner's 1893 account of "The Significance of the Frontier in American History" depicts the progressive story as a traditional hero narrative. European immigrants are the heroes who settled the land, crossing the Atlantic and the eastern United States in successive frontier lines. According to Turner, the frontier transforms the settler. The encounter with wilderness "strips off the garments of civilization." It puts the colonist "in the log cabin of the Cherokee and Iroquois." But as the victor gains control, "little by little he transforms the wilderness." The outcome, however, is not the old European émigré, but a new American, formed by the frontier. Democracy and American civilization, in a perennial rebirth, fill the land. Democracy, according to Turner, is "born of free land."

The fur traders, who constituted Turner's first frontier, explored the West during the 1820s through the 1840s, employed by trading companies such as John Jacob Astor's (1763–1848) American Fur Company. The competing company, the Rocky Mountain Fur Company, employed some of the world's most famous traders, such as William Ashley, Jim Bridger, Milton Sublette, and Kit Carson. Competing with these companies from the north was the Hudson's Bay Company, which merged with the Northwest Fur Company and held a monopoly in Canada. Traders scoured the West, capturing and shipping eastward the pelts of fur-bearing animals, in particular the beaver. Mountain men fanned out over the Rockies along Indian trails, coming together at a given time and place each year — a "rendezvous" — to which they brought the year's collection of pelts. Held between 1825 and 1840, rendezvous sites included camps along the tributaries of the Green and Snake Rivers. French, Scottish, and German traders, along with Native Americans and African Americans participated in the trade as hunters, trappers, voyageurs, and entrepreneurs. African Americans were often used as go-betweens in negotiations with Indians, to reduce friction between the parties.

In Turner's progressive narrative, the rancher's, miner's, and farmer's frontiers followed that of the fur traders. Turner writes: "Stand at Cumberland Gap and watch the procession of civilization, marching single file — the buffalo following the trail to the salt springs, the Indian, the fur-trader and hunter, the cattle-raiser, the pioneer farmer — and the frontier has passed by. Stand at South Pass in the Rockies a century later and see the same procession with wider intervals between. The unequal rate of advance compels us to distinguish the frontier into the trader's frontier, the rancher's frontier or the miner's frontier, and then finally the farmer's frontier."[11]

South Pass, on a gradual rise at 7,375 feet at the Continental Divide in Wyoming, was on the trail most settlers followed west. Here they could take the Sublette Cutoff (named after the fur trader, Milton Sublette) across Colorado and north to Fort Hall in Idaho, where they followed the Oregon Trail along the Snake River, crossed into the Columbia watershed, and traveled south to Oregon's fertile Willamette Valley. Alternatively, they could take the California trail westward along the Humboldt River, crossing the Sierras and heading down to California's Gold Country and Central Valley. Some dipped south into Salt Lake City, Utah, settling in the region developed by Mormons. On arriving at their destinations, the emigrants pursued mining, ranching, and farming, settling the far West and the Great Plains.

The Rancher's Frontier

Historian Walter Prescott Webb's *The Great Plains* (1931) builds on Turner's progressive narrative, describing the rancher's and farmer's frontiers on the Plains in terms of two formative factors — environment and technology. Webb states: "New inventions and discoveries had to be made before the pioneer farmer could go into the Great Plains and establish himself."[12] Technologies allowed settlers to subdue a forbidding environment that had three main characteristics not found in the eastern United States. First, as pioneers moved west of the 100th meridian, the environment of the Plains became increasingly arid, lacking the minimum twenty inches of rainfall per year that would support agriculture reliably. Second, the Plains were treeless, and therefore did not provide the timber for fuel and building materials readily available in the East. Third, the Trans-Mississippi Plains were level, rising only gradually westward, which meant that rivers were shallow and lacked the power to operate mills or float ships.

The rancher's frontier, according to Webb, was initiated by longhorn cattle drives north from Texas. "The Cattle Kingdom," he wrote, "was a world within itself, with a culture all its own, which though of brief duration, was complete and self-satisfying."[13] Between 1866 and 1886 an open range, characterized by cattle drives and nomadic cowboys, prevailed from north to south on the short grass Plains. Longhorn cattle, released into Mexico with the conquistadors, were raised in the lower Rio Grande Valley of Texas and herded north to markets in the Midwest. Abilene, Kansas, in 1867, was the first depot on the east-west Kansas Pacific Railroad to which cattle were driven. Soon cattle towns, such as Sedalia, Missouri and Wichita, Ellsworth, Dodge City, and Ellis, Kansas sprang up across the Plains as additional east-west railroads were constructed. The open range created cattle barons and itinerant cowhands during a two-decade expansion. In addition to whites, black and Mexican cowboys rode the Chisholm and other trails as highly skilled cowboys.

The drought of 1883 put the first crimp into the open range. Then, the nationwide panic of 1884, part of America's continuing cycle of boom and bust capitalist expansion, wreaked havoc in the cattle industry. The blizzards of 1885 dealt the final blow, ending the heyday of the open range. During the next decade, tensions and competition ensued as ranchers, sheepherders, and homesteaders all vied for access to the range. The overgrazed Plains were depleted of the perennial grasses that had supported one steer on every two acres and were seeded with less nutritious annuals that supported one steer per 5 to 10 acres. As perennials declined, wind and water erosion increased and topsoils were lost. Donald Worster calls the results an ecological disaster. "The 'tragedy of the laissez-faire commons'" he notes, "was one of the greatest in the entire history of pastoralism."[14]

The Farmer's Frontier

The farmer's frontier that followed the rancher's frontier was enhanced by the Homestead Act of 1862, which allowed people to claim title to 160 acres of land. For Webb, six technologies made it possible for farmers to control and manage the Plains environment. Following the Colt six-shooter (1835), which had subdued the Plains Indians, the second piece of technology that, from Webb's perspective, transformed the Plains was barbed wire. Invented in 1874 by Joseph Glidden on an old homestead in DeKalb, Illinois, double-stranded barbed wire was significant because it made the homestead possible. Fields could be fenced in and purebred stock could be kept isolated. The homestead and the vegetable garden could be fenced to keep out cattle, sheep, or deer. Barbed wire made homesteading possible, but spelled the end of the open range. The farmer's frontier of sedentary life and power, however, could now replace the nomadic pastoral frontier of the Indian and rancher.

A third piece of technology that subdued the Plains environment, according to Webb, was the windmill. Developed in Europe in the Middle Ages, the windmill was a part of the technological complex associated with the domestication of animals and crops. It accompanied the European advance onto the Plains, giving settled agriculture a stationary power-base compared to the shifting campsites of nomadic Indians. The windmill kept its vanes pointed into the wind, and the gear changed the rotary motion of the vanes into a vertical motion to pump subsurface water, which could then be stored in a watering trough or pond. Wherever a well could be dug and a windmill situated, water for cattle and crops could be controlled, steers could be fattened, and wheat harvested. The windmill, like barbed wire, made the homestead possible because water for vegetables and other crops could be stored and channeled to gardens and fields.

The fourth technology that subdued the Plains was the John Deere plow. Like the mill, the plow was one of the important pieces of technology that changed Western history. The earliest plows of southern Europe, pulled by oxen, successfully scratched the dry shallow soils of the Mediterranean region. But in northern Europe, the heavy, wet soils required wrought iron plows hitched to several horses, shared communally. Those plows were increasingly refined and brought to Puritan New England and the East Coast. Iron plowshares worked well in the East because the iron could scour the gritty, wet soil. But on the Great Plains, prairie grasses with matted roots and deep growing points resulted in very thick sod, with no grit to scour the moldboard. In 1846, John Deere took a piece of steel off a broken mill saw and attached it to a plow. It worked so well that he ordered high-grade steel from the Pittsburgh steel mills and began manufacturing the John Deere plow. By 1858 his factory was producing 13,000 steel plows.

The fifth technology that changed the Great West was the railroad. The first transcontinental railroad was completed in Utah in 1869 when the Central Pacific and the Union Pacific were joined. Windmills along the tracks drew up water for the steam engines. Railroads not only provided a means for immigrants to reach the Plains, but transported cattle and crops such as wheat and corn to eastern markets.

The sixth technology was the harvester. Mechanized harvesters, cultivators, threshers, and tractors became increasingly sophisticated, allowing the production of monocultures of wheat and corn. In Webb's narrative, technologies — such as barbed wire, windmills, plows, harvesters, and irrigation systems — enabled people to gain some measure of control over the arid, treeless, level environment on the Plains.

The progressive story of the Great Plains reveals that, from the mid-nineteenth century to the present, many cowboys and settlers successfully ranched, homesteaded, and gradually urbanized the Plains. Human production began to move away from subsistence-oriented homesteading and towards capitalist ranching and large-scale agribusiness. As corporate ranching replaced the free range and absentee landowners bought out small farmers, attitudes toward nature became increasingly profit-oriented, managerial, and scientific. Nature was subdued by technology; an ethic of human domination controlled development.

Narratives of Blacks and Women

The progressive story of the Great Plains often ignores the roles and contributions of minorities and women. African Americans and women contributed both ingenuity and skilled labor to transforming the Plains environment, but they were also victims of discrimination or were left out altogether in progressive histories, such as those by Turner and Webb. Including their stories adds complexity to the ways in which the Great Plains environment presented both challenges and opportunities to those disadvantaged by existing attitudes toward race and gender in American society.

African-American homesteaders were among those who moved to the Great Plains, escaping post–Civil War discrimination in the South. In 1877, a committee of five hundred black men documented beatings, violence, and continuing oppression, but were unable to achieve redress in the U.S. Senate. Former Georgia slave and Civil War veteran Henry Adams and others organized the "Exodus of 1879" to Kansas. During that year, some 20,000 to 40,000 poor blacks migrated into Kansas along the Chisholm Trail, seeking land and work on the western prairies. More than 5,000 African Americans worked the Chisholm Trail as skilled riders and cowhands, while other blacks became homesteaders. Some towns became primarily black, while in others, blacks mingled with whites at work, in schools, and in sports. In the early 1880s, a period of heavy rainfall induced large influxes of people, including many families of black homesteaders, westward to the High Plains of Nebraska and Kansas and then into eastern Colorado. But the drought of the mid-1880s forced large numbers to give up and leave. By 1910, however, approximately one million blacks lived in the western states.

Women were also important actors in Great Plains history. They were portrayed as reluctant pioneers in literature such as Ole Rolvaag's *Giants in the Earth* (1929), but as self-reliant homesteaders in other narratives, such as Willa Cather's *O Pioneers* (1913) and *My Antonia* (1918). The 1862 Homestead Act did not disqualify women and was not a gender-biased or discriminatory act. Women, like men, qualified for homestead entry if they were single and over 21 years of age, or as heads of households if they were widowed or divorced and had a family.

Homesteaders included single, divorced, married, and widowed women. Sheryll Patterson-Black described the experiences of women in each these categories in *Western Women in History and Literature* (1978). Single women, such as the five Chrisman sisters, sometimes homesteaded together on the plains. The life of divorcée and homesteader Mary O'Kieffe, mother of nine children, was described by her son Charley: "Mother herded cattle all day long in the broiling hot sun, so the children could attend a fourth of July celebration in a nearby community. The next morning around two A.M. I was born. No doctor, no nurse, no midwife, just Mother and God."[15] In 1884, Mary O'Kieffe left her lazy husband, packed up the wagon, hitched up the work horses, and, with milk cows and children in tow, headed out to western Nebraska to homestead the Plains. Wives were often homesteading partners, rather than reluctant pioneers. In some cases, a woman pretended to be single, but was actually married, placing her own sod house next to that of her husband, each filing claims for their respective 160 acres.

Widows were also successful homesteaders. One widowed woman living with her four children went to Nebraska and filed a claim. She supported herself, went to school with her children, learned to read and write, and began teaching school herself. She remarried, lost her second husband to a blizzard, and made final proof on both her and her husband's claims. When the Kincaid Act was passed in 1904, allowing larger claims, she filed for an additional 320 acres. A final example is that of Laura Cruz, who was part of the Oklahoma rush for Cherokee lands in 1893. She reconnoitered for a year to find the site she wanted. When the shot was fired and the race began, 100,000 participants mounted horses and charged across the land. Laura rode seventeen miles in fifty-nine minutes astride her horse, and staked out a piece of bottom land near a creek. She worked her claim for years, built a dugout house, dug a well by hand, lived on corn bread and rabbit stew, and made a living from selling eggs. Later, oil was discovered on her land. She was still living there at the age of 105.

Not only did women help to settle the Plains, often succeeding where men had failed, their technologies transformed the Plains environment. Women were skilled at building sod houses, digging wells, planting and harvesting crops, collecting dried buffalo dung for fuel, preserving fruit and vegetables in Mason jars, and herding cattle and sheep. Like men, they appreciated "nature" on the Plains, while also harnessing its unique features to make life both possible and comfortable.

The Dust Bowl of the 1930s

Donald Worster's *Dust Bowl* (1979) analyzes the factors that led from the Euro-American settlement of the Plains to the disastrous Dust Bowl of the 1930s. Worster's story, unlike the triumphal stories of Turner and Webb, begins with the ecologically evolved buffalo-grass biome, details capitalist expansion on the Plains, and ends with the defeat of the exodusters — refugees from the Plains, many of whom migrated west to California in the 1930s. How did the yeoman farmer become a Dust Bowl refugee?

The ecological cycles of wet and dry years west of the 100th meridian were alternately favorable and disastrous for ranchers and homesteaders. People settled and then retreated from the Plains, fooled by the abundance of crop growth during rainy years. The move to monocultures of wheat and corn, mechanized farming equipment, and land tenancy created a situation that was increasingly untenable from an ecological point of view. A disastrous drought in the 1890s devastated ranchers and farmers on the Plains. In the 1900s, however, when the rains returned, another major planting boom ensued.

By the early 1900s, most of Kansas, Oklahoma, the Texas panhandle, and eastern Colorado were settled. Wheat yields took off. In the eight Great Plains states, in 1899, 54 million acres of crops were harvested; in 1919, 88 million acres; and in 1929, on the eve of the depression, 103 million acres were harvested. Then the stock market fell, initiating the Great Depression. Wheat prices dropped to a dollar a bushel. Crops were left standing in the fields. The Depression was followed by the Dust Bowl of the 1930s, another major disaster for Great Plains' inhabitants.

According to Worster, the Dust Bowl derived from a 30-year history of monocultures that depleted the prairie grasslands of fertility, eroded the topsoil, and evaded the rainfall limits imposed by nature on the region. Its effects on both the land and its settlers were devastating. Families went on welfare and tried to survive by any means possible. Husbands and older children left home to search for work. Women took in washing, but tubs hung empty because no one had money to pay for laundry. Desperate for jobs, Dust Bowl refugees migrated to California in old Ford trucks with a trailer on the back and mattresses on the roofs. They came in battered-up cars, with chicken coops tied on the trunk and goats on the running board, camping with infants and young children beside the road. Many settled in the Bakersfield and Fresno areas of California's Central Valley, where they could work as migrants in the fields. Some 6,000 poverty-stricken people poured into California every month. Between 1935 and 1939 there were 300,000 Dust Bowl refugees, poignantly described by John Steinbeck in *The Grapes of Wrath* (1939). The result was a human and ecological disaster of the highest magnitude. The economic yardstick of growth at all costs began to be questioned, and the idea was promoted that human agriculture had upset the natural balance of the evolved grasslands.

The declensionist story of the Great Plains, in contrast to the progressive story, begins with a pristine grassland and ends in economic and ecological disaster. It organizes its data according to a different beginning, middle, and end than does the progressive story. Like the progressive story, it captures a real picture of society and nature, but its plot is downward rather than upward. As such, it exemplifies a different way to write history, one that is compelling from an environmental point of view.

Conclusion

Narratives of the settlement of Western frontiers, focusing on the Gold Rush and the Great Plains, can be told from several different perspectives. Stories of progress relate the triumph of humanity over a recalcitrant nature, ending with the extraction of wealth in the form of gold, cattle, and wheat from the land. Stories of decline emphasize the ecological impacts of gold-mining, ranching, and large-scale agriculture. Alternative stories challenge these two linear plots, inserting the complexities of the lives of Indians, women, blacks, Chinese, Mexicans, and buffalo into a larger picture of both failure and triumph, revealing the many dimensions of ecological transformation and potentials for resource use.

Notes

1. George H. Phillips, *The Enduring Struggle: Indians in California History* (San Francisco: Boyd and Fraser, 1981), 13.

2. John L. O'Sullivan, "Manifest Destiny," *United States Magazine and Democratic Review* 17 (July–August 1945): 215–20.

3. Walter Pigman, quoted in "Gold Diggers: Indian Miners in the California Gold Rush," by James J. Rawls, *California Historical Quarterly* 55 (Spring 1976): 28–45, at 41.

4. Samuel Ward, quoted in Rawls, "Gold Diggers," 33.

5. Quoted in Rudolph Lapp, *Blacks in the Gold Rush* (New Haven: Yale University Press, 1977), 21.

6. "A Great Gravel Mine," *The Nevada City (Calif.) Daily Transcript,* 30 July 1879, as excerpted in *Green Versus Gold: Sources in California's Environmental History,* ed. Carolyn Merchant (Washington, D.C.: Island, 1998), 110–11.

7. "A Great Gravel Mine," 111.

8. Robert Kelley, *Gold vs. Grain: The Hydraulic Mining Controversy in California's Sacramento Valley* (Glendale, Calif.: Arthur H. Clark, 1959), 57–58.

9. Joaquin Miller, *My Life Amongst the Indians* (Chicago: Morril, Higgins, 1892 [1890]), 55.

10. William Cronon, "A Place for Stories: Nature, History, and Narrative," *Journal of American History* 78 (March 1992): 1349.

11. Frederick Jackson Turner, "The Significance of the Frontier in American History" (1893), in American Historical Association, *Annual Report for the Year 1893* (Washington, D.C.: Government Printing Office, 1894), 199–227.

12. Walter Prescott Webb, *The Great Plains* (Boston: Ginn, 1931), 205.

13. Ibid., 206.

14. Donald Worster, "Cowboy Ecology," in *Under Western Skies: Nature and History in the American West* (New York: Oxford University Press, 1992), 41.

15. Charley O'Kieffe, *Western Story: The Recollections of Charley O'Kieffe, 1884–1898* (Lincoln: University of Nebraska Press, 1960), 3–4, quoted in *Western Women in History and Literature,* by Sheryll Patterson-Black and Gene Patterson-Black (Crawford, Nebr.: Cottonwood, 1978), 20.

6 Urban Environments, 1850 – 1960

Late nineteenth-century industrialization was a second major phase in the development of the United States economy, following the market revolution of the early to mid-nineteenth century. Urbanization is a core topic for environmental history because the increasing density of industry, transportation, and housing transformed both the land and the lives of urban dwellers. After the Civil War, industrialization — particularly in large Eastern and Midwestern cities — was accompanied by air, refuse, noise, and water pollution. This chapter deals with the evolution of environmental problems brought about by and associated with industrialization, urbanization, suburbanization, and the efforts of engineers, citizens, and legislators to deal with them.

Urbanization, Industry, and Energy

Living nature disappeared from everyday experience for most Americans by the mid-twentieth century. Amidst the urban world's masonry, steel, and asphalt, manicured remnants of greenery camouflaged the built environment's dependence on the natural world of forests, waters, air, and wildlife. Urbanization and industrialization, the twin processes propelling this shift, had moved into high gear following the market revolution of the early nineteenth century.

In 1800, only 320,000 people — or 6 percent of the population of the United States — lived in urban areas of more than 2,500 persons. Except for the northeastern seaboard, the country was largely rural. By the advent of the Civil War sixty years later, urban dwellers had jumped to some six million (20 percent of the total population); by 1920 to almost 54 million (about 50 percent); and by 1970 to more than 149 million (72 percent).

The rapid growth of urban population was fostered by the growth of manufacturing on an ever-larger scale, using ever-more potent methods for extracting energy from the natural environment. Human muscle and animal power were first augmented by the energy of falling water, as stored behind milldams and transmitted by waterwheels, drive shafts, and belting to the grindstones of gristmills in colonial and rural America. In the last decade of the eighteenth century, waterwheels drove the small textile mills that arose along the fall line of coastal New England. This early hydropower peaked in the 1820s and 1830s along the Merrimac River in northeastern Massachusetts and southern New Hampshire. Here, large mills, which had improved on British machinery for spinning and weaving, vibrated with energy drawn from White Mountains snowmelt rushing to the sea. Similarly, in western New York, the thundering falls of the Genessee River energized a clattering assemblage of large, mechanized flourmills.

The renewable but all too variable energy of falling water was soon supplanted by the nonrenewable energy of ancient plant life, as fossilized in coal and transmitted reliably to machines by steam engines. Steam, generated more efficiently by fossil fuel than by firewood, drove both production and propulsion. The stationary steam engine freed energy-hungry industries from scarce water-power sites to seek inland locales with better access to raw materials, labor, and markets, while the steamboat and the steam locomotive brought inputs to their doors and delivered outputs to their customers. Coal mining itself became a major industry, exploiting at first Pennsylvania's limited supplies of clean-burning anthracite (or hard) coal, and soon moving west to mine abundant supplies of sooty-burning bituminous (or soft) coal, from West Virginia to the Rocky Mountains. Coal quickly replaced wood for heating and cooking in urban homes.

Iron and steel superseded the infrastructure of timber and masonry when coal combined with iron ore to usher in a new industrial order. Coal-fueled ironworks and open-hearth steel mills, as well as the foundries and machine shops that fabricated their output into rails, cars, and locomotives for railroads, girders for bigger buildings and bridges, a new generation of heavy machinery for factories, pipe for water and sewers, and ironwork for a multitude of other uses. After British inventor Henry Bessemer's converter was introduced to America in the 1870s, coal-fueled production of high-grade steel spread from the upper waters of the Ohio River in western Pennsylvania to the shores of the Great Lakes. By 1890, this mode of exploiting a bounteous natural endowment enabled the United States to outstrip British production and claim industrial leadership of the world. While "the old nations of the world creep on at a snail's pace," exulted steel magnate Andrew Carnegie, "the Republic thunders past with the speed of the express."[1]

As the age of coal and iron culminated at the end of the nineteenth century, two new forms of energy — petroleum and electricity — promised even greater productivity in the twentieth. A cleaner, nonrenewable fossil fuel became available when Samuel Drake struck oil at western Pennsylvania's Oil Creek in 1859. Petroleum was used initially for oil lamps and stoves in homes, for street lamps, and to lubricate machinery. After new oil strikes in Texas, Oklahoma, and California at the turn of the century, production increased dramatically, and fuel oil, which needed little refining, began to compete for coal's major uses. In California, the leading oil-producing state between 1900 and 1926, fuel oil supplemented the Sierra's hydropower to fuel furnaces, factories, ships, and locomotives. By mid-century, pipelines radiated from the oil fields, carrying crude oil to tidewater refineries, and natural gas for home heating and factories to cities.

The growing popularity of automobiles in the early twentieth century enormously expanded the demand for petroleum, both as gasoline for fuel and as asphalt for paved roads. The automobile itself was a revolutionary economic force, spawning new industries to mass-produce cars, trucks, rubber tires, and road building machinery, and multiplying demand for steel, glass, and cement. Its environmental effects were also profound, fouling the air with noxious emissions from vehicles and refineries and paving over much of the landscape with highways and shopping malls.

Meanwhile, energy's portability and versatility had been greatly enhanced by its conversion to electricity. Whereas coal and petroleum had to be transported laboriously from mine or wellhead to user, electrical energy whizzed effortlessly, instantly, and far less expensively through transmission lines to both industrial producers and individual consumers. The remarkable Thomas A. Edison (1847–1931) developed an array of ingenious electrical applications, most notably a practical incandescent lamp that dispelled "night and its darkness ... from the arena of civilization" by lighting streets and homes. The "wizard of Menlo Park" became a popular hero by also inventing the multiplex telegraph, an improved telephone transmitter, the mimeograph machine, the phonograph, the microphone, and the motion picture.[2]

New industries arose to transform urban life by mass-producing these and countless other electrical applications for home, office, transport, and factory. Electric power propelled elevators and trolleys. Electrical smelting of aluminum, developed by Oberlin undergraduate Thomas Hall in 1888, became a major factor after 1930 in making modern aviation possible. Americans increasingly experienced the world through the electrically transmitted sounds and images of radio, motion pictures, and television. As the twentieth century ended, a digital computer revolution opened vistas of instant, universal access to electronically transmitted information and to un-limited production by electronically programmed robots.

Hydroelectricity was hailed as "clean" energy because it resuscitated re-newable and nonpolluting water power to drive its generators. This marked a shift, according to historian of technology Lewis Mumford (1895–1990), from the paleotechnic age of coal to the neotechnic age of electricity. The shift came at the cost of constricting great rivers by serial dams, from the Chattahoochee, Catawba-Santee, and Tennessee in the East to the Colum-bia and Colorado in the West. With hydroelectric sites fully exploited by the late twentieth century, an insatiable appetite for electricity was increasingly supplied by coal-fired generating plants that spewed contaminants into the atmosphere across many states, degrading Appalachian forests, dangerously depleting the earth's ozone layer, and threatening global warming.

At the dawn of a new millennium, the environmental costs of existing sources of energy were becoming all too apparent, and prospects for new and cleaner sources were clouded. The extravagant hopes held out for nu-clear power were dying under repeated assaults — near disaster at Three Mile Island, Pennsylvania; total disaster at Chernobyl, Russia; leaking or abandoned generating plants; and inability to dispose of nuclear wastes safely. Nor had nonpolluting and renewable forms of energy, such as solar power and wind power, been able as yet to compete substantially in the market. Meanwhile the citizenry, who would ultimately determine Ameri-can society's relation to its natural base, had been experiencing the environ-mental stresses that impinged most directly on urban-industrial life.

Industrial Cities and Labor

During the nineteenth century, industrialization transformed the urban landscape. Originally the urban system had been devoted mainly to trade, from the wharves and warehouses of the great Atlantic ports, to the whole-sale merchandising houses of interior towns along major transportation routes, to the country stores around which rural villages straggled. Inter-spersed among these commercial facilities were skilled and independent "mechanics" or artisans, who owned their small home workshops and spe-cialized in producing by hand the various commodities in demand. The in-troduction of large, mechanized factories gradually forced mechanics into unskilled wage labor, created great disparities of wealth, and segregated ur-ban geography into separate sectors — offices for business, mansions for the affluent, factories and tenements for workers, and slums for a new urban underclass of the dispossessed.

Industrialization also gave rise to different types of cities. In the old Atlantic ports, various new industries mixed with established commercial activities. New York City became a center for shipbuilding, steam printing, publishing, and clothing manufacture; Philadelphia turned out machinery and locomotives; and Richmond supplemented tobacco processing with a major ironworks, which proved an indispensable linchpin of the Confederate war effort. Although most cities and towns spawned several industries, others specialized in some newly mass-produced commodity. In Massachusetts, Lowell and Lawrence sprang up as cotton-mill cities, while rural Lynn evolved into a center of mechanized shoemaking. Elsewhere, Rochester milled flour, Louisville distilled whiskey, and Cincinnati became the "Porkopolis" of the West. In areas devoted to resource extraction, towns and villages were variously dominated by mines and smelters, by lumbering and sawmills, or by mechanized fishing and canneries. The growth of some large cities was propelled by particular industries: stockyards and meatpacking in Chicago, steel in Pittsburgh, automobiles in Detroit, and tires in Akron.

Demand for industrial labor attracted a mass migration of ethnically diverse peoples, both from the rural countryside and from overseas. Throughout the nineteenth century, sons and daughters deserted foundering subsistence farms in droves to seek jobs in rising industrial centers. At mid-century, large influxes of Irish immigrants provided much of the labor for canals and railroads while establishing their largest enclave in Boston. Germans settled heavily in Cincinnati, St. Louis, and Milwaukee. Later in the century, southern and eastern European peoples arrived in New York City and other northeastern communities. Although many boarded railroads to claim homesteads on the Great Plains, 80 percent of the "new immigrants" settled in the Northeast, primarily in urban mosaics of polyglot ethnic neighborhoods. The ethnic and racial diversity of urban America was rounded out in the twentieth century by Asian immigration to the West Coast, a massive flight of African Americans from southern oppression to northeastern cities, and an influx of Mexicans to the Southwest.

Working in cities posed living and laboring hazards for the urban poor. Immigrants to cities often lived in poorly constructed, drafty, unsanitary tenements and worked in unhealthy, hazardous environments. In Chicago, Jane Addams (1860–1935) spearheaded a middle-class movement to improve living conditions for poor tenants by living and working among them. She founded Hull House, documented in her *Twenty Years at Hull House*

(1910), where members instituted childcare, built playgrounds, and initiated cooperative housing arrangements for working women. She and others pressured the city to extend garbage and refuse collection to poor neighborhoods, to halt animal slaughtering in tenements, and to remove dead carcasses. Addams's colleague, bacteriologist Alice Hamilton (1869–1970) publicized the spread of infectious diseases resulting from piles of garbage and untreated sewage and lobbied to alleviate workplace hazards. She promoted the field of industrial medicine in America and worked with labor leaders to identify "industrial poisons," such as carbon monoxide in steel mills, and work-related hazards, such as "phossy jaw" that disfigured match factory workers using white phosphorous. Her work extended to the effects of toxic chemicals found in petroleum, solvents, and heavy metals and resulted in her classic text *Industrial Poisons in the United States* (1925), opening up the field of occupational health.

The City as Wilderness

In reaction to all these changes, a new perception of the city as wilderness emerged. In 1898, writer Robert Woods described *The City Wilderness* as a dark, dismal, depressing place filled with squalid alleys and poverty-stricken inhabitants. Novelist Upton Sinclair's *The Jungle* (1905), written to expose the unsanitary conditions and working-class misery in Chicago's meatpacking industry, depicted the urban environment as dingy, smoky, and rancid with stockyard odors. Booth Tarkington's *The Turmoil* (1914) characterized the city as a locale dedicated to growth and the production of wealth at any cost to its inhabitants. Evil wilderness was embedded in the city, by contrast with the good wilderness of clean air and pristine, unblemished nature found in the Appalachian and Rocky Mountains.

In *American Hunger* (1944), novelist Richard Wright recalled vividly his impressions of the city wilderness on reaching Chicago in 1927 as an African-American refugee from southern racism: "My first glimpse of the flat black stretches of Chicago depressed and dismayed me.... Chicago seemed an unreal city whose mythical houses were built of slabs of black coal wreathed in palls of gray smoke.... The din of the city entered my consciousness.... I looked northward at towering buildings of steel and stone. There were no curves here, no trees, only angles, lines, squares, bricks and copper wires.... I was learning already ... the strain that the city imposed upon its people."[3]

For many white urbanites, negative feelings about the city wilderness were fueled by racist reactions, not only to blacks like Wright, but also to dark-haired, dark-eyed immigrants from southern Europe. Newspapers and cartoonists portrayed impoverished refugees from Mediterranean farms and villages as living in pestilent squalor with the animals they brought. Newcomers, who could have squatted on public lands and staked out a homestead farm in an earlier era, became urban squatters while they looked for housing and jobs. When they camped by necessity alongside their poultry and goats in city parks and on streets, they were stigmatized as uncivilized threats to the safety and health of the community.

Urban businesses also seemed threatening to much of the citizenry as they increased in size and economic power. Editorial cartoons portrayed corporations, monopolies, and trusts as gigantic octopuses with tentacles reaching out to envelop the entire world. The turn-of-the-century Progressive movement, which looked to efficiency in industry and government, tried to restrain corporations by laws enforcing competition. Congress passed the Sherman Antitrust Act outlawing monopolies and conspiracies in 1890, and a decade later President Theodore Roosevelt built a popular mandate against abuse of power by the American oil industry. In 1911, the Supreme Court found Standard Oil in violation of the Sherman Antitrust Act and ordered its dismantlement.

Yet the problems of the industrial city had deeper roots than corporate greed and ethnic prejudice. Its health was entwined, as few understood, with the health of the natural environment it was steadily despoiling and polluting. "The idea of the city as animate — if not 'natural' — " according to urban historian Martin Melosi, "is essential for an understanding of urban growth and development. Cities are not static backdrops for human action ... : they are ever mutating systems." Cities both depend on and modify the physical environment, he continues: "Urbanization removes much of the filtering capacity of soil and rapidly channels precipitation into available watercourses, thus encouraging flooding. City building affects the atmosphere by increasing airborne pollutants and also creating 'heat islands' where temperatures are greater than the surrounding area. Various urban activities produce huge volumes of waste products that require complex disposal mechanisms."[4] Unfortunately, as geographers Thomas R. Detwyler and Melvin G. Marcus note, the air and water extracted from the ecosphere by cities, "are not returned to the ecosphere in the same condition in which they were received."[5]

By the turn of the nineteenth century, the intertwined ecological effects of population growth, industrialization, urbanization, and their concomitant pollution called forth new technologies, citizen movements, and legislation. According to environmental historian Joel Tarr, "The construction of technological networks in American cities for the transmission of water, wastewater, power, communications, freight, and people dramatically altered the context of city life and the effect that urban centers had upon their surrounding environments."[6] Beginning with the recognition of air pollution, the environmental problems of the industrial city soon extended to garbage, noise, and water pollution.

Air Pollution

Concern about quality of life issues originated, Melosi points out, not with the environmental movement of recent decades, but "with the more obscure antismoke campaign of the 1910s."[7] The history of air pollution moves from acceptance of smoke to campaigns against it. Early in the age of coal and iron, urban industrial centers often took pride in black smoke as symbolizing progress and the triumph of "civilization." At worst, smoke was an inescapable nuisance that had to be endured, a necessary cost of the creation of jobs, the rise of the middle classes, the improvement of living standards.

The symbolic meaning of smoke turned negative by the late nineteenth century in areas where industries abounded and coal was used for domestic heating. Barrels of cinders and ashes lined streets and walkways, where they might remain for weeks owing to inadequate collection methods. People with rain barrels on their roofs found the water contaminated with soot; city trees began dying. Breathing polluted air had disastrous effects on the human respiratory system. Smoke and soot irritated the nose, throat, and lungs, giving rise to or exacerbating a multitude of ailments ranging from the common cold to asthma, bronchitis, pneumonia, black lung disease, tuberculosis, diphtheria, and typhoid fever. The poor, immigrants, and blacks who lived and worked in crowded inner cities were especially prone to diphtheria and pneumonia.

Smoke was most oppressive in towns dominated by coal-burning industries. Rebecca Harding Davis described life in the iron mill town of Wheeling, West Virginia, in 1861: "The idiosyncrasy of this town is smoke. It rolls sullenly in slow folds from the great chimneys of the iron-foundries, and settles in black, slimy pools on the muddy streets.... clinging in a coating of greasy soot to the house-front, the two faded poplars, the faces of the passersby. Smoke everywhere! ... From the back-window I can see ... the slow stream of human life creeping past, ... breathing from infancy to death an air saturated with fog and grease and soot, vileness for soul and body."[8]

Women expressed concern over air pollution because it affected their abilities to uphold standards of cleanliness. A Milwaukee woman reported, "It's impossible for me to have my laundry work done at home because of the smut that falls on clothes while they're drying," while another noted, "It is bad for the furniture, for the clothing, for the health, and for the temper."[9] Psychological consequences of living in smoky areas were also noted. A doctor asserted that "women living in sunless rooming houses and attired in somber, dirty clothes are prone to be irritable. They scold and whip their children and greet their husbands with caustic speech."[10] Domestic deterioration led to excessive drinking, the doctor stated, a situation that caused men to flee to bars, seeking "the cup that cheered." Physical dirt was considered to be moral dirt.

A major clean air reform movement ensued. In the late nineteenth and early twentieth centuries, civic groups took up the issue. Chambers of Commerce in large cities such as Cleveland, Chicago, St. Louis, Toledo, and Pittsburgh began to study the problem. Pittsburgh industrialist Andrew Carnegie (1835–1919) urged the city and state to deal with smoke pollution. A Committee on Smoke Prevention recommended that legislation be passed. "The way to abate smoke is to abate it," trumpeted an inspector in Toledo. "There are three remedies. Fine the violators, second fine them again, third, keep fining them until they are bankrupt or repentant."[11] But even with the passage of new laws, the courts imposed only meager fines.

Women were especially motivated to clean up cities. According to historian Suellen Hoy, "Women were a significant force in the movement to improve living conditions in the industrial cities of the United States. Motivated by a desire to protect their homes and families, large numbers of women believed that they, the nation's homemakers, could make a special contribution to the housekeeping practices of their communities."[12] They turned their energies to abating air, water, noise, and garbage pollution. During the late nineteenth and early twentieth centuries, women's groups were instrumental in passing abatement ordinances that regulated smoke outputs from factories. In St. Louis, the Women's Organization for Smoke Abatement pressed for compliance with city smoke ordinances. In Cincinnati, women's groups appealed to the mayor to act on smoke abatement. In Chicago, an antismoke league was formed that was chaired by women. The league worked with smoke inspectors, helping to coordinate efforts to locate and halt the worst polluters.

Along with increased pressure from civic reformers and women's groups, engineers sought technical solutions. Engineers and smoke inspectors were active in every major city by 1912, and engineers themselves frequently made rounds to inspect factories. They pushed for the use of better equipment, such as stokers and banked and down-draft furnaces, and urged that the height of chimneys be raised so that smoke was carried higher into the atmosphere. The results, however, were limited by the technological options available. While some success was achieved in smoke abatement, and while technological advances continued, the problem continued in most urban industrialized areas. During World War I, with the country's efforts focused on increasing industrial output, progress was especially limited.

Garbage

A second problem for urban areas was garbage disposal. On farms the practice was often to throw buckets of gray water or garbage out the door. But when that method of disposal was transferred to cities, garbage and rubbish rapidly piled up in alleys, where it caused odors and created health hazards. In large cities, such as New York and Los Angeles, garbage and refuse accumulated faster than it could be collected. Collection was haphazard and primitive. Street teams collected garbage, and large barges carried it out to sea. Or refuse collectors simply dumped it in fields or lots beyond city limits.

A major concern for cities was the removal of refuse from horse-drawn trolleys. Miles of rails laid through cities were polluted by horses whose equine wastes created health hazards and odors, as each horse discharged several gallons of urine and 20 pounds of manure per day. By the mid-1880s, 100,000 horses and mules were pulling 18,000 horse cars on 3,500 miles of track. By 1900, there were 3.5 million horses in cities. In Chicago alone 82,000 horses produced 600,000 tons of manure per year. At night horses had to be stabled or kept in fields and barns. During the day, when working the streets, they deposited piles of manure along the rails. Such sights and smells contributed to characterizations of urban areas as wilderness.

What was done to combat garbage? The American Public Health Association undertook a major study of refuse, looking to Europe for experience and guidance in dealing with waste. Women's groups formed health protective associations in New York and other cities. The General Federation of Women's Clubs (G.F.W.C.) advocated urban cleanup and promoted better methods of street cleaning. In Chicago, Jane Addams's Hull House lobbied for clean streets and the excavation and removal of years of accumulated waste. Cities also created cleaner, healthier recreation spaces such as urban parks. Such efforts, combined with new laws regulating sanitary conditions, contributed to civic improvement. By 1900, ninety-four per cent of cities had refuse collections.

Street cleaning was initiated in major urban areas. Blacks and immigrants were employed to push refuse carts on wheels and clean up streets, but early attempts were sporadic. Some parts of the city remained laden with trash, owing to lack of coordination. Another method was to institute traveling garbage burners, a type of a steam engine on wheels with a long smokestack. Workers moved through the streets, shoveling garbage into the furnaces, burning it as they went. Horse-drawn water carts went through city streets flushing them of refuse even as the horses themselves added to the problem. Such systems were employed in New York, St. Louis, Cincinnati, and Los Angeles, among others.

A major effort in garbage and refuse reform was undertaken in New York City by George Waring, a cleaning commissioner known as "The Apostle of Cleanliness," who assumed office in 1895. He advocated source separation. Every household had containers in the kitchen and basement for garbage, paper, and ashes that were taken to the street on pickup days. Waring hired young engineers to design the collection system and employed urban workers to clean up neighborhoods. He created a cadre of "White Wings," 2,000 white-uniformed adult cleaners, who patrolled the streets with brooms, buckets, and can carriers. He raised their pay and improved their working conditions. He also formed a juvenile street-cleaning league, in which he enlisted 500 young people to disseminate information on sanitation to citizens.

Additionally, Waring created municipal refuse-sorting plants staffed by immigrant workers, the poor, blacks, and other minorities. They removed salvageable materials, sorted paper, and picked over goods that could be sold or recycled. Workers experienced a mixed blessing, caught between the advantage of employment and the consequences of unsanitary working conditions. Exposed to smoke and dust while working in sorting plants, they were also condemned to living in substandard apartment buildings.

By the 1890s, engineers began to address problems of garbage with increasing effectiveness. A Society for Street Cleaning and Refuse Disposal was formed. Numerous books and professional pamphlets were written on the topic of pollution; analyses and inventories were compiled; and citizen groups were formed, all of them reflective of early responses to the problem of polluted cities. Engineers constructed the first American garbage furnace in New York City on Governor's Island in 1885, and other cities, such as Montgomery, Alabama, followed suit in 1911. Carts loaded with refuse and garbage ascended a ramp and deposited their contents on an unloading deck at the top of the building. The refuse was then lowered and burned. Engineers also invented better garbage scows. Ashes and rubbish were loaded onto barges and taken outside New York Harbor, where they were dumped into the ocean, a method still used illegally to dispose of garbage and medical wastes. The rationale was, "dilution is the solution to pollution."

Noise Pollution

A third problem for urban development was noise from new forms of transportation. Industrial cities continually generated huffings, clankings, screechings, and smokings that went on for 12 to 14 hours a day and sometimes around the clock. Brakes, gears, motors, and bells mingled with the noise of street traffic and elevated railroads. Noise was a major issue for those in inner cities. Poor women in Philadelphia complained: "What we can't stand is the noise. It never stops. It is killing us. We work hard all day and need sleep and rest at night. No one can sleep till midnight and all the noise begins again at five. Many of us have husbands who work all night and must get their sleep during the day." They told wealthy Civic Club members, "You can get away from the noise during the summer, but we cannot.... Now what can your civic club do for us?"[13]

What was done about noise? Noise reform was already under way in the early twentieth century. Civic clubs formed committees to deal with unnecessary noise and worked to prepare legislation. Antinoise groups cooperated with local boards of health. They argued that sound endangered life and health, asserting that if people could be prevented from polluting water supplies, they could be prevented them from congesting the air with noise. Police departments assisted in abating unnecessary noise. In Atlanta, the Chamber of Commerce's antinoise committee, appointed in 1918, received 140 letters of complaint and secured the cooperation of the police department. In Baltimore, the city's medical society formed an antinoise commission, which compiled a noise inventory and drew up legislation to abate the noise.

In New York City, a special concern was mounted about whistles from tugboats and barges on the rivers that outlined the five boroughs. Julia Bar-

nett Rice, after having raised six children, began a campaign against shriek-
ing tugboat whistles and riverboat noise. She collected 3,000 signatures
from rich and poor, from hospitals and boards of health, and persuaded her
congressman, William Bennett, to push a law through Congress. Known as
the Bennett Act of 1907, it regulated boat whistling in harbors and prohibited unnecessary whistling. But enforcement of laws regulating ephemeral,
nuisance noises was almost prohibitive. Despite laws and police enforcement, compliance with antinoise regulations was virtually impossible, and
noise continued to be a major urban problem.

Water Pollution

A fourth major problem faced by burgeoning urban areas was access to
clean water. Historian Theodore Steinberg notes the strain put on rivers and
streams by the polluting effects of industry: "Industrial transformation ...
created a new ecology of its own with far-reaching effects on the water quality of [a] region's rivers.... Rivers were used to generate energy for production, to carry off human and factory waste, and ultimately to supply cities
with water for domestic use."[14] Contaminants such as sulphuric acid, soda
ash, muriatic acid, lime, dyes, wood pulp, hair, flesh, and hides polluted the
early mills of the Northeast. During the nineteenth century, prior to the
construction of municipal water and sewer works, keeping water clean and
disease-free for drinking was a major problem. Contaminated water flowed
into rivers and oceans and raw sewage was pumped into waterways. "Years of
abuse and neglect," states Steinberg, "had by the 1870s left a rather disturbing mark on the waterscape. In that decade, as concern emerged over public health ... some state officials began to wonder about the costs of water
pollution."[15] After 1880, many cities began to install sewer systems, but raw
sewage continued to be dumped into bays and rivers.

Water-related infections occurred throughout the nineteenth century.
Cholera outbreaks devastated the urban economies of cities such as New
Orleans between 1846 and 1850. In Memphis, in the 1870s, epidemics led
to improvements in sanitation and sewage systems. By 1873, it was discovered that the vibrio that caused cholera could be killed if sufficient sunlight
was allowed to fall on the water supply. If reservoirs were opened and exposed to light, the vibrio was prevented from reproducing. Typhoid fever
was another waterborne disease, spread by flies lighting on fecal contaminants, by impure, unpasteurized milk, and by contaminated raw fruits and

vegetables. A rate of typhoid above 15 to 20 per 100,000 indicated a polluted water source.

Another water-related disease was yellow fever, caused by the mosquito *Aedes aegypti*. It was especially dangerous in warm, humid cities and rural areas, but reached into northern cities as well. The Massachusetts Bay Colony experienced an outbreak of yellow fever in 1647, while in Philadelphia in 1793, 10 percent of the population died from the disease. In New Orleans in 1853, some 8,000 people died, with another 4,000 fatalities in 1878. In Memphis, Tennessee, more than 5,000 people in a population of 38,000 died during the 1878 outbreak. By 1905, however, the virus was eradicated from the United States.

Municipalities began to build clean water supplies so that drinking water could be kept free of contamination. Large cities created gravity-flow reservoirs above the city so that water could flow into distribution systems. In 1801, Philadelphia built the Schuylkill Waterworks, the first large municipal system that used steam to push water through its pipes. In 1835, New York City constructed the Croton River Dam forty miles north of New York, a gravity aqueduct consisting of tunnels that conducted fresh water into the city. In California, water works supplying Los Angeles and San Francisco were constructed toward the end of the century, initially by leading water from coastal mountain ranges. In the early twentieth century, they were followed by larger projects, such as the Owens River Valley project to supply Los Angeles and the Hetch Hetchy Valley dam for San Francisco, both enormously controversial at the time. Water projects were thus a major part of the great era of publicly financed municipal water and power systems for large and growing cities in the name of the public good.

The Sanitary City

The links between clean water as an ecological input to the city and sewage and garbage as outputs became regional concerns after World War I as the economy moved from an industrial to a consumer base and populations burgeoned. "Sanitary services," Melosi notes, "provide water for domestic and commercial uses, eliminate wastes, protect the public health and safety, and help to control many forms of pollution."[16] Public waterworks grew from more than 10,000 in 1932 to more than 14,000 in 1940, and regional water districts formed to deliver service to customers. The Public Works Administration (PWA) during the New Deal years financed some

2,500 water projects. Public health was of paramount concern. Water treatment facilities improved through better filtration, control of chlorination, and aeration. With better treatment, deaths from typhoid fever declined from 33.8 per 100,000 in 1920 to 3.7 by 1945. Water outputs also received more consideration, and regional water and sewage treatment facilities were increasingly linked under joint management and administration.

Sewer systems, like water systems, received government support during the New Deal era. Between 1933 and 1939, the PWA funded approximately 65 percent of the nation's new sewage disposal plants. Early methods of disposal by dilution moved toward a combination of bio-physico-chemical treatments. Sewage was treated by processes of clarification, oxidation, and disinfection. Technologies included filtration through sand and gravel beds, sludge digestion and drying, chemical precipitation, and the use of septic tanks and contact beds. By 1945, about 63 percent of the U.S. population lived in communities with sewage treatment plants.

During the interwar period, the concept of pollution expanded from sewage and domestic wastes to include industrial effluents. Untreated industrial wastes were discharged into sewers, which flowed into rivers, lakes, and oceans. Increasingly, public health officials raised doubts over whether dilution was sufficient to dissipate polluting chemicals. Pollutants included mineral matter in coal and iron industries; acids and salts from mines and oil wells; lead from slag; benzene, toluene, and naphtha from oil distilleries; sulphites from pulp mills; grease and oils from manufacturing plants; arsenic from paints; animal refuse from meatpacking plants; and hot liquids and inflammable wastes from chemical processing plants. Effects included waterfowl and fish kills, corroded sewer pipes, gas leaks, explosions in pumping and treatment plants, and reduction of dissolved oxygen in water. Questions arose over whether industries should be allowed to discharge such wastes into municipal sewer systems and how to assess costs and levy taxes.

After World War II, thousands of new chemicals entered the waste stream. Petroleum distillates, detergents, and pesticides produced burning rivers, foaming streams, and dying lakes. In 1959, the Cuyahoga River between Cleveland and Akron burned for eight days. In 1963, nationwide pollution from municipalities, industry, and agriculture killed 7.8 million fish. In 1965, Lake Erie was so polluted with coliform bacteria, phenols, iron, ammonia, chlorides, and phosphates from industrial processes and municipal sewage that it was ecologically dead.

In the postwar period, federal legislation began to address industrial effluents. The Water Pollution Control Act of 1948 provided loans to local interstate agencies, municipalities, and states to reduce water pollution. The act was amended in 1956 and 1961 to provide grants for municipal sewage treatment plants for domestic and industrial wastes and to continue a cooperative approach between the federal government and the states. Post–World War I pollution problems, however, led to widespread unease over the nation's ability to resolve the interconnected problems of inputs and outputs to urban water supplies, setting the stage for the emergence of the age of ecology in the 1960s through 1990s.

From City to Suburb

The sense of the city as a polluted arena helped push people into suburbs, beginning at the turn of the nineteenth century and accelerating in the post–World War II period. Mass transit made the suburbs a new option for the middle class. People could now work in city business districts but live outside of town. Horse-drawn omnibuses and commuter railroads in the 1830s had been followed by electric streetcars (1880s), elevated trains (1890s), and subways (1900s), creating tentacle-like suburban patterns in which people lived within a mile of a mass-transit station. After World War I, the automobile's use for commuting diffused those early patterns, allowing workers to live anywhere that roads and later freeways reached, while trucks made possible the location of commercial distribution centers on urban edges. Suburbs created safe, sanitary environments for those who could afford the homes and travel costs, but condemned those who could not to crowded slums. The big boom in housing expansion occurred after World War II. Single-family housing starts ballooned from 114,000 in 1944 to post–World War II highs of 937,000 in 1946; 1,183,000 in 1948; and 1,692,000 in 1950. Both highways and automobiles supported the growth in suburban living. The interstate highway system enacted by President Dwight D. Eisenhower's signing of the Federal-Aid Highway Act in 1956 created multilane highways accompanied by automobile-generated sprawl, traffic congestion, and environmental pollution.

The automobile suburb fostered privately owned, single-family homes built on larger lots, wider streets, and with greener lawns than in urban centers, giving rise to what Kenneth Jackson has called *The Crabgrass Frontier* (1985). "The dream of a detached house in a safe, quiet, and peaceful place," he argues, "has been an important part of the Anglo-American past and a potent force in the development of the suburbs."[17] Suburban homes followed the designs of architects such as Andrew Jackson Downing (1815–52), who stressed the need for affordable functional designs, and Frank Lloyd Wright (1867–1959), who focused on single-story ranch homes with street-accessible carports and garages.

In addition to affordable housing, private property, and the automobile, race was a decisive factor in the evolution of American cities and suburbs. Underlying prejudice, fear of urban crime, and crowded conditions, along with court-mandated school desegregation in 1954, resulted in a flight to the suburbs by white affluent families. Tract developments, single-story schools with large playing fields, and suburban shopping malls characterized post–World War II construction patterns. "Racial prejudice," states Jackson, "and a pervasive fondness for grass and solitude, made private and detached houses affordable and desirable to the middle class, and they produced a spread-out environment of work, residence, and consumption."[18] Low-density housing and the internal-combustion engine, however, combined to foster waste in land and fuel and to increase air and water pollution. Racial prejudice, wage and housing discrimination, and the lack of mass-transit systems kept minority workers in inner cities, where pollution and hazardous wastes threatened both the natural environment and human health.

Minorities and Pollution

"The age of ecology," according to environmental historian Andrew Hurley, "coincided with changing forms of environmental degradation that discriminated along racial and class lines."[19] In his study of environmental inequalities in Gary, Indiana, Hurley shows how patterns of waste disposal differentially affected black and white, rich and poor, white collar and blue collar, urban and suburban populations. The vast industrial output during World War II drastically increased inner city pollution. People in minority and poor communities, adjacent to industrial areas, were affected by pollution to a far greater extent than middle-class and well-to-do whites. The case of Gary's Steel industry illustrates post-War pollution problems.

By the end of World War II, Gary had become one of the nation's most polluted cities. Steel manufacturing was located along the southern shore of Lake Michigan. Here, U.S. Steel Corporation used the Grand Calumet River, which flowed through Gary and into Lake Michigan, for water in the production of coke and finished steel and for waste disposal. Most of the post–World War II population was centered in the area just south of the plant, where the Grand Calumet flowed through the city and where most black steel workers lived.

U.S. Steel emitted 100,000 tons of particulate dust into the air every year, much of which settled over north Gary. Coke plants released small carbon particles, while open hearth furnaces, used for converting the iron into steel, emitted iron particles. The air was polluted with sulfur dioxide, nitrogen oxide, and hydrocarbons. Winds carried the particles southward into the areas where workers lived.

North Gary likewise bore the brunt of the factory's water pollution. U.S. Steel dumped toxic liquid wastes into the Grand Calumet River, including ammonia, cyanide, phenols, and oils that flowed into Lake Michigan. In northwestern Gary, children played in the river, while the water utility, located just east of the waste stream, supplied well water polluted by seepage to several thousand families in northwest Gary.

Hazardous waste was also a major pollutant. U.S. Steel used its own vacant land for disposal sites, and by the 1970s was burying 600,000 barrels a year of hazardous waste. After the 1960s and 1970s, private scavenger companies began to buy up vacant lots throughout the area, burying the waste in lots, marshes, and trenches left over from highway construction. Wastes were also disposed of within the city limits in fifteen private dumps. Chemicals, pesticides, and solvents were introduced into the waste stream — PCBs (polychlorinated biphenyls); insecticides such as DDT; and solvents, such as benzene and toluene — chemicals that were subsequently reclassified as carcinogens.

Waste sites were differentially located in poor areas. Wealthier people, some of whom managed the plants and industries, lived in Miller, located east of the downtown area, and in Aetna and Glen Park to the south. These areas were least affected by contaminants. Glen Park and Miller each had one hazardous site; Aetna had none. By contrast, many sites were located in the inner city and in West Gary, where wage workers and the poor made their homes.

By the 1950s, Miller and West Gary had become the two fastest-growing areas. Just after World War II, blacks had constituted less than 30 percent of the population, but by the 1980s they made up 71 percent. They began to move into white areas, initiating white flight and racial discrimination, such that western and central Gary became primarily black while other areas became white. By 1980, blacks comprised 90 percent of the population and Hispanics made up 8 percent. They earned good salaries from the steel works, and therefore most of them chose to remain and live close to their jobs.

The history of urban environmental problems going back to the turn of the century and their differential impact on minorities and the poor illustrates the increasing complexity of environmental problems as populations moved from rural to urban areas and from other countries into the United States during the era of industrialization.

Conclusion

As the United States became increasingly urbanized by the turn of the nineteenth century, problems of urban pollution became widespread. Smokestack industries polluted the atmosphere, garbage piled up on city streets, the noise from cars, trucks, trains, and boats rent the air, and domestic and industrial effluents caused odors and illnesses. Minority populations experienced these effects on a larger scale than did white communities, which were able to marshal legislators and technologists to create solutions. Efforts to stave off the worst effects of urban pollution met with marginal success at first, but increased over time. The greatest successes, however, would come during the era of environmentalism in the latter half of the twentieth century.

Notes

1. Andrew Carnegie, *Triumphant Democracy* (New York: Scribner's, 1886), 1, as quoted in *Civilizing the Machine: Technology and Republican Values in America, 1776–1900*, by John F. Kasson (1976), 183.

2. Ibid., 183.

3. Richard Wright, *American Hunger* (New York: Harper and Row, 1944), 1–3.

4. Martin V. Melosi, *The Sanitary City: Urban Infrastructure in America from Colonial Times to the Present* (Baltimore: Johns Hopkins University Press, 2000), 3–4.

5. Thomas R. Detwyler and Melvin G. Marcus, eds., *Urbanization and Environment: The Physical Geography of the City* (Belmont, Calif.: Duxbury, 1972), 21, quoted in Melosi, *The Sanitary City*, 4.

6. Joel Tarr, *The Search for the Ultimate Sink: Urban Pollution in Historical Perspective* (Akron, Ohio: University of Akron Press, 1996), xxx.

7. Martin Melosi, "Hazardous Waste and Environmental Liability: An Historical Perspective," *Houston Law Review* 25, no. 4 (July 1988): 741.

8. Rebecca Harding Davis, "Life in the Iron Mills," *Atlantic Monthly* (April 1861): 430–51, quotation at 430.

9. *Milwaukee Sentinel*, 10 November 1903, quoted in Dale Grinder, "The Battle for Clean Air: The Smoke Problem in Post–Civil War America," by Dale Grinder, in *Pollution and Reform in American Cities, 1870–1930*, ed. Martin Melosi (Austin: University of Texas Press, 1980), 87.

10. J. B. Stoner, "The Ill Effects of Smoke on Health and Comfort," *Military Surgeon* 32 (1913): 373, quoted in Grinder, "The Battle for Clean Air," 86.

11. Frederick Upham Adams, in the *Toledo Blade*, 23 January 1906, quoted in Grinder, "The Battle for Clean Air," 95.

12. Suellen M. Hoy, "'Municipal Housekeeping': The Role of Women in Improving Urban Sanitation Practices, 1880–1917," in Melosi, ed., *Pollution and Reform*, 193–94.

13. Imogen B. Oakley, "The Protest Against Noise," *Outlook* 90 (17 October 1908), 351–55, quoted in "Toward an Environmental Perspective: The Anti-Noise Campaign, 1893–1932," by Raymond W. Smilor, in Melosi, ed., *Pollution and Reform*, 139.

14. Theodore Steinberg, *Nature Incorporated: Industrialization and the Waters of New England* (New York: Cambridge University Press, 1991), 206.

15. Ibid., 211–22.

16. Melosi, *The Sanitary City*, 1.

17. Kenneth T. Jackson, *Crabgrass Frontier: The Suburbanization of the United States* (New York: Oxford University Press, 1985), 288.

18. Ibid., 296.

19. Andrew Hurley, "The Social Biases of Environment in Gary, Indiana, 1945–1980," *Environmental Review* (Winter 1988): 19.

7 Conservation and Preservation, 1785–1950

A core topic for environmental history is the formation of land, water, and conservation policies: how land was allocated as the country was being settled; how land use policy developed; and what laws allowed people to gain title to land as private property. By the late nineteenth century, most of the unsettled land had been allocated and people began to press for the conservation of natural resources for efficient use and to join a growing national movement to set aside wilderness areas for recreation. This chapter looks at the history of land use policy as it developed in the colonial and national periods, and at the emergence of resource conservation and wilderness preservation at the turn of the nineteenth century.

Colonial Land Policy

American land was the lure for the greatest human migration in modern history. Nowhere else was such a bounteous environment — fertile in soil and temperate in climate, well-watered, amply wooded, and richly stocked with minerals — so ready to wrest from its native inhabitants. Until the conservation movement of the late nineteenth century, an exploding populace of European immigrants and their descendants reveled in unfettered exploitation of this favored land as private property. The New World's abundance of cheap land held out opportunities unimaginable in an Old World where land was too scarce and expensive to meet the needs of the many for subsistence or the ambitions of the few for wealth, status, and power. The American Eden offered the independence, comfort, and security of subsistence farming for most colonists, and the wealth of large-scale staple production for many others. Yet the opportunities that drew a flood of voluntary immigrants from Europe also entailed a flood of involuntary immigrants from Africa. Because the ready availability of land for private ownership made Euro-Americans reluctant to work for others, many soon discovered that the wealth of large-scale production could be realized only through the enslaved labor of Africans.

In all the English colonies, however varied in other respects, abundant American land quickly subverted the traditional European notion that people should accept their inherited class positions and communal obligations in a rigidly stratified society. The Virginia Company that founded Jamestown discovered that profits could not be made from the labor of settlers sent over as company servants. The colony began to flourish only when the company gave settlers land to work for themselves. Immigration soared when a "headright" of 50 acres for every new settler was given to whoever paid for their ocean passage. The immigrant who could pay his family's passage was guaranteed an ample farm on arrival, while planters could swell their landholdings by buying indentured servants or slaves off arriving ships. As slavery grew, this system produced enormous concentrations of land in the hands of great planters.

In the next colony, Plymouth in New England, the Pilgrims similarly discovered that not even their intense religious commitment to communal labor could sustain them, and shifted to private property. Like the larger Massachusetts Bay Colony that soon arose next door, they were trying to establish a holy commonwealth as a refuge from English oppression of Puritans. Both of these colonies were composed of self-governing farming towns that parceled out free land to families according to traditional norms of need and status. Each family was given a large lot for a house, garden, and orchard near the church in the town center, as well as ample strips for cultivation in outlying communal fields, and grazing rights on the town common. But as population thickened in the eighteenth century, communalism declined, land became a marketable commodity, and new towns were promoted as land speculations.

The lessons of these early examples was not lost on the royal favorites who competed to attract settlers as the proprietors of later colonies. Lord Baltimore of Maryland, the eight Lords Proprietors of Carolina, and William Penn of Pennsylvania all tried to make their colonies attractive by offering generous land policies as well as religious freedom. In the eighteenth century, the rising power of representative assemblies in all the colonies ensured an accelerating transfer of lands to private hands on liberal terms. Exploitation of the land as private property became the basis of Anglo-American society, and the British effort, through the Proclamation of 1763, to restrict land-hungry colonists from encroaching on western Indian lands helped to precipitate the American Revolution.

Federal Land Policy

With independence from Britain assured as the Revolutionary War drew to a close, American leaders conceived a visionary land policy to realize their conception of the new republic's destiny. In 1780, the Continental Congress called for converting the lands west of the Appalachian Mountains into a great public domain, to be parceled out to citizens under federal authority and eventually erected into states fully equal with the original thirteen. This plan required much hard bargaining to secure cessions from the seven states that claimed territory extending westward to the Mississippi River under their colonial charters. By 1786, New York, Virginia, Connecticut, and Massachusetts ceded their overlapping claims to the territory northwest of the Ohio River, with Virginia reserving much of present southcentral Ohio for grants to its Revolutionary veterans, and Connecticut retaining land rights to a broad strip along Lake Erie for sale to its citizens. In 1792, the rest of Virginia's western territory entered the Union as the state of Kentucky, and by 1802, the Carolinas and Georgia had ceded the western territory south of Kentucky. The southwestern cessions were obtained only by federal promises to honor the extensive claims already awarded to land speculators by the ceding states, and to clear Indian claims from Georgia's remaining territory.

In the Land Ordinance of 1785, the Confederation Congress provided for the orderly conversion of this vast public domain into private property. A rectangular system of surveying the public lands as settlement advanced prepared them for sale to the highest bidders at periodic auctions. Federal surveyors superimposed on the topography a grid of east-west and north-south lines at one-mile intervals. These divided the landscape into numbered townships that were six miles square, and each township into 36 numbered sections that were one mile square (640 acres). Section 16 at the center of each township was reserved to support a public school. In the Northwest Ordinance two years later, the Congress provided for temporary territorial governments north of the Ohio, which would evolve into full statehood when a sufficient population was reached. Slavery was prohibited here, but not in the later territories established farther south. In 1812, the General Land Office was established within the federal government to administer the disposition of federal lands, to conduct land surveys, and to process and record sales, entries, withdrawals, reservations, and leases on the public lands.

The regime of freehold property and representative government — an empire for liberty, Thomas Jefferson called it — spread all the way across the continent in three generations of aggressive expansion. It was extended west to the Rocky Mountains by Jefferson's Louisiana Purchase from France (1803); south to the Gulf of Mexico by the purchase of Florida from Spain (1819); and southwest to the Rio Grande River by the annexation of Texas (1845). The continental domain was rounded out to the Pacific under President James K. Polk by the Oregon Treaty with Great Britain (1846) and the conquest and purchase of the Southwest from Mexico (1846–48). A final chunk of Mexico was acquired by the Gadsden Purchase (1853) to obtain a favorable route for a transcontinental railroad through present southern Arizona.

Although the basic system of selling public lands through rectangular survey and auction continued unchanged throughout the period of expansion, constant conflict occurred over the minimum amount that had to be bought, and the minimum price that had to be paid. High minimums were favored by wealthy land speculators, by eastern interests averse to the westward migration of cheap labor, and by politicians and financial interests anxious to raise revenues for paying off Revolutionary war debts and restoring American credit. These high-minimum forces prevailed at first. The Ordinance of 1785 required a minimum purchase of an entire mile-square section (640 acres) at a minimum price of $1 per acre, and the Land Act of 1796 raised the minimum price to $2 and the minimum purchase for half the tracts on sale to 5,760 acres.

But the tide turned in favor of small purchasers with the rise of the more democratic and pro-western forces that elected Jefferson to the presidency in 1800. Although the Harrison Land Act of that year kept the minimum price at $2, it lowered the minimum purchase to a half-section (320 acres) and spread the payment over four years. In 1804, Congress reduced the minimum price to $1.64 and further reduced the minimum acreage to a quarter-section (160 acres), conventionally regarded as enough for a viable family farm. When the minimums were further reduced to 80 acres and $1.25 per acre in 1820, Congress also ended the credit system, since a financial crisis had forced so many purchasers into default.

Meanwhile, poor settlers had increasingly taken matters into their own hands by squatting on the public lands, building a cabin, clearing fields, and pasturing livestock without a formal title. When the land came up for sale at auction, squatter associations intimidated potential bidders, arguing that people who had put their labor into improving the land should have a right to hold it. After Congress granted temporary relief to such squatters in the 1830s, the Preemption Act of 1841 legalized squatting by any citizen who was the head of a family (of any age), a widow, or a single man over the age of 21. He or she, under this "Log Cabin Law," could obtain up to 160 acres of any surveyed public land by settling on it, building a house, and paying the minimum price of $1.25 per acre when it came up for sale. With this act, the settlement of the public domain by family farmers became the primary object of public land policy.

The implications of the Log Cabin Law were not fully implemented until southern opponents of a liberal land policy disappeared from Congress during the Civil War. The Homestead Act of 1862 made it possible to preempt a farm on public lands free of charge, and extended the privilege from heads of families to unmarried women as well as unmarried men. Any citizen "who is the head of a family, or who has arrived at the age of twenty-one years," in the language of the act, could occupy 160 acres of public land "for the purpose of actual settlement and cultivation," after filing for entry at a cost of $10. Full title to the land would issue without further cost when it could be "proved by two credible witnesses that he, she, or they have resided upon or cultivated the same for the term of five years."[1] At last, it seemed, anyone could own a farm. But by the time railroads carried immigrants from Europe and the East Coast to settle the Plains, the best land had already been claimed under the Log Cabin Law, and that available for homestead entry lacked the fertility, water, timber, or other resources necessary for farming.

Land Law in the Arid West

A land policy shaped by the agrarian ideal in the well-watered and wooded East often proved disastrous when settlement pushed onto the arid plains beyond the 100th meridian. Traditional farmers lured too far west by wet years in the climatic cycle were devastated when more normal drought years followed. Moreover, the cattle ranching, dry-land wheat production, and mining that could be practiced in the arid West required far more land to support a family than the 160-acre norm of federal land policy. During the cattle boom of the late nineteenth century on the high plains, ranchers typically grazed their stock without cost on vast reaches of public lands by controlling the only water sources where settlement was possible. Range wars and intimidation of intruders by armed cattlemen's associations were not uncommon, but where possible ranchers and miners sought title to strategic sites.

Symptomatic of this problem were the extensive frauds perpetrated under both the Log Cabin Law and a provision of the Homestead Act allowing a homesteader who did not want to wait five years for his farm to buy it after six months for $1.25 per acre. Ranchers and others needing large tracts hired people to file claims with the local land office, enter and cultivate the land, and live there for six months. The "homesteader" would then pay for the land with the rancher's money and turn over the title in return for a fee. The meanings of "settlement," "cultivation," and "house" were abused. A "house," for this purpose, could have four walls but no chimney, or lack a floor, or even measure 12 by 14 inches in size. In addition, many land offices in western townships were staffed with people who accepted bribes to write fraudulent titles and file fraudulent claims.

In the 1870s, Congress sought to adapt land policy to the special circumstances of the arid West through a series of special land acts. The Timber Culture Act of 1873 allowed a person to claim title to 160 acres of land by planting trees on one quarter, or forty acres, of it. Most of that land was on the prairies and plains of the arid West, and although planting trees was a worthy objective, most settlers lacked information on planting, cultivating, and watering trees. The government soon reduced the requirement to ten acres, but even with that modification, the law was largely unsuccessful.

In 1877, the Desert Lands Act made 640 acres of desert land available for $1.25 per acre if it was irrigated in three years. But irrigating arid land was very labor-intensive. One had to find land near a river and dig irrigation ditches to divert the water to the soil. Again the results were not often successful.

In 1878, the Free Timber Act gave people the right to cut timber on public lands reserved for mineral use in order to obtain timber for building farmhouses and towns. A related act of the same year, the Timber and Stone Act, stated that 160 acres of land valuable for minerals or stone as well as timber could be purchased from the federal government. Both these acts were bonanzas for timber companies, who hired people to stake claims to forest lands and then turn the lands over to the lumber company.

The land acts by which people acquired property for homesteading were differentiated from acts for acquiring mineral lands, beginning with the General Mining Act of 1872. The law stated that "all valuable mineral deposits in lands belonging to the United States, both surveyed and unsurveyed, shall be free and open to exploration and purchase, and the lands in which they are found to occupation and purchase, by citizens of the United States." On mineral lands, one could simply stake a claim — that is, put stakes or monuments in the ground, and then extract the minerals. While this gave one claim to the gold, silver, or other minerals on the land, the title to the land itself had to be obtained through the land laws. One could obtain title to the land around a placer, lode, or gravel mine by paying $2.50 an acre or, in the case of hard-rock mining, $5 an acre, prices still in effect. Although the law was meant to encourage development of the land and its minerals for the public good, environmentalists today consider it questionable policy, while mining companies consider it essential to their success.

Lands for Railroads and Education

Federal lands were granted not only to individuals, but also to states and railroads. In addition to the federal government, most of the western states deeded lands for railroad development. Railroads were granted a right of way of 100 yards and then 20 square miles on either side of the track in alternate one-square mile sections, creating checkerboard patterns. The sections could then be sold to homesteaders. Agents for the railroads went to Europe and the East Coast to encourage people to travel west by rail and settle the alternate checkerboards they had acquired. They offered land to immigrants by giving them incentives on transportation costs for specific crops, such as corn or wheat, to be grown on the Plains. This practice dovetailed with the federal government's policy of encouraging the transportation, marketability, and internal development of the country.

Another purpose of the disposal of the public domain was for education. Each state acquired school lands to establish agricultural and technological colleges for farmers and mechanics — the yeoman and the artisan, mainstays of the democratic ideal. The goal was to educate them for good citizenship and to continue to develop the crops and technologies that would advance the country. The Morrill Act of 1862, which established land grant colleges in each state, was followed by the Hatch Act of 1887, which created the agricultural experiment stations. The mission of the experiment stations was "aiding and acquiring and diffusing among the people of the United States useful and practical information on subjects connected to agriculture." Research done by agricultural colleges would benefit the people of each state. Departments such as soils, plant pathology, entomology, and nutrition hired professors to conduct research on solving practical problems for the benefit of family farmers. Cooperative extension agents then took the results and disseminated them to farmers.

The Conservation Movement

By the late nineteenth century, most of the land and natural resources of the western United States had been entered, claimed, and developed under the federal government's liberal land laws. Historian Frederick Jackson Turner, in his 1893 essay "The Significance of the Frontier in American History," argued that the frontier had come to a close. "In a recent bulletin of the Superintendent of the Census for 1890," he wrote, "appear these significant words: 'Up to and including 1880 the country had a frontier of settlement, but at present the unsettled area has been so broken into by isolated bodies of settlement that there can hardly be said to be a frontier line.'" Only small areas of land throughout the West remained to be claimed, and however significant the frontier had been for the emergence of American democracy, "never again will such gifts of free land offer themselves."[2]

The perception of abundant unexploited lands teeming with wildlife and fertile soils began to turn to one of wasted resources and inefficient use. Timber companies cut the best trees and moved on without reforestation, leaving slash and litter behind. Ranchers exploited the perennial grasses of the open range, leaving sagebrush and eroded soils. Settlers discovered that periods of drought made farming in the arid West unreliable and that dams and canals for irrigation were costly to construct and maintain. Market hunters and fishers depleted supplies of fish and game and decimated bird populations for feathers for women's hats. Rivers, lakes, and air were polluted from industrial development.

George Perkins Marsh (1801–82), one of the earliest advocates for the wise use of land, warned that American lands might suffer the fate of the ancient civilizations of the Mediterranean world, where once abundant forests had turned to denuded slopes and eroded soils. His book *Man and Nature*, published in 1864, sold 100,000 copies in a few months, testimony to the level of concern over the waste of resources in the United States. "Man," Marsh wrote, "has too long forgotten that the earth was given to him for usufruct alone, not for consumption, still less for profligate waste.... [M]an is everywhere a disturbing agent. Wherever he plants his foot, the harmonies of nature are turned to discords." Marsh proposed that the pioneer settler "become a co-worker with nature in the reconstruction of the damaged fabric.... He must aid her in reclothing the mountain slopes with forests...."[3] Marsh's concerns were echoed by a growing number of eastern elites who hiked in the Appalachian Mountains, western sportsmen who hunted and fished in the Rockies, women's groups, birders, nature writers, artists, scientists, and foresters who saw wild nature vanishing under ax and plow and valuable resources wasted under exploit-and-move-on policies.

By the late nineteenth century, *laissez-faire* capitalism, which had supported the free market's development of natural resources, began increasingly to be scrutinized and curtailed. The initial period of disposition of the public domain was followed by a second period of withholding lands for forest reserves, game refuges, national parks, and wilderness areas. During the Progressive Era of the early twentieth century, the government passed laws that regulated both corporations and land use, initiating a shift away from the policy of unregulated development, characterized by the term, *laissez-faire*, a French phrase meaning "let it be." The Division of Forestry's first chief, Bernhard Fernow (1851–1923), in his 1902 book *Economics of Forestry*, proposed instead a policy of *faire-marcher*, another French term, meaning "to make it work," or to give direction or guidance to development through resource conservation.

Utilitarian conservation began to displace the earlier policies of "free" disposition of federal lands between the 1880s and the beginning of World War I. Utilitarianism, an ethic developed by philosophers Jeremy Bentham (1748–1832) and John Stuart Mill (1806–73) in England, was based on the idea of "the greatest good for the greatest number." "Utility" meant putting the land to work to promote people's happiness. The conservation ethic of the early twentieth century added the idea of time, captured in the phrase of conservationist WJ McGee (1853–1912), "the greatest good of the greatest number for the longest time." [4]

The concept of conservation was promoted in 1908 by forester Gifford Pinchot and President Theodore Roosevelt at the White House Conference on Conservation. It brought under one rubric the various strands of the movement toward resource efficiency in the use of forests, water, and rangelands that had developed separately during the prior three decades. Resource conservation came to mean the wise and efficient use of natural resources. "Conservation above all," states environmental historian Samuel Hays, "was a scientific movement.... Its essence was rational planning to promote efficient development and use of all natural resources."[5]

Pinchot followed Fernow as the Chief of the Division of Forestry (now the Forest Service). The Division, established in 1886, managed the forest reserves, established in 1891 through the Forest Reserve Act. During this period, certain lands within the public domain were reserved from homestead entry as forest reserves. In 1905, Pinchot transferred the forest reserves out of the Department of the Interior and into the Department of Agriculture, on the grounds that forests ought to be managed as if they were a crop. He also advocated allowing controlled grazing in the forest reserves. He instituted a sustained yield process in which timberlands must be reforested after cutting. With the work of Fernow and Pinchot, the twentieth-century forestry movement, devoted to the efficient use of natural resources and sustained timber yields, was launched.

Reclamation and Water Law

Water reclamation and development were likewise of paramount importance in settling the arid American West. "In the East," states water historian Marc Reisner, "virtually every acre received enough rainfall.... [M]uch of the West ... suffers through months of habitual drought." [6] In his 1878 *Report on the Lands of the Arid Region of the United States,* John Wesley Powell (1834–1902) identified the amount of annual average rainfall as constituting a fundamental difference between the lands of the western and eastern United States. Lands west of the 98th meridian were characterized by erratic rainfall in amounts from 5–20 inches per year. Less than 20 inches of rain a year was insufficient to grow crops reliably. Powell classified western lands into irrigable lands and arid lands. For an irrigated farm, 80 acres, or one-half of the quarter section that the Homestead Act had set out as the basic unit, would be enough for a homestead. Fertile land existed in the river valleys and adjacent lowlands, where water could be easily diverted, for use by irrigated farms.

But Powell also recognized that many lands throughout the West did not have easy access to water. Here pasturage farms, devoted not to crops but to grazing, would need 2,560 acres, or four sections of land. A third classification was forested land in the high mountains as far as the timberline, valuable for commercial purposes and for buildings, housing, and fencing. Timberlands, he proposed, should be set aside and reserved for commercial use.

Powell's 1878 report was not received with enthusiasm by members of Congress, and he retreated to pursuing science, as head of the Bureau of Ethnology (1881–1902) and as head of the U.S. Geological Survey from 1881–1892. But toward the end of the century, in 1902 — the year that Powell died — his ideas began to gain favor. Francis Newlands (1848–1917) of

Nevada, after winning a seat in the House of Representatives, introduced Powell's recommendations into Congress. At first his bill was couched in terms too radical for Congress to accept, but President Theodore Roosevelt (1858–1919), who had by then gained office, urged Newlands to rewrite his bill, and in 1902 the Reclamation Act was passed.

The Reclamation Act of 1902 established a fund from the sale of public lands to be used for the construction and maintenance of irrigation works in the western states. The Act set up the Bureau of Reclamation (BuRec) to oversee the construction of federal water systems, and continued the tradition of 160 acres as the basic farm unit. The irrigated farm supplied with water from federal projects would be the key to farming in the West. Single people could obtain 160 acres for an irrigated farm, married couples 320 acres, with the acreage limitation intended to privilege the family farmer. But farming interests found ways to subvert the Reclamation Act's 160-acre limitation law, just as speculators had found ways to undermine the 1862 Homestead Act.

Western water law had developed differently from that in the eastern states, which initially was based on the riparian doctrine, stemming from English common law. Under riparian law, farms that abutted a river had the right to use the water. Owning the adjacent land conferred the right to use the stream, a right that inhered in the property. The water could be used for drinking, farming, or watering animals, but there was no obligation to use it. Because one could simply sit beside the stream and contemplate its flow, riparian law was known as the "Rocking Chair Law." The owner, however, could not build a dam or construct a mill that would alter the river's flow for those downstream.

With industrialization, eastern water law began to change toward a policy of appropriation, a doctrine that developed in the nineteenth-century mill towns of New England and New York. According to legal historian Morton J. Horwitz, "When American judges first attempted to resolve the tension between the need for economic development and the fundamentally antidevelopmental premises of the common law, the whole system of traditional rules was threatened with disintegration."[7] In the late eighteenth century, courts in Massachusetts and Connecticut had refused to recognize the right to interfere with the flow of the water, but by the early nineteenth century, a number of cases in New York and New England began to challenge the common-law doctrine of using, but not altering the river, and instead allowed capitalist development to take place if the flow was altered for a beneficial purpose. The case of Palmer v. Mulligan (New York, 1805), followed by a second case in New York in 1818 and another in New Hampshire in 1823, modified riparian law to allow for the appropriation of water for beneficial use. If factories needed water, or dams were constructed to operate mills, looms, and spinning jennies for the textile industry, it was permissible to alter the downstream flow on the grounds that the "natural rights of all" would prevail.[8]

In the arid West, appropriation became established in Colorado by 1882, and henceforth was known as the Colorado Doctrine. The eight mountain states of Arizona, Colorado, Idaho, Montana, Nevada, New Mexico, Utah, and Wyoming rejected riparian rights in their constitutions in favor of state control over water. As applied to surface waters, the doctrine of prior appropriation is the concept of "first in time, first in right." The first person who begins to use the water for a beneficial purpose, such as growing crops, has the right to divert it to their land. The senior appropriator, or the first individual to use the water, has the first right to use or divert it. The person who follows in time is the junior appropriator, and even if located upstream from the senior appropriator, he or she cannot divert all the water from the river, leaving nothing for the first user. The water may be used for growing crops or watering cattle, but in times of drought is also allocated on the basis of the rule, "first in time, first in right." Second, in contrast to the riparian doctrine, appropriation is based on the concept of "use it or lose it." If the water is not used, the appropriator loses the right to the water. Third, water use is quantified, and appropriators can sell water among themselves.

The California Doctrine, a dual riparian-appropriation system, resulted from a famous water law case, Lux v. Haggin, filed in 1879 and decided by the California Supreme Court in 1886. According to environmental historian David Igler, "Lux v. Haggin served as a public forum for a wide range of ongoing conflicts in California's arid San Joaquin Valley — 'land monopolists' against 'small farmers,' San Francisco control of hinterland resources, and perhaps most notably, 'riparian' landowners versus water 'appropriators.'"[9] Henry Miller (1827–1916) and Charles Lux (1823–87), immigrants engaged in the San Francisco butchering business, needed to maintain an adequate supply of beef cattle to serve their customers. They therefore formed the Miller-Lux Land Company, eventually becoming the wealthiest landowners in the West, with holdings in California, Oregon, and Nevada. In the process, they acquired land along the Kern River. Because they were ranchers, they could drive their cattle to the river and water them at the stream itself. They claimed their water on the basis of the California state constitution, which granted water rights under riparian law. But they soon came into conflict with upstream farmer Ben Ali Haggin (1821–1914), who claimed water rights on the basis of prior appropriation. Haggin argued that he had appropriated the water and was diverting it to his crops through irrigation. The conflict that developed was between ranching under riparian law and irrigated farming under appropriation.

The court case that established the California Doctrine was resolved in 1886 when the California Supreme Court ruled that both sides had rights to the water. Riparian rights prevailed if a person had private land along the river, but appropriation prevailed over riparian rights if the appropriator was the first user. Both the riparian and appropriation laws were, therefore, valid, but it was timing — when the river was first used to draw off water versus when the land abutting the river was purchased — that determined who had first right to the water in the event of a conflict. With the legal issues settled, Miller and Lux went into business with Haggin and the parties built a dam together. The California Doctrine, however, became the basis for water law in nine western states — California, Kansas, Nebraska, North Dakota, Oklahoma, Oregon, South Dakota, Texas, and Washington.

The Preservation Movement

As development proceeded throughout the country, pristine areas of wilderness vanished. The movement to preserve wilderness arose during the second half of the nineteenth century and commanded national attention in the first two decades of the twentieth century. Environmental historian Roderick Nash writes: "All the nineteenth-century champions of wilderness appreciation and national parks in the United States were products of either urban Eastern situations or of one of the West's most sophisticated cities, such as San Francisco."[10]

During the last half of the nineteenth century, many upper- and middle-class Americans came to view wilderness as a threatened national asset. Wild nature was a treasure to be cherished and preserved, rather than an evil to be eliminated. Mountains, waterfalls, valleys, and even deserts took on characteristics of the sublime, associated in the public mind with the awesome power of God. Nature writer John Muir (1838–1914), among others, viewed mountains as God's cathedrals.

Wilderness was also associated with the formation of the American character, especially the male character. It was an arena in which men could reassert and reaffirm their masculinity. According to Nash, "wilderness ... acquired importance as a source of virility, toughness, and savagery — qualities that defined fitness in Darwinian terms."[11] The more wealth people acquired and the more a middle-class lifestyle seemed attainable by large numbers of people, the more that way of life came to be considered soft. As an antidote, the outdoor movement arose, exemplified by groups such as the Appalachian Mountain Club (1876), the Boone and Crockett Club (1885), the Mazamas of Portland, Oregon (1894), the Sierra Club (1892), and the Campfire Club of America (1897), whose members camped in the outdoors and hiked in the mountains in the summertime. As more people acquired access to wilderness, they came to appreciate its value. Books celebrating the virtues of the wilderness sold in increasing numbers, and people displayed pictures in their homes of beautiful mountains, lakes, and meadows.

Among those who promoted the idea of wilderness was landscape architect Frederick Law Olmsted (1822–1903) who visited California's Yosemite Valley in 1863–4 as a commissioner appointed to manage the grant of the Yosemite Valley to the state of California. His report, presented in 1865 to his fellow commissioners, argued that exposure to the wilderness was an important complement to city life, especially as therapy for the mind. As people became caught up in the pressures of ordinary urban life, they developed "softening of the brain, paralysis, palsy, monomania, or insanity, mental and nervous excitability and moroseness, melancholy and irascibility."[12] These psychological and physical problems could incapacitate people and prevent them from exercising their "intellectual and moral forces." The antidote was to get out into the wilderness. Olmsted was especially adamant that poor people, unable to afford vacations in the mountains, should have access to city parks through public transportation. Olmsted's 1866 design for New York City's Central Park reflected this vision by rejecting formal design in favor of wooded hills, lakes, and undulating pathways.

The rapid destruction of the enormous redwoods, which had graced California's coast ranges and the Sierra Nevada range for thousands of years, heightened interest in wilderness preservation. At the time, redwood lumbering was on the rampage throughout California, and the plight of the redwoods became known worldwide, as images and even bark from the trees were displayed in cities as far away as London.

Many mountain lumber companies did business on the basis of the Timber and Stone Act of 1878, which permitted individuals to file claims to land that had timber of commercial value. Agents for the lumber companies looked for people they could hire to enter forests and stake claims to 160 acres of land. They then took them to a local land office to file a claim. After the deed was paid off in six months, the title would be transferred to the timber company for a sum of money paid to the participant. Thus much of the western wilderness was disappearing under axe and saw, as lumber companies harvested the giant trees.

The preservation of California's coastal redwoods (*Sequoia sempervirens*) became a cause célèbre by the turn of the nineteenth century, as lumbering proceeded unhindered along the northern California coast from Santa Cruz to Humboldt County. Because there were few roads and railroads, logs and lumber were hauled to docks on the Pacific Ocean and sent down shoots or dragged onto wharves, whence by chains and pulleys they could be loaded onto schooners offshore. They were then shipped to San Francisco Bay and eastward by train.

A major issue that mobilized redwood preservationists was the proposed cutting of the Calaveras Big Trees (*Sequoia gigantea*) in California's Stanislaus River Valley, which Europeans discovered in 1850. In 1900, women in San Francisco — members of the California Federation of Women's Clubs, spearheaded by Laura White — began a campaign to save the trees. The result was the passage of a bill curtailing tree-cutting in 1900, followed by an authorization to exchange the Calaveras Grove for lands of equal value in 1909. Although cutting was halted, the preservation project took many more years to complete. Additionally, in 1901, the Sempervirens Club, named for the coastal redwood, *Sequoia sempervirens*, was successful in persuading the state to set aside Big Basin, south of San Francisco, as a park. In 1918, the club became the Save the Redwoods League, which spearheaded the creation of state and national redwood parks in California throughout the twentieth century.

In addition to hiking and outdoor clubs, women's clubs, such as the General Federation of Women's Clubs (1890), the Women's National Rivers and Harbors Congress (1908), and the Daughters of the American Revolution Committee on Conservation (1909) promoted the wilderness ideal and the preservation of parks. Other preservation organizations included the American Civic Association (1900), the National Audubon Society (1905), and the railroads, which immediately saw that tourism would be to their financial benefit.

Creation of the National Parks

The movement to preserve areas for their aesthetic and recreational benefit led to the establishment of state and national parks. Between the creation of Yellowstone National Park in 1872 and the passage of the National Park Service Act in 1916, thirteen National Parks were set aside to "conserve the scenery and the natural and historic objects and the wild life therein and to provide for the enjoyment of the same in such a manner and by such means as will leave them unimpaired for the enjoyment of future generations."[13] The states also began to preserve lands for watershed and recreation. In 1885, the state of New York approved a 715,000-acre "Forest Preserve" in the Adirondacks, to be "kept forever as wild forest lands."[14]

The first national park to be created was Yellowstone in Wyoming. In 1870, an expedition to the region was led by Henry D. Washburn (1832–71), surveyor general of Montana, with a second in 1871 authorized by Ferdinand Hayden (1829–87), head of the United States Geological Survey. Soon thereafter a movement to preserve the area as a park was mounted. At the forefront of the process was A. B. Nettleton, spokesperson for Northern Pacific Railroad financier, Jay Cooke. The bill was shepherded through the legislative process by Hayden, and the Yellowstone Park Act was passed in 1872. With the country in depression at the time, a proposed railroad extension from Cooke's Northern Pacific Railroad into the park was not completed until 1883. But by 1915, almost 45,000 people were arriving by train in Yellowstone each season and staying in hotels financed by the railroad.

Yosemite National Park in California was created through a sequence of acts. First, in 1864, during the Civil War, President Lincoln ceded Yosemite Valley to the state of California for the purpose of its preservation. The mountains around the valley were designated a National Park in 1890, and then the valley was re-ceded to the federal government to become part of Yosemite National Park in 1905. The first tourists had entered Yosemite Valley in 1855, four years after the Mariposa Indian War resulted in the removal of the Yosemite Indians. Soon thereafter, some 400 visitors came by steamboat and stage. By 1869, when the transcontinental railroad was completed, tourists arriving in San Francisco were offered trips into the park. By 1875, wagonloads of supplies were taken to park hotels to supply the needs of tourists. The great champion of Yosemite was naturalist and writer John Muir (1838–1914), who helped to found the Sierra Club in 1892 and was its first president. He encouraged people to visit Yosemite and to hike in the mountains as he had done.

The preservation of Yosemite, however, became the source of national controversy over the proposal to build a dam within the park in Hetch Hetchy Valley. After the 1906 earthquake, San Francisco needed a more reliable supply of water than that available in the coastal mountain ranges and turned to the Sierra Nevadas for new sources. San Francisco mayor, James Phelan, hired agents to survey all suitable water sources. The use of Cherry Lake or Lake Eleanor were two of the most viable alternatives to damming the Tuolomne River in Hetch Hetchy Valley. The issue came to national attention in 1909, the year in which Muir split with Gifford Pinchot over the latter's advocacy of the Hetch Hetchy site as a dam for public power. The conservation movement's ethic of "the greatest good of the greatest number for the longest time" reinforced Pinchot's goal of providing water for the public good.

The debate over whether or not to dam Hetch Hetchy was spearheaded by Muir, who wrote in florid terms about the valley as "sublime rocks whose walls glow with life" and described "the birds, bees and butterflies [that] would stir the air into music." Muir's ethic, in contrast to that of the public power advocates, was couched in theocentric terms. The mountains were God's cathedrals, and Hetch Hetchy Valley ought to be preserved on those grounds alone. People who had never visited Yosemite National Park or the Hetch Hetchy Valley wrote to Congress, urging that the valley be saved. Muir described the valley in a number of articles, and women's clubs throughout the country rallied to his side, denouncing Pinchot, whom they had supported on most conservation issues. The battle over Hetch Hetchy was lost, however, with the passage of the Raker Act in 1913. The O'Shaughnessy Dam across Hetch Hetchy's Tuolomne River, built between 1915 and 1920, became the centerpiece of San Francisco's effort to bring water and electricity to the city.

In 1916, after Muir's death, the National Park Service began to administer the thirteen parks established by that date and formerly managed by the U.S. Army. The list included well-known parks in the western United States, such as Yellowstone (1872), Yosemite (1890), Sequoia (1890), General Grant (1890), Mt. Rainier (1899), Crater Lake (1902), Mesa Verde (1906), Glacier (1910), and several smaller parks. Henceforth, the National Park Service administered all national parks, monuments, and historic sites set aside for their natural and scenic value, a list that grew to more than 350 sites by the late twentieth century.

Grand Canyon National Park (1919) was the first to commemorate the stark beauty of the desert. John Wesley Powell, on his expeditions through the Grand Canyon in 1869 and 1871–72, described the canyon's sheer walls, magnificent cliffs, and panoply of colors. Geologist Clarence Dutton's *Tertiary History of the Grand Cañon District* (1882) subsequently brought the science, aesthetics, cartography, painting, and photography of the canyon to public attention. Environmental historian Steven Pyne, in *How the Canyon Became Grand* (1998), writes: "By the time Dutton finished his inquiry, the Grand Canyon, essentially unknown twenty years before ... had become ... an exemplar of landscape aesthetics."[15] Pyne attributes the park's popular appeal to descriptions by scientists such as Dutton and Grove Karl Gilbert (1843–1918) and to paintings by artists such as Thomas Moran (1837–1926) and Gunnar Widforss (1879–1934). The Grand Canyon became a national monument in 1908 and a national park in 1919. While the Grand Canyon is perhaps the country's most spectacular national park, it is in desert country, and as such was a departure from the magnificent forests and snow-capped mountains that characterized the Adirondacks, the Rockies, and the Sierras. The preservation of the Grand Canyon was an acknowledgment that not only could the southwestern rocks and canyons be beautiful, but also that the arid landscape of the deserts was worthy of preservation.

Over the years, the railroads built trunk lines to the national parks. The Union Pacific and Northern Pacific had lines going into the Yellowstone area. The Great Northern Railroad crossed the top of the country through Glacier National Park. In Utah, Bryce and Zion National Parks were served by the Union Pacific. The Southern Pacific had a route into Yosemite National Park, and the Yosemite Valley Railroad reached nearby El Portal. Mount Rainier was served by both the Northern Pacific and Great Northern railroads, and Crater Lake, in southern Oregon, was accessed by the Southern Pacific. The Grand Canyon could be visited by taking the Atchison, Topeka, and Santa Fe railroad's trunk line into the canyon just off its main route. Abandoned in 1969, it was restored in 1989 to became the Grand Canyon Railway. The idea of seeing the Great West, in the company of the railroads, has been a significant part of the preservation of America's scenic heritage.

Throughout the twentieth century, debates over the preservation of wilderness areas versus the need to exploit water and forests for the public good continued. Controversies over whether to dam the Colorado River in Glen Canyon and the Grand Canyon and the Green River in Dinosaur National Monument aroused the passions of the nation. Whether to cut the ancient forests of the Northwest for timber or to preserve the Headwaters redwood forests of California have spawned citizen demonstrations and legal battles. Such issues keep alive the concerns raised by early twentieth-century conservationists and preservationists over the vanishing West and the vanishing wilderness.

Conclusion

Federal land policy moved from the acquisition of territory during colonial times through the mid-nineteenth century to the encouragement of rapid settlement between the American Revolution and the closing of the frontier in the 1890s. With much of the best land allocated by the turn of the century, the nation's focus shifted to resource conservation and wilderness preservation. The conservation movement stressed the efficient use of natural resources for "the greatest good of the greatest number for the longest time." Forests, rangelands, and water were managed for productivity, sustained yield, and year-around conservation. The preservation movement aroused the nation to the need for setting aside sublime and picturesque areas for aesthetic appreciation and recreation. Supported by citizen's groups, hiking clubs, and the railroads, 13 national parks were created by the time the National Park Service Act was passed in 1916. Throughout the twentieth century, additional parks were added to the system, many amid controversy over dam construction for water conservation versus preservation for wilderness appreciation.

Notes

1. "An Act to Secure Homesteads to Actual Settlers on the Public Domain," 20 May 1862, in The Statutes at Large, Treaties, and Proclamations of the United States of America, ed. George P. Sanger (Boston: Little, Brown, 1865), 12:392–93.

2. Frederick Jackson Turner, "The Significance of the Frontier in American History," American Historical Association, Annual Report for the Year 1893 (Washington, D.C., 1894), 199.

3. George Perkins Marsh, Man and Nature (New York: Scribner's, 1864), 29–37.

4. Gifford Pinchot, Breaking New Ground (Washington, D.C: Island, 1947), 326.

5. Samuel P. Hays, Conservation and the Gospel of Efficiency (Cambridge: Harvard University Press, 1959), 2.

6. Marc Reisner, Cadillac Desert: The American West and Its Disappearing Water (New York: Viking, 1986), 43.

7. Morton J. Horwitz, The Transformation of American Law, 1780–1860 (Cambridge: Harvard University Press, 1977), 36.

8. Platt v. Johnson, 15 Johns. 213, 218 (N.Y. 1818), quoted in Horwitz, Transformation of American Law, 37.

9. David Igler, "When is a River Not a River? Or, Reclaiming Nature's Disorder in Lux v. Haggin," *Environmental History* 1, no. 2 (April 1996): 52–69, at p. 52.

10. Roderick Nash, "The Value of Wilderness," *Environmental Review* 3 (1977): 14–25, at 16.

11. Roderick Nash, *Wilderness and the American Mind*, rev. ed. (New Haven: Yale University Press, 1973), 145.

12. Frederick Law Olmsted, "The Yosemite Valley and the Mariposa Big Trees" (1865), reprinted in *Landscape Architecture* 43 (1952): 17–21, at 17.

13. *U.S. Statutes at Large*, H.R. 15522, Public Act 235.

14. New York Laws, 1885, chap. 238, p. 482.

15. Steven Pyne, *How the Canyon Became Grand: A Short History* (New York: Viking, 1998), 69, 70.

8   Indian Land Policy, 1800 – 1990

Indians are central actors in American environmental history.
They were the primeval users, shapers, and stewards of the land.
The natural resources of the lands they occupy — soil, timber,
grasses, water, and minerals — have played major roles in how the
environment was developed in the past and is used today. Indians,
together with the national parks and recreational resources
created from their lands, figure preeminently in the evolving
attitudes toward nature and the wild that underlie environmental
policy. The policies of the United States government toward
Indians evolved over time from land acquisition by conquest and
treaties, to the removal of tribes to western reservations, to
expulsion from the national parks, and finally toward Indian
sovereignty, civil rights, and co-management of resources. This
chapter surveys conflicts over Indian land and water rights,
Indians and the creation of the parks, and the role of the United
States government in establishing and administering policies
toward Indians.

Indian Land Treaties

The lands of the eastern states, beginning with the thirteen
original colonies, were ceded to English proprietors by Indians who could
neither read nor fully appreciate the implications of the agreements they
signed. In many cases, tribal members did not completely comprehend the
size of the ceded lands, or that — in many cases — they were giving up long-
established hunting and fishing rights. In other cases, whites negotiated
with chiefs from whom they could receive the most land, and used fraud
and bribery to accomplish their desired ends. The proclamation line of
1763, established at the end of the Seven Years' War between the French
and British, temporarily protected Indians from further settlement and reserved lands west of the Appalachians as hunting grounds. The British
Crown retained the exclusive right to negotiate cessions from Indians, and

settlement was forbidden to colonists. Between 1768 and 1783, Indians ceded areas west of the proclamation line, but, although the lands were officially ceded, Indians still occupied them, making settlement difficult for Americans and European immigrants.

After the American Revolution, the new United States government continued the British policy of regarding Indian tribes as sovereign nations and regulating relations with them through treaties. In the Northwest Ordinance of 1787, the Confederation Congress declared: "The utmost good faith shall always be observed toward the Indians, their lands and property shall never be taken from them without their consent."[1]

Government policy became less favorable to Indians, however, in the context of westward expansion and American land hunger in the early nineteenth century. During the War of 1812 against Great Britain, the Indians' power to resist American demands was shattered by their defeats at the hands of General William Henry Harrison in the Old Northwest and General Andrew Jackson in the Old Southwest. The devastated tribes were forced to accept treaties confining them to reservations on the least desirable lands, thus opening most of the country for white settlement westward to the Mississippi. Indian cessions in the Old Southwest were extorted by General Jackson's threats that those who resisted his demands would suffer the fate of the Creeks he had annihilated during the war. Although the tribes' new lands were guaranteed to them "forever," the government soon began urging them to resettle west of the Mississippi, on reservations created in an "Indian territory" comprising present-day Oklahoma and parts of Arkansas. In 1823, the U.S. Congress changed the legal status of the tribes from sovereign nations to domestic dependent nations, and the next year President James Monroe called for removing all remaining Indians beyond the Mississippi.

Indian Removal

American demands for removing Indians mounted as settlement thickened around their reservations, cotton boomed across the South, and settlers began to resent the federal government's treaty obligation to protect the reservation Indians from white intruders. The extensive reservations still held in the South by the so-called five "civilized" tribes, the Choctaw, Chickasaw, Cherokee, Creek, and Seminole, were particularly coveted by settlers and cotton planters. Their opportunity came when the Indians' nemesis, Andrew Jackson, won the presidency in 1828. Promptly, Georgia, Alabama, and Mississippi extended their jurisdiction over Indian reservations in defiance of federal authority. State jurisdiction outlawed tribal governments and subjected Indians instead to local officials for whom they could not vote and to local courts in which they could neither sue nor testify, leaving them helpless to prevent seizure of their lands by white intruders. The Indians' only hope for survival was the federal government's obligation to protect them, as promised in the Constitution, which stated that "all Treaties made, or which shall be made, under the Authority of the United States, shall be the supreme Law of the Land."[2]

President Jackson quickly made it clear that states' rights, in his singular understanding of the Constitution, forbade federal interference with state policy toward Indians. Striking a magnanimous pose, he offered removal as a generous government's only means of rescuing Indians from state encroachments. "My red Choctaw children, and my Chickasaw children," he told two of the tribes, "must" move beyond the states, where they would be guaranteed "land of their own ... as long as the Grass grows or water runs."[3] In 1830 Jackson persuaded Congress to pass an Indian Removal Act, authorizing the president to move Indians to public lands beyond the Mississippi in exchange for their eastern reservations.

The Supreme Court responded by pronouncing the states' invasion of Indian reservations and rights to be unconstitutional. This 1832 case, Worcester v. Georgia, involved two white missionaries who were jailed under a Georgia law forbidding their presence on Cherokee lands without a permit from the governor. The Court ordered their release, because, in the words of Chief Justice John Marshall: "The Indian nations had always been considered as distinct, independent political communities, retaining their original natural rights as the undisputed possessors of the land from time immemorial."[4] But without support from the executive branch, the Court pronounced and ordered in vain. "John Marshall has made his decision," a defiant Jackson reportedly scoffed: "now let him enforce it!"[5]

During the 1830s, the U.S. Army forcibly corralled members of the "civilized" tribes and marched them westward. In 1831, the Choctaw were removed, many dying en route. When the Creek were removed in 1836, many went in chains, and 3,500 died along the way. Then, in 1837, the Cherokee, who — in white terms — had been the most civilized of the Indian nations, were also sent west. They had adopted a constitution, taken up private property, grown grain, and built mills and stores. But the army herded them into stockades and removed 15,000 people, about one quarter dying along their "Trail of Tears." The army faced dogged resistance before it managed to extract some of the Seminole from Florida's swamps and herd them along the same route in 1842. Subsequently, the prairie tribes of the Sac and the Fox were sent from Iowa and Illinois to reservations in Indian territory.

As Euro-Americans invaded the Trans-Mississippi plains after the 1840s seeking range, farm, and mineral lands, battles between settlers and the western Indian tribes stimulated a new round of removals to a new set of reservations. During the 1850s, the government prevailed on fifty-two tribes to relinquish large acreages of land that settlers wanted in return for protection on smaller tracts of reservation lands that settlers did not, as yet, want. In 1871, the status of Indians changed again. Congress declared that "hereafter no Indian nation or tribe ... shall be ... recognized as an independent ... power with whom the United States may contract by treaty."[6] Indian nations were reclassified as non-national entities, and weaker legal "agreements" replaced formal treaties.

In 1864, minerals were discovered on western lands that Indians retained, and new battles ensued between whites and Indians. Black Kettle of the Cheyenne was tricked and more than one hundred members of his band killed in 1864, while the tribe was in the process of surrendering to the governor of Colorado, and in 1868 Black Kettle himself met the same fate while leading his band to join tribes friendly to the government. In 1866, the Sioux War was fought when miners who had invaded Sioux lands wished to remove the tribe from the mineral-rich areas. The Sioux, in retaliation, ambushed and destroyed the soldiers who entered their land. In 1875, similar problems occurred with gold prospectors in the Black Hills of South Dakota, resulting in the Battle of Little Big Horn. Then in 1886, the Apaches — the last Indian tribe in the Southwest to hold out against the government — surrendered; and in 1887, the Nez Perce, under Chief Joseph, after a thousand-mile flight through Idaho, Wyoming, and Montana, were captured just before reaching a haven in Canada.

Defeated and in despair, the Plains Indians, in 1888, began a revival they called the Ghost Dance movement. The Ghost Dance was a tribal effort to renew Indian ways of life and preserve Indian traditions. The dance itself revealed a paradise in which Indians could live free of pain and suffering. Most wanted to live harmoniously, but a radical faction wanted to eliminate whites. The religion represented a threat to white expansion, and so, in 1890, the government banned the Ghost Dance on the Sioux reservations. This act precipitated the so-called "final battle" of the Indians against the whites, the 1890 Battle of Wounded Knee, in which Sioux leader Sitting Bull (1831–90) was killed and more than three hundred Indians died.

Underlying the removals to reservations was a policy of assimilation by turning Indians into farmers. The progress of civilization, which was the major narrative of Western culture, required that Indians should become civilized and move up from the level of "savagery" and "barbarism" (hunting/gathering and pastoralism) to that of farming and commerce. The idea of progress, prevalent in the work of John Locke, Adam Smith and other philosophers of the Enlightenment, was reinforced by Social Darwinist ideas that the white race was the highest of all the races. The goal was to "civilize" Indians by converting them to farmers. First in time came the supposedly "wild" Indian, then the partially "civilized" Indian who had donned a few western clothes and habits, and finally the totally educated, "civilized" Indian who had been incorporated into and made a full part of American society. Reformers such as Richard Henry Pratt, with government funding, started boarding schools (e.g., the Carlisle Indian School in Pennsylvania, founded in 1879), where Indian children were taught English and Anglo-American culture in the belief that obliteration of Indian beliefs and practices would bring about assimilation.

Between 1877 and 1881, under Interior Secretary Carl Shurz, the Bureau of Indian Affairs (BIA, created in 1824) moved toward greater centralized control over Indian culture through appointments of professional bureaucrats loyal to Washington rather than to local agents or Indian tribes. In the 1880s, the Bureau banned such Indian practices as medicine-making, the Lakota Sun Dance, and ritual bundles associated with spiritual release of the dead. Indian education and the banning of tribal rituals made up two parts of a three-pronged program of "civilizing" the Indian. The third component was the transformation of communal reservation lands into private property.

The Dawes Act

In order to "civilize" Indians and turn them into settled farmers, Euro-Americans believed that private property should replace the communal property Indians held on reservations. In 1887, the General Allotment (or Dawes) Act, sponsored by Massachusetts Senator Henry Dawes, authorized the allotment of reservation lands to individual Indians in 160-acre parcels, the official homestead unit. The objective was to create independent Indian

farmers and to dissolve the tribes. The Secretary of the Interior could negotiate sale of the remaining tribal lands, and after an initial period of federal
trusteeship, Indians could sell their 160-acre allotments. So much land
passed to non-Indians through both processes that Indian landholdings of
155 million acres in 1881 had shrunk to 56.6 million acres by the 1990s.
Only five percent of the lands distributed by the Dawes commission remain
in Indian hands.

"[T]he ... most effective means of encouraging the tribes to meld into
white society," as legal historian Lloyd Burton explains the Dawes Act, "was
simply to relieve them of much of their well-watered, arable land."[7] In the
process, Indian population declined steadily to 237,196 individuals by 1900.
Western travel writer Samuel Bowles stated bluntly the underlying rationale
for this drastic change in Indian policy. Arguing for dispossessing Indians of
their lands, he wrote, "We know they are not our equals; we know that our
right to the soil, as a race capable of its superior improvement, is above
theirs." The Indian did not belong in the West, Bowles maintained. "Let us
say to him, you are our ward, our child, ... ours to displace, ours to protect.... We want your hunting grounds to dig gold from, to raise grain on,
and you must 'move on.'"[8]

The Dawes Act meshed strategically with the conversion of Indian lands
into national parks during the late nineteenth century. For white settlers and
exploiters, as historians Robert Keller and Michael Turek explain, the
"opening of reservations for settlement compensated for millions of acres
withdrawn from the public domain by national forests and parks."[9] Whites
gained at both poles of this Euro-American bargain, as demand for recreational resources joined demand for natural resources in motivating expropriation of Indian lands. The saga of Indian removal from the parks, as revealed by a flood of recent scholarship, is the dark side of environmental
history's most celebrated creation story.

Indians and the Creation of the National Parks

That Indians, always and everywhere, had to "move on" was demonstrated most forcefully when Henry David Thoreau's dream of preserving
wildness achieved its greatest triumph in the creation of the national parks.
The very idea of a national park was first articulated by the painter of Indian
life George Catlin (1796–1872), who wanted, like Thoreau, to save both
vanishing wildness and the vanishing Indian. Proposing a national park on

the Great Plains in his 1844 book *North American Indians*, Catlin imagined the vanishing buffalo "as they *might* in future be seen (by some great protecting policy of government) preserved in their pristine beauty and wildness, in a *magnificent park*, where the world could see for ages to come, the native Indian in his classic attire, galloping his wild horse, with sinewy bow, and shield and lance, amid fleeting herds of elks and buffaloes. What a beautiful! and thrilling specimen for America to preserve and hold up to the view of her refined citizens and the world, in future ages! A nation's park containing man and beast, in all the wild and freshness of their nature's beauty!"[10]

But when parks came, Indians were expelled from lands they had long inhabited and ranged to create recreational resources for whites. Except as tourist attractions, Indians vanished from memory as well as from view. Wilderness was redefined as untainted by human presence, and parks were conceptualized as places where white tourists could be inspired by the sublimity of depopulated natural beauty. "Generations of preservationists, government officials, and park visitors have accepted and defended the uninhabited wilderness preserved in national parks as remnants of a priori Nature (with a very capital N)," writes environmental historian Mark David Spence. "Such a conception of wilderness forgets that native peoples shaped these environments for millennia."[11]

Indians were expelled from the national parks by techniques of removal perfected over three centuries of Euro-American settlement — military force, treaties of cession, and violation of treaty stipulations. The victims were often dehumanized in the process, states Philip Burnham, as tourism businesses in and around parks "used Indians as advertising icons" or isolated them "in 'model' settlements."[12] The varied processes and rationales by which Indians were displaced from parks are illustrated by four cases — at Yosemite and Yellowstone by military force and intimidation in the nineteenth century, and at Glacier and Mesa Verde by legal subterfuge and protracted negotiation in the twentieth.

Indian Removal from Yosemite and Yellowstone

Especially rich in pathos is the saga of Indian removal from California's magnificent Yosemite region. When Congress and President Abraham Lincoln ceded Yosemite Valley to the state for preservation in 1864, it had already been cleared of Indians. In the decade following the 1849 Gold Rush, a hundred thousand Indians disappeared from California's population. When miners flooding the foothills of the Sierra Nevada met resistance to staking their claims on Indian lands, a state-armed Mariposa Battalion of volunteering miners invaded Yosemite in 1851. Determined "to sweep the territory of any scattered bands that might infest it,"[13] this force systematically burned the Yosemite Indians' villages and food caches to starve them into accepting removal to a reservation on the Fresno River in the Sierra foothills. Dr. Lafayette Bunnell, a member of the battalion who published his account of the *Discovery of the Yosemite and the Indian War of 1851* (1880), was both enthralled by Yosemite's sublime beauty and sympathetic to its Indians. His account left no doubt, however, that the Indians had to go.

After the last defiant Yosemite were routed from their high-country refuge on a stunningly beautiful alpine lake (now the scenic jewel of the park's Tioga Road), Bunnell tried to console their despairing Chief Tenaya by telling him that "we had given his name to the lake and river. At first he seemed unable to comprehend our purpose, and pointing to the group of glistening peaks near the head of the lake, said, 'It already has a name; we call it Py-we-ack [Shining Rocks].' Upon my telling him that we had named it Ten-ie-ya, because it was upon the shores of the lake that we had found his people, who would never return to it to live, his countenance fell.... indicat[ing] that he thought the naming of the lake no equivalent for his loss of territory."[14]

Many years after Yosemite's ancient inhabitants were "removed," some of the refugees charged in a petition to Congress that whites were despoiling "our beloved" Yosemite Valley by "gradual destruction of its trees, the occupancy of every foot of its territory by bands of grazing horses and cattle, the decimation of the fish in the river, ... [and] the rapacious acts of the whites in the building of their hotels and stage lines." Complaining that "all seem to come only to hunt money," they claimed that "this is not the way in which we treated this park when we had it."[15] Meanwhile, a remnant of Yosemite had left the reservation for a camp in their ancestral valley, where they were tolerated for several decades as a tourist attraction, performing dances, posing for a fee before visitors' Kodaks, and selling the women's beautifully woven willow baskets. These "unwelcome guests," the petition alleged, were "the object of curiosity ... to the throngs of strangers who yearly gather in this our own land."[16] Eventually the Army expelled these showcase Indians, and park managers burned their camp as an eyesore. The goal was to reaffirm the national park ideal of a wilderness sublimely uninhabited, except for hikers and campers.

Yellowstone, by contrast, presents a saga of coerced expulsion of Indians after the park was created, justified by tendentious claims that Indians were threats to tourists. Here, after several expeditions publicized the area's natural splendors, and after a campaign spearheaded by a railroad hungry for tourist traffic, Congress authorized the country's first national park in 1872. The economies and cultures of Crow, Bannock, Shoshone, Salish, Nez Perce, and Northern Paiute depended on seasonal visitation to the Yellowstone high country for hunting, harvesting the dietary staple camas, lodgepole cutting, and ceremonial or intertribal gatherings. Because of the Yellowstone plateau's high altitude, however, averaging 8,000 feet, its only year-round residents were small bands of Shoshone, called Sheepeaters because they lived by hunting bighorn sheep in the loftiest and most remote areas. The reclusive Sheepeaters fled Yellowstone to escape management by the U.S. Army, which administered the early national parks until the National Park Service was created in the twentieth century. They departed under the assurances of a treaty negotiated in 1868, which insured their right to hunt in Yellowstone in return for ceding their lands. But once they were gone, the government refused to ratify the treaty or to recognize the claims of the other tribes who had long depended on seasonal use of the area.

The park then banned all Indian entry, allegedly to protect tourists, on the basis of several incidents. In 1877, nine white visitors were briefly held captive and one of them was wounded when they encountered 750 Nez Perce Indians, who were in a thirteen-day flight across Yellowstone while being pursued by two thousand U.S. troops through Idaho, Wyoming, and Montana. Then, in 1878, a group of Bannock alarmed park Superintendent Philetus Norris by crossing the park boundary in pursuit of cattle that had destroyed gathering areas on their reservation. The final provocation was the so-called Sheepeater War of 1889 in central Idaho, when the U.S. cavalry hunted down 51 Sheepeaters, mainly women and children with less than a dozen firearms, who were suspected of involvement in a murder of Chinese miners. Although this incident occurred far from Yellowstone, Norris erected a fort to protect the park from further encroachment by Indians. The result was another chapter in the cultural construction of wilderness as an area where visitors could view sublime nature unimpeded by Indian presence.

Legal Maneuvers in Glacier and Mesa Verde

Congress and the courts replaced military force as instruments of Indian removal when western Montana's Glacier National Park was created in

1910. On the Blackfeet reservation where the park came to be located, the Indians had traditionally gained their sustenance by moving freely between the mountainous western part of their territory and their homes on the more easterly plains. By the 1880s, warfare, disease, and disappearing bison had reduced the Blackfeet to some two thousand individuals. After several hundred starved to death in the harsh winter of 1883 – 84, they yielded sweeping land cessions, including their prized Sweet Grass Hills, to the United States in order to obtain provisions.

Meanwhile the Blackfeet were befriended by George Bird Grinnell (1849 – 1938), naturalist, admirer of Indian cultures, conservationist editor of the widely read *Forest and Stream*, and founder of the Audubon Society. Dreaming of preserving the spectacular "Backbone of the World" along the Rocky Mountain crest in the western part of the Blackfeet reservation as a national park, he cooperated with potential exploiters of the area in persuading the Indians to sell it to the government. In 1895, they finally agreed to a price of $1.5 million, but only when the government promised that Blackfeet would have the right "to go upon any portion of the lands ... and to cut and remove wood ... and to hunt upon said lands and to fish in the streams thereof so long as the same shall remain public lands of the United States."[17]

When valuable minerals believed to be in the ceded area failed to materialize, a campaign led by Grinnell, along with the Sierra Club and the Great Northern Railroad, finally persuaded Congress to create Glacier National Park in 1910. But the act creating the park invalidated the Blackfeet's 1895 agreement, government lawyers maintained, giving rise to a running conflict in the courts, and contestation that continues to this day along the park's hundred-mile boundary with the Blackfeet reservation. Meanwhile the Great Northern Railroad built a large hotel beside its tracks at the park's southern entrance and extensive tourist facilities within the park. Both the railroad and the park developed an extravaganza of Indian display and entertainment by recruiting "some good type Indians ... who ... have good costumes, put on a good show, and live in peace and harmony" to participate in this early ecotourism.[18]

Mesa Verde, in the red-rock canyon country of southwestern Colorado, presents a final example of expropriation, this time to memorialize ruins left by Indians long dead by taking land from living Indians who were absent from the park. The lands were part of a reservation assigned to the Mountain Ute in return for larger tracts they had ceded in 1874 and 1880. Settling down here and utilizing gifts of wagons, tools, and cattle in compliance with white reformist hopes, they provided food and clothing for themselves through farming, constructed sawmills and gristmills, and set up schools. All this went for naught when, in the 1870s, whites discovered on their lands the multistoried cliff dwellings occupied by the Anasazi (Ancient Ones) until drought drove them out around A.D. 1300. Preservationists sought cession of these ruins, which the Ute seldom visited, not wishing to disturb the spirits of the dead.

The Ute were understandably reluctant to surrender part of their reservation for a park, fearing that they might be removed in the process. While white women, organized as the Colorado Cliff Dwelling Association, collected petitions from women's clubs and other groups for preserving the Anasazi ruins, their leader, Mrs. Virginia McClurg, met nothing but rebuffs through ten years of negotiation with the Indians. Finally in 1906, the Ute ceded 42,000 acres, and President Theodore Roosevelt signed a bill creating Mesa Verde National Park. Only then did the government discover that its faulty survey had left most of the ruins outside the Ute cession. After Congress unilaterally placed another 175,000 acres of mostly tribal land under park administration, a further cession was demanded from the Ute. A long deadlock was broken only when the government sent in its most skilled troubleshooter in Indian affairs, and the Indians were bluntly informed that "the government is stronger than the Utes.... [and] Congress is going to have that land."[19] Finally the Ute consented in 1911 to trade 10,000 acres in the ruins area for almost twice that acreage elsewhere, only to discover later that, due to government error, the land they received in exchange was land they already owned.

Mesa Verde enabled the government to commemorate an ancient pueblo people, while obscuring the violent wars against the Indians of the recent past and ignoring the present Indians upon whose lands the park was created. As Philip Burnham assesses the outcome: "First, Indian cultures could now be perceived to have come and gone long before the days of Manifest Destiny.... Second, the government was now cast in the role of resurrecting the memory of Indian people, not destroying them."[20] Unlike Yosemite, Yellowstone, and Glacier National Parks, Mesa Verde memorializes Indian archeological remains, but like them it celebrates a cultural landscape in which living Indians no longer dwell.

These four parks exhibit a variety of rationales for removing Indians: protection of parks to create tourism, protection of tourists from Indians, protection of sublime canyons and mountains at the expense of Indians, and protection of archaeological sites delineating ancient Indian cultures. Variations on these themes are exhibited by most national parks and monuments — including the Grand Canyon, Mount Ranier, the Everglades, Death Valley, the Great Smoky Mountains, the Badlands, the Black Hills, and Grand Teton — as well as most of the national forests and outdoor recreational sites administered by the Forest Service, the Bureau of Land Management, the Bureau of Reclamation, and the Fish and Wildlife Service. However varied in other respects, relations between park lands and Indians almost everywhere exemplify two consistent themes. Indians had to "move on," unless they could be used to entertain tourists. And the Indians' ancestral lands had to be repackaged for consumption by tourists as wild Nature, uncontaminated by human presence.

The dialectical relationship between the parks and Indian lands had been evident from the inception of the national park system. Within a year of Yellowstone's creation in 1872, Indian tribes lost their sovereignty, and the shift from formal treaties to agreements made land cessions from Indians more available and concessions to Indians more revocable. Indian lands were ideal raw material for parks, offering the remote and unprofitable scenery least in demand for private development, and therefore available in large parcels at low cost. While the parks commemorated the grandeur of American scenery, the reservation system removed unwanted people to unwanted lands. While the parks depended on attracting tourists, reservations restrained Indians from bothering tourists. After 1916, the parks and the reservations vied for the same resources within the Department of the Interior. The former succeeded at the expense of the latter.

The Winters Decision

In the twentieth century, government policy moved sporadically toward greater autonomy for Indians, retroceding at times before moving forward again. The first important movement in this direction occurred in the courts and involved the hotly contested issue of western water rights. Although historical events had usually conspired against Indians when land was being appropriated by speculators, in the case of the Blackfeet Indians' appeal for water rights, the courts actually ruled in favor of Indians. "In January 1908," writes Lloyd Burton, "the U.S. Supreme Court issued its seminal decision in *Winters v. United States*, the first case in which the federal courts explicitly affirmed the water rights of Indian reservations."[21]

The Blackfeet Indians, soon to be barred from the Glacier National Park created on part of their reservation, claimed they did not have enough water on the reservation for watering cattle or buffalo, or even for farming, at a time when the 1887 Dawes Act was promoting Indian agriculture. In the resulting Winters Doctrine, Indians were designated as the senior appropriators of the water on their reservation. The date a reservation was established, in most cases before white settlers arrived and began diverting water for irrigation, determined the date of water appropriation. As a result, in many cases, reservation tribes became the senior appropriators with the right to use millions of acre-feet of water in the western United States. Moreover, their rights were not subject to quantification — i.e., the "use it or lose it," principle of western water law — but they were instead entitled to as much water as the reservation required. From 1908 to the present, the Winters decision has spawned numerous water conflicts in the West and numerous court cases between the states and private companies who wished to divert water despite the rights of Indian reservations.

The Indian New Deal and Civil Rights

The Bureau of Indian Affairs had been responsible for protecting Indian rights since the nineteenth century, but by the New Deal era of the 1930s, Indian lands were reduced to sixty-five million acres. In 1934, under BIA director John Collier, Indian policy was reconfigured through the Indian Reorganization Act (IRA). Collier sought to preserve Indian cultural heritage and to give Native Americans more power to manage their lands and natural resources. Although controversial among both Indians and non-Indians at the time, the IRA, known as the "Indian New Deal," reconstituted the tribes as sovereign entities. It stopped the Dawes Act land allotments to individual Indians, prohibited the sale of non-allotted lands to non-Indians, and stopped the subjection of allotted trust lands to the states. Until then, the states had been able to intervene in Indian affairs, a situation that had weakened Indian control over their remaining lands. The act allowed Native Americans on reservations to establish self-governance and permitted tribal corporations to develop reservation resources. It established programs for forest management, health services, and education, including teaching tribal culture. On Collier's recommendation, President Roosevelt also abolished the entrenched Board of Indian Commissioners, which had overseen the BIA since 1869, thereby allowing the reforms to proceed.

The Depression of the 1930s, combined with drought and record-breaking summer temperatures, was especially hard on Indians, who already had the deepest poverty levels of any ethnic group in the United States. Collier pressed Congress to pass the Pueblo Relief Act in 1933, and established the Indian Emergency Conservation Work program, which created seventy-two camps on thirty-three reservations and employed eighty-five thousand Native Americans between 1933 and 1942. Under the program, Indians worked to improve roads, protect forests, construct dams, dig wells, and erect fences. Collier worked with Secretary of Agriculture Henry Wallace to assist Native Americans in purchasing purebred cattle to improve the quality of their herds, with Harry Hopkins, director of the Federal Emergency Relief Administration, to allocate relief funds to reservations, and with the Department of War to distribute surplus clothing and shoes. The Public Works Administration (PWA) employed Indians to construct hospitals, schools, and sewer systems on reservation lands, while the Resettlement Administration assisted them in improving water distribution. Collier also worked with Secretary of the Interior Harold Ickes and President Roosevelt to cancel Indian debts for roads, bridges, and irrigation systems. Although sweeping in scope, Collier's programs were not always positive for Indians. In some cases the reforms drew Native Americans as individuals more tightly into wage-dependent work, while weakening existing tribal economies.

The pendulum swung briefly but sharply the other way in the 1950s, when the "termination" policy of the Eisenhower administration sought to end federal responsibility for tribes deemed able to fend for themselves. A total of 109 tribes were terminated, and lost title to 1.4 million acres of tribal land. Federal trust responsibility and supervision were halted on terminated reservations, along with federal health, education, and welfare programs and funds. The terminated tribes were assessed state taxes and subjected to state laws, as well as state criminal and civil jurisdictions. They lost sovereignty over all land that had been sold to both Indians and non-Indians. Most Indians reacted with alarm, and some of the terminations were reversed during the ensuing Kennedy administration of the early 1960s. Congress stopped the terminations in 1962, and in 1963 it reaffirmed the Winters' decision that considered Indians the first appropriators of water on reservation lands.

Tribal autonomy was further protected by the Indian Civil Rights Act of 1968, which declared that the states could no longer extend civil and criminal jurisdiction over Indian lands without the tribes' consent. Indian policy generally seemed to unfold politically along the same lines Lloyd Burton describes for water policy. "From the 1908 *Winters* decision up through the

1960s," he writes, "the federal courts for the most part have tended stead-
fastly to uphold Indian water rights as preemptive federal obligations, while
Congress and the executive — because of the Indians' relative lack of political clout — have tended either to ignore the Indian right or to subvert it directly."[22] When the Congress and the Executive Branch had power, especially under Republican administrations, they gave Indian jurisdiction to
the states, and the states then undercut Indian power. Throughout American history therefore, the judiciary has often been the strongest defender of
Indian rights.

Despite this history, however, the Republican administration of Richard
Nixon brought the protection of Indian autonomy, as well as the environment, to a high-water mark in the early 1970s. "Termination of tribal recognition will not be a policy objective and in no case will it be imposed without Indian consent," Nixon had promised as a candidate. " ... The right of
self-determination of the Indian people will be respected and their participation in planning their own destiny will actively be encouraged."[23] Indeed,
after his election, the termination policy was decisively repudiated in the
Menominee Restoration Act (1973), reversing for a Wisconsin tribe one of
the largest terminations. The Alaska Native Claims Settlement Act (1971)
assigned 40 million acres and nearly one billion dollars to village and regional corporations managed by Alaska's native peoples, laying the basis for
autonomous development of indigenous economies and societies on an unprecedented scale.

Perhaps the most important Nixon policy in the long run was to encourage active participation by Indians in managing the federal health, education, and welfare programs provided for their benefit. Originally the administration offered a "takeover" bill, transferring to Indians full responsibility
for managing services, with the federal government providing the funds
along with technical assistance. But Indians considered the proposal too
radical, fearing that it was but another step toward termination. Instead, they
opted for a substitute bill providing for Indian management by contract under the Indian Self-Determination and Education Assistance Act (1975). By
sharing responsibility with federal agencies, most tribes quickly came to ex-
ercise substantial control over the governmental activities that most affected
their daily lives and welfare. Such "co-management" soon became a model
for all government agencies, and even the National Park Service has begun
to promote active participation in policy formation and management by Indians within its purview.

Indian Lands and Environmental Regulation

From the 1970s to the present, American Indians have engaged in a variety of natural resource development and co-management projects. Both Indian and non-Indian development on reservation lands is subject to the environmental protection regulations of the National Environmental Policy Act of 1969 and the Surface Mining Control and Reclamation Act of 1977. The 310 Indian reservations include 6.3 million acres of commercial forests, 43 million acres of rangeland, and 3 million acres of farmland. They contain 20 percent of U.S. oil and natural gas reserves, one-third of the western states' low-sulfur coal reserves, and one-half of all uranium reserves. Tribes have gathering, hunting, and fishing rights on their own lands and certain off-reservation rights on lands ceded to the United States. Tribal councils have the power to lease land to non-Indians for farming, ranching, logging, and mining.

Indian tribes are adversely affected by environmental pollution and degradation from adjacent lands. Shoshone-Bannock tribal lawyer Jeanette Wolfley states: "Due to the unique nature of tribal land tenure and tribal culture, tribes cannot simply relocate to new areas when their lands become contaminated, their water polluted, or their wildlife resources decimated...."[24] Rivers and airsheds that abut or cross reservations may be contaminated from industrial pollution, logging debris, nuclear wastes, and weapons testing, while dams may impede fish runs and affect water levels and water quality. Off-reservation effects on wild animal and plant habitats can decrease the availability of foods and medicines necessary for cultural survival, affecting long-term tribal health. Both private corporations and federal agencies that regulate adjacent public lands are responsible for such impacts. In 1994, President Clinton instructed all federal agencies to implement programs in a manner that respected tribal sovereignty. They were directed to operate in a government to government relationship with the tribes, engage in the best practicable consultation with them, assess the impacts of federal decisions and programs on tribal lands, and review and remove procedural impediments to dealing effectively with tribes.

The tribes have the authority to regulate, manage, and protect resources on their own homelands. Many tribes have developed resource management programs to conserve and restore fisheries, forests, and waters on reservation lands. While some tribes have used their lands to increase profits through forestry and fishing, others, such as the Colville Confederated tribes of Washington, have engaged in sustainable forestry practices that harvest trees only as they are replaced in the growth cycle and have enacted fishing and hunting practices that conserve wildlife. Some tribes, such as the Blackfeet, Sioux, and Assiniboine, have enacted water quality management programs and others, such as the Cheyenne of Montana, have set stricter air pollution regulations than federal or state standards require.

Mineral resources constitute an important component in tribal land use and regulation. In 1975, tribal leaders formed the Council of Energy Resources Tribes (CERT) to coordinate leasing policies for oil, gas, uranium, geothermal, and mineral resources on Indian lands. In 1982, the Indian Mineral Development Act was passed, allowing tribes to enter into joint ventures with non-Indians. The tribes receive royalties on the leaseholder's income. Uranium mining on Indian lands, begun in the 1940s and 1950s, resulted in deadly consequences to Native Americans who worked in the mines, owing to continued exposure to radioactivity. Although too late to benefit those who died from radiation-induced cancers in those years, in 1990 Congress passed the Radiation Exposure Compensation Act, entitling survivors and their beneficiaries to receive $100,000 in compensation if they worked in the mines between 1947 and 1971 and received doses of radiation higher than federally specified limits or contracted cancer. The act, however, did not compensate for the long-term effects of uranium mining on reservation lands and waters and their cumulative effects on human lives.

Coal and oil extraction is a controversial dimension of energy resources development on Indian lands. In 1969, the Peabody Coal Company began mining coal in accordance with leases on Hopi and Navajo reservation lands. While some Indians benefited from the profits in the form of jobs, homes, and schools, others complained that their sacred lands were being desecrated in pursuit of power for white society's cities and lifestyles. Similarly, in Alaska, native peoples, conservationists, and developers disagree frequently over such issues as drilling for oil in the Arctic National Wildlife Refuge, clear-cutting of timber in the Tongass National Forest, and development in the Copper River delta. Such issues illustrate ongoing environmental and natural resource conflicts among private industry, the federal government, the states, and Indian sovereignty. Anishinabe Indian advocate Winona LaDuke concludes: "People of color and indigenous people who have been marginalized historically ... may offer leadership for the kind of structural change that we need to make. We need to expand the vision of the social movements in this country to allow for that voice to come forth."[25]

Conclusion

The native peoples of the United States resisted and survived a series of destructive American policies — military subjugation, removal to reservations, dependency on government bureaucracies, assimilation to majority norms, and termination of tribal organization and government support. Gradually, in the twentieth century, shifting government policies permitted a reassertion of Indian identity and interests, both cultural and political, a regeneration of tribal organization and activity, and growing autonomy in co-managing and managing their daily life and natural resources. In recent years, Indians have worked toward a balance between tribal welfare and environmental conservation. Their expulsion in the creation of national parks dramatized an ambiguous relation to mainstream conservationism that still persists. When the Wilderness Act of 1964 brought Thoreau's dream of preserving wildness to its greatest triumph since creation of the national parks, the act's authors defined wilderness as an area "where the earth and its community of life are untrammeled by man, where man himself is a visitor who does not remain." Once again Indians were erased from the wilderness.

Notes

1. "An Ordinance for the Government of the Territory of the United States, North-West of the River Ohio," July 13, 1787, *Columbian Magazine* (Philadelphia) 1 (1787): supplement, sec. 14, art. 3.

2. U.S. Constitution, art. 6, sec. 2.

3. Michael P. Rogin, *Fathers and Children: Andrew Jackson and the Subjugation of the American Indian* (New York: Knopf, 1975), 216–17.

4. *Worcester v. Georgia*, 5 Peters 17–18.

5. Ronald N. Satz, *American Indian Policy in the Jacksonian Era* (Lincoln: University of Nebraska Press, 1975), 61.

6. *United States Statutes at Large*, 16: 566.

7. Lloyd Burton, *American Indian Water Rights and the Limits of the Law* (Lawrence: University Press of Kansas, 1991), 18.

8. Samuel Bowles, *The Parks and Mountains of Colorado: A Summer Vacation in the Switzerland of America* (Springfield, Mass.: Bowles, 1868), 145–46.

9. Robert H. Keller and Michael F. Turek, *American Indians and National Parks* (Tucson: University of Arizona Press, 1998), 27.

10. George Catlin, *North American Indians* (Philadelphia: Leary, Stuart, 1913 [1844]), 1:294–95.

11. Mark David Spence, *Dispossessing the Wilderness: Indian Removal and the Making of the National Parks* (New York: Oxford University Press, 1999), 5.

12. Philip Burnham, *Indian Country, God's Country: Native Americans and the National Parks* (Washington, D.C.: Island, 2000), 13.

13. Lafayette Houghton Bunnell, M.D., *Discovery of the Yosemite and The Indian War of 1851 Which Led to That Event*, 4th ed. (Los Angeles: G.W. Gerlicher, 1911 [1880]), 110.

14. Bunnell, *Discovery of the Yosemite*, 240.

15. Ed Castillo, ed., "Petition to Congress on Behalf of the Yosemite Indians," *Journal of California Anthropology* 5, no. 2 (1978): 271–77, quotations at 273, 275. The petition was apparently submitted ca. 1890–91.

16. Castillo, "Petition to Congress," 273.

17. Keller and Turek, *Indians and National Parks*, 49–50.

18. Ibid., 57.

19. Ibid., 37–38.

20. Burnham, *Indian Country, God's Country*, 50.

21. Burton, *American Indian Water Rights*, 6.

22. Ibid., 34.

23. Quoted in Francis Paul Prucha, *The Indians in American Society: From the Revolutionary War to the Present* (Berkeley: University of California Press, 1985), 83.

24. Jeanette Wolfley, "Ecological Risk Assessment and Management: Their Failure to Value Indigenous Traditional Ecological Knowledge and Protect Tribal Homelands," in Duane Champagne, ed., *Contemporary Native American Cultural Issues* (Walnut Creek, Calif.: AltaMira, 1999), 303.

25. Winona LaDuke, "From Resistance to Regeneration," *The Nonviolent Activist* (September–October, 1992): 5.

9   The Rise of Ecology, 1890–1990

The development of ecology as a science is an important theme in environmental history because the scientific analyses of human surroundings provide a basis for resource management and land-use development. Environmental historians have delineated a number of different approaches to the evolution of scientific ecology in twentieth-century America. They include human ecology, organismic ecology, economic ecology, and chaotic ecology. This chapter looks at the historical development of and implications for managing the human environment inherent in each of the four approaches.

Ernst Haeckel and the Origins of Ecology

Ecology derives from the Greek word *oikos*, meaning "household," and is the study of the relationships among organisms and their surroundings. The science was named by German biologist Ernst Haeckel (1834–1919), who introduced the term in several works in the 1860s and 1870s, first in German and then in English, inspiring others to develop the science. In his *Generelle Morphologie* (*General Morphology*, 1866), Haeckel included a section entitled "Oecologie und Chorologie," in which he defined ecology as the study of the organic and inorganic conditions on which life depends. "By ecology, we mean the whole science of the relations of the organism to the environment including, in the broad sense, all the 'conditions of existence.' These are partly organic, partly inorganic in nature; both, as we have shown, are of the greatest significance for the form of organisms, for they force them to become adapted."[1]

Haeckel further designated the inorganic conditions as the physical and chemical properties of the habitat, including climate, nutrients, and the nature of water and soil. The organic conditions included "the entire relations of the organism to all other organisms with which it comes into contact, and of which most contribute either to its advantage or its harm." He noted that each organism had friends and enemies that favored or harmed its existence, but that science had so far mainly investigated the organism's functions for preserving itself through nutrition and reproduction. Instead, a more comprehensive approach was needed. Haeckel employed the influential idea of nature as a household, stating that biologists had neglected "the relations of the organism to the environment, the place each organism takes in the household of nature, in the economy of all nature. . . ." Darwin's theories of descent and natural selection, he argued, filled these gaps by showing "all the complicated relations in which each organism occurs in relation to the environment."[2] In this, his first use of the term, Haeckel placed the new concept squarely within the framework of evolution introduced just a few years earlier by Charles Darwin in *The Origin of Species by Means of Natural Selection* (1859).

Haeckel presciently connected two words, both deriving from *oikos* — "ecology," or the study of the household, and, "economy," or the management of the household. Haeckel's idea of ecology as the study of the household, or home, would inspire those who later developed the field of human ecology, chemist Ellen Swallow Richards (1842–1911) and ecologist Eugene Odum (1913–). His idea of ecology as investigating the economy of nature would inspire those who later developed an economic approach to ecology, such as Charles Elton (1900–91), Arthur Tansley (1871–1955), and Raymond Lindeman (1915–42).

In 1869, Haeckel elaborated on his definition of ecology in his inaugural lecture to the philosophical faculty of the University of Jena, Germany. Here he drew specifically on Darwin's theory of the struggle for existence and again included the idea of nature's economy in his definition. "By ecology we mean the body of knowledge concerning the economy of nature — the investigation of the total relations of the animal both to its inorganic and to its organic environment; . . . in a word, ecology is the study of all those complex interrelations referred to by Darwin as the conditions of the struggle for existence."[3] This definition is significant, because in English translation it was included in the introductory epigraph (dated 1870) of the Chicago school of ecology's classic 1949 treatise, *Principles of Animal Ecology*, thereby validating Haeckel's preeminence as founder of the science. The authors were zoologists Warder C. Allee, Alfred E. Emerson, Orlando Park, Thomas Park, and Karl P. Schmidt.

The term "ecology" found its way into English in the translation of Haeckel's two-volume work, *The History of Creation* (1873), a popular rendition of his 1866 *General Morphology*. In the preface "oecology" is listed in a series of sciences that explicate Darwin's theory of natural selection, but in volume two, he elaborated on it further as one of several "facts" proving Darwin's Theory of Descent: *"The oecology of organisms, the knowledge of the sum of the relations of organisms to the surrounding outer world,* to organic and inorganic conditions of existence; the so-called *'economy of nature,'* the correlations between all organisms living together in one and the same locality, their adaptation to their surroundings, their modification in the struggle for existence. . . ." Here Haeckel argued in opposition to the "unscientific" explanation based on "the wise arrangements of a creator acting for a definite purpose," that phenomena could be explained as "the necessary results of mechanical causes."[4] The creationist versus evolutionary explanation of the origins of the natural world generated ongoing conflict in later years, sparking a debate over human origins that continues to the present.

In the English translation of *The Evolution of Man* (1879), drawn from a set of lectures he gave at the University of Jena, Haeckel provided a more detailed discussion of ecology. Here again, he drew on the root of the term *oikos* as a household and argued that Darwin's "Theory of Descent" would explain "all the remarkable phenomena which, in the 'Household of Nature,' we observe in the economy of the organisms." He elaborated further on the role of adaptation in the new science of ecology, stating, "all the various relations of animals and plants, to one another and to the outer world, with which the Oekology of organisms has to do . . . admit of simple and natural explanation only by the Doctrine of Adaptation and Heredity." Finally, he reiterated his own historically significant assessment that the older theory of creation by a benevolent deity was far weaker in explanatory power than was Darwin's new scientifically derived theory of evolution. "While it was formerly usual to marvel at the beneficent plans of an omniscient and benevolent Creator, exhibited especially in these phenomena, we now find in them excellent support for the Theory of Descent; without which they are, in fact, incomprehensible."[5] In these books and articles, Haeckel defined and elaborated the concept of ecology, but did not explicate the science itself. A handful of followers, who read Haeckel's work in its original German and later in English, began to use ecology as an explanation for the role played by the environment in maintaining biotic, including human, life and the ways humans could use the science to manage the economy of nature.

The first approach to ecology that developed in the United States was that of human ecology, put forward by chemist Ellen Swallow Richards in the 1890s. The second was the organismic approach developed by Frederic Clements (1874–1945) in 1916 and elaborated by scientists who worked on the Great Plains in the 1930s and 1940s. The third was the economic approach, emerging out of the nineteenth-century science of thermodynamics as elaborated by British ecologists Charles Elton and Arthur Tansley and many American ecologists during the 1940s through the 1960s. The fourth, or chaotic, approach rose out of work in population ecology as influenced by chaos theory in mathematics during the 1970s to 1990s, and was explicated by scientists such as Robert M. May, S.T.A. Pickett, and P.S. White. There are major differences in the assumptions that each of these approaches makes about nature itself, its management, and the human ethical relationship to it. In Swallow's human ecology, people are part of nature and work within it. In the organismic approach, humans are separate from nature and should follow nature as a teacher. In the economic approach, humans are managers of nature and assume control over it, while the chaotic approach implies that humans need to relinquish the hubris implicit in attempts to control the natural world, accept the disorderly order of nature, and work within nature's limits.

Human Ecology

Chemist Ellen Swallow Richards (1842–1911) developed the concept of human ecology and applied it to the human home and its surroundings. In his 1973 book *Ellen Swallow: The Woman Who Founded Ecology*, Robert Clarke attributed the introduction of ecology in the United States to Swallow. "As early as the late 1860s, [Haeckel] is credited with suggesting a science be developed to study organisms in their environment....Then leaving "Oekologie" for others to develop, he concentrated on other sciences.... Fluent in German, Swallow traced the German word Oekologie to its origin — the Greek word for house."[6] Following Clarke's lead, ecologist Robert P. McIntosh, in *The Background of Ecology* (1985), wrote: "Ellen Swallow (Mrs. Richards) recognized many problems of the environment in an era when industrialization and modern technology were just developing some of the bases of air and water pollution and rural and urban decay which are the bane of modern environmentalists. She was a crusader for establishing a scientific basis for bettering human life."[7] In 1892, at a meeting of the Boston Boot and Shoe Club, she introduced the term to the United States as "oekology," or the science of right living, "perhaps," notes McIntosh, "the first appearance in the public press of Haeckel's word."[8]

For Swallow, ecology had its roots in the biology and chemistry of life. Chemistry, which had been her focus as a special student at the Massachusetts Institute of Technology, was the science of the environment. The human home, or community, was one point in a larger environmental framework. Water flowed into and out of that home, air surrounded it, and fertile soils within it produced nutritious food.

In 1907, Swallow used the term "human ecology" in her book *Sanitation in Daily Life*. She stated: "Human ecology is the study of the surroundings of human beings in the effects they produce on the lives of men."[9] The features of the environment came both from climate, which was natural, and from human activity, which was artificial and superimposed on the environment. The sanitation movement of the post–Civil War era had raised public consciousness of problems caused by humans living in the cities — for example, garbage, noise, smoke, and water pollution. Swallow brought all this under the rubric of what she called municipal housekeeping. She argued that fresh air and clean water free of pollutants from factories were as necessary to human health as was good nutrition. Swallow's approach was quite different from that of scientific ecology, which was developing about the same time. She viewed humans as part of nature, while scientific ecology, as it later developed, separated humans from nature in order to study the environment prior to human influence.

The Organismic Approach to Ecology

The history of scientific ecology and its various schools of thought in the United States have been described by environmental historian Donald Worster in his foundational work, *Nature's Economy* (1978). Frederic Clements (1874–1945), a major founder of ecology, developed an organismic approach to the field. "Two interrelated themes dominated Clements' writing," states Worster, "the dynamics of ecological succession in the plant community and the organismic character of the plant formation."[10] Clements was influenced by the work of U.S. ornithologist C. Hart Merriam (1855–1942) and Danish botanist Eugenius Warming (1841–1924), both of whom looked at relationships among plants, animals, and their populations. Warming had written a treatise on the foundations of ecological plant geography in 1895 entitled *Plantesamfund: Grundträk af den okologiske Plantegeografafi*, and Clements followed in the footsteps of the evolving science of ecology by publishing "Research Methods in Ecology" in 1905 and "Plant Physiology and Ecology," in 1907. Working on the Great Plains at the University of Nebraska, Clements approached the evolution of plant relationships in terms of plant communities. He investigated how those communities changed over time and were dependent on climatic factors, such as temperature, rainfall, and wind. If dry climate with little rainfall prevailed, deserts developed, whereas moderate rainfall produced grasslands, and heavier rainfall resulted in forests. The type of plant community in a particular locale developed through stages, initiated by soil formation. Soils determined which plants could thrive in a particular type of climate. Over time a stable equilibrium evolved through succession, a process he characterized in his 1916 book *Plant Succession.*

For Clements, the plant community developed like a living organism. As he stated in *Plant Succession* (1916): "The unit of vegetation, the climax formation, is an organic entity. As an organism, the formation arises, grows, matures, and dies.... The climax formation is the adult organism, the fully developed community...."[11] The vegetative community found on the Great Plains was the mature grassland, developing from an early youthful period into a mature adult period and then finally dying back as the individual organisms died. In *Bioecology* (1939), Clements integrated animals into the plant community, using a concept he called the biome. Plants determined which animals could be supported in a given place and mediated between the habitat and animal populations. They provided a buffer against the extremes of climate and supplied food for animals that established themselves there.

Clements believed that there were several types of climax communities, but they evolved in two ways, both starting from bare areas denuded of vegetation. He differentiated between primary and secondary areas: "Primary areas, such as lakes, rocks, lava-flows, dunes, etc., contain no germules at the outset, or no viable ones other than those of pioneers. Secondary areas, on the contrary, such as burns, fallow fields, drained areas, etc., contain a large number of germules, often representing several successive stages."[12] Primary succession was the process by which vegetation began to cover an area that had never before borne vegetation, proceeding from an initial stage to an intermediate stage and then to a mature climax formation. A pond or a lake would be invaded by the marshy lands around it as plants began to cover the waters. Then it would become muck, supporting herbaceous species that began to come in from the edges. Eventually an entire forest would cover the land, obscuring the water. Succession was thus the development of the life history of the climax formation. Terms such as "pioneers," "invaders," "migration," "communities," and "societies" recalled stages in the developmental history of the United States, from pioneering through farming and finally urbanization. Clements identified several types of mature climax formations in North America: the deciduous forest climax, the prairie-plains climax, the Cordilleran (or mountain range) climaxes of the Rocky Mountains, and the desert climaxes of the Southwest.

Secondary succession resulted from either natural disturbances, such as hurricanes, tornadoes, fires, or floods, or from human activities, such as clearing land or draining swamps. In each case there was a reinvasion of vegetation through successional stages on the disturbed or abandoned land. For example, in the first stage of forest succession, weedy species, such as crabgrass, began to cover an abandoned field. In the second stage, after two or three years, grasses and herbaceous plants began to come in. After several more years, trees, such as the pines of the eastern forests, seeded themselves among the grasses, and in ten to thirty years, the pine forest matured. Finally, in about seventy years, hardwoods, which seeded themselves beneath the pine trees, produced the eastern climax forest.

In the context of the dry, windy conditions of the 1930s, known as the Dust Bowl, Clements brought humans into the equation. The grasslands of the Great Plains represented the climax community, or the vegetative organism. Humans were outsiders to that biome — disrupters of the naturally evolved grasslands. The eastern forests had been settled by Europeans, whose axes gave them victory over nature, but in the grasslands, nature began to get the upper hand, defeating human efforts to subdue the land. Clements wrote two papers, one on "Plant Succession and Human Problems" (1935) and the other on "Environment and Life in the Great Plains" (1937), in which he identified harvests of monocultures, such as wheat and corn, as the transformers of the natural grasslands of the Plains that produced the Dust Bowl.

The Dust Bowl of the 1930s influenced the organismic approach to ecology. In his book *Deserts on the March* (1935), ecologist Paul Sears (1891–1990) stated that the Great Plains were turning into a desert much like the Sahara. The pioneers had not appreciated or understood the grassland biome native to the Plains. A land-use policy modeled on the climax theories of ecology was needed. A resident ecologist on the Plains could promote ecological education for the benefit of the people and could advise farmers on how to conserve the soil.

Scientists working at the Universities of Kansas and Nebraska in entomology and soil science further developed the organismic approach during the Dust Bowl years. Entomologist Roger Smith of Kansas State University considered the ways in which insect pests began to increase and take over during periods of drought, again helping to upset the balance of nature.

John Weaver and Evan Flory, ecologists at the University of Nebraska, proposed that the country should create a grassland conservation program based on Clements' concept of the climax ideal. Clements himself began to look at regional management plans from an ecological point of view. He proposed saving relict prairies adjacent to areas where species had disappeared. Relicts remained in cemeteries, along railroad tracks, and along fencerows, where one could identify native species and then perhaps replant and restore them. Ecological historian Ronald Tobey writes: "Clements envisioned ... that a plan to save the plains would require the federally regulated use of the midcontinental grasslands on scientific principles and the resettlement of the region's rural population.... The ecologist was to be a social planner and manager, deriving his authority from the highest level of government."[13]

Another branch of the organismic school of ecology developed at the University of Chicago after World War I through the collaboration of a group of animal ecologists: Warder C. Allee, Alfred E. Emerson, Orlando Park, Thomas Park, and Karl P. Schmidt. The group rejected Social Darwinist assumptions of a nature characterized by Thomas Henry Huxley as "red in tooth and claw," for a nature of cooperation among individuals in animal and human communities. According to historian of science Gregg Mitman, this school of ecologists developed "a focus on the organism-environment relationship as an interactive process, in which the organism, through its behavior and activity, continually restructured its environment to meet new demands.... It was a developmental picture that was goal directed and progressive, lending credence to one of ecology's earliest cherished principles — succession."[14] The group published a foundational volume in 1949 entitled *Principles of Animal Ecology*, in which they elaborated their theories of community organization, succession, and development. For Allee and Emerson, the ecologist functioned as a social healer, applying knowledge gained from the study of the natural world to repair the human social organism damaged by depression and war, aggression and competition. They believed, states Mitman, that "the task of the biologist was to discover nature's moral prescriptions and thereby serve as savior of society."[15]

While Allee and Emerson sought to use ecology to heal human society, ecologist Aldo Leopold (1887–1948) wanted to use society to heal the land. Leopold's work became the basis for restoration ecology and for a new kind of ethical relationship between the human being and nature, one that derived from Clements' organismic approach. By following nature as teacher, people could restore the land as doctors healed their patients. Leopold's famous manifesto, "The Land Ethic," written about 1948, was published posthumously in *A Sand County Almanac* (1949). Here, Leopold differentiated between two forms of ethics that he called the A-B cleavage, that is, the land used for human benefit, a homocentric view, versus the land as a biotic pyramid, an ecologically centered view. In the A portion of the A-B cleavage, forestry, wildlife, and agriculture were managed for human benefit; in the B portion, they were seen as parts of a larger ecological system. In *Game Management* (1933), Leopold had viewed game — such as deer or elk — as crops, just as trees or foods were crops. But by 1948, he was differentiating between the land as a managed system and the land seen as a natural process that reproduced and maintained itself. The organismic approach thus saw people as separate from nature, but capable of both disrupting its successional systems and having the scientific and ethical tools to follow nature and heal it.

The Economic Approach to Ecology

Donald Worster contrasts the organismic approach to ecology with the competing economic approach, or "new ecology," in which organisms are characterized as producers and consumers and nature maximizes efficiencies and yields, ultimately to the benefit of man, the consumer at the apex of the food chain. "The New Ecology," Worster notes, was "an energy-economic model of the environment that began to emerge in the 1920s and was virtually complete by 1950."[16] The economic approach grew out of critiques of Clements' organismic theories. Among the early critics of Clements was botanist Henry Gleason (1882–1975), who in 1926 published a paper entitled "The Individualistic Concept of the Plant Association." Gleason argued that the type of vegetation that occurred in a particular place was more accidental than Clements' theory allowed. "The vegetation of an area is merely the resultant of two factors, the fluctuating and fortuitous immigration of plants and an equally fluctuating and variable environment." Plants migrated from one place to another as seeds, carried by winds, storms, and other disturbances. Some of those seeds took root and thrived. The environment itself was likewise more variable than Clements had held. Gleason believed that if two areas had the same vegetation, it was by chance alone. "Every species of plant is a law unto itself," he wrote. "The species disappears from areas where the environment is no longer endurable."[17] Migration and environmental selection operated independently in each area, no matter how small. Clements' climax community was not the result of a determinative law, but was a mere association or mosaic of plants. Succession was haphazard, not a harmoniously ordered developmental process.

Another critic of Clements, British ecologist Arthur Tansley (1871–1955), developed ideas that led directly to the economic approach to ecology. In a 1935 paper entitled, "The Use and Abuse of Vegetational Concepts and Terms," he introduced the term "ecosystem" to the field. He challenged the use of anthropomorphic concepts from human development that Clements had imposed on nature when in the process of plant succession, nature developed as did a living organism. Tansley, schooled in the laws of thermodynamics developed in nineteenth-century studies of heat transfer in steam engines, began to apply those concepts to nature. Rather than considering a plant community as greater than the sum of its parts, as in organismic theory, Tansley held that the whole was simply the sum of its parts, as in Euclidean geometry. The ecosystem was made up of individual components that were physical entities, like the parts of a machine. The components were quantifiable and the method reductionist and analytic. Tansley applied terms from thermodynamics, such as "energy," "systems," and "fields." He described the structure of the ecosystem in terms of abiotic factors — such as oxygen, nitrogen, carbon dioxide, calcium, and phosphorous — and biotic factors — such as plants, animals, and bacteria. With the ecosystem concept, ecology moved toward an economic or managerial approach, in which humans considered themselves above nature and able to control it. The goal was to manage nature, using utilitarian values to enhance human life. If nature was managed in a beneficial way, the productivity and yields of forests, soils, and crops could be increased.

The economic approach to ecology drew on such concepts as producers, consumers, efficiencies, and yields. As early as 1926, August Thieneman had begun to designate producers, consumers, and reducers. Producers are organisms, such as trees and grasses that collect energy from the sun by photosynthesis and create complex organic compounds from simple chemicals. Consumers, such as animals, use these chemicals for growth and for metabolism. In a terrestrial ecosystem, the primary consumers are herbivores, such as rabbits or deer, which eat the grasses. Then secondary consumers — predators, such as mountain lions or wolves — eat the herbivores. At the top of the scale are tertiary consumers, such as humans. Completing the cycle are reducers or decomposers — micro-organisms that break down animal dung and detritus, such as falling leaves and dying grasses — into simple chemicals to be taken up by the roots of trees. A pond ecosystem is similar to the terrestrial ecosystem. Here, the producers are aquatic plants, such as marsh grasses growing along the edge of the pond. Phytoplankton are plant-like producers in the water, while the primary consumers are zooplankton, or animal-like plankton. A fish is a secondary consumer and a turtle a tertiary consumer. In the pond's sediment are Thieneman's reducers, or what are now called decomposers.

The economic approach to ecology uses other terms derived from agriculture, such as annual yields. The total organic structure formed per year at any one of the levels of producers — or primary, secondary, and tertiary consumers — can be calculated, along with the efficiency of energy transfer from one level to another.

In 1942, Raymond Lindeman (1915–42) introduced into ecology the idea of the food chain, or trophic levels, in a paper called "The Trophic-Dynamic Aspect of Ecology." Trophic means nutrition, and the food chain represents the process of food metabolism at each trophic level. Metabolism produces energy, so that when one organism is eaten by one at the next level, its energy is transformed into food, and in turn into that organism's own energy. Each organism thus converts energy from the trophic level below it for its own survival. From measurements of the energy expended at each level can be calculated the efficiency of energy transfer and hence the productivity of each level.

English ecologist Charles Elton (1900–91) of Cambridge University integrated animal populations into the system of trophic levels. He counted the numbers of various species in the ecosystem and studied the ways the numbers of those species changed, developing the concept of the pyramid of numbers. As he formulated it, the combined weight of predators must be less than that of all the food animals, and these in turn must be less than plant production. The Eltonian pyramid can be illustrated by the number of individuals at each trophic level. If at the bottom of the pyramid, there are a million producers, or phytoplankton, then at the next higher level, the number of primary consumers, or zooplankton, would be much smaller, for example 10,000. Secondary consumers, such as fish in the aquatic ecosystem, would then be 100, and the apex organism at the top, such as a human being who consumes the fish, would only be one. Thus one person might be supported by a million phytoplankton. If people are vegetarians, more people can eat at a lower level of the Eltonian pyramid, and therefore the plant community can support a larger human population than if they are consumers who eat beef.

According to Elton, at each level there is an inefficiency in energy transfer owing to the second law of thermodynamics. The energy available for work declines at each level of the energy-producing process. As in the case of the steam engine, one hundred percent efficiency of energy output is never possible. No steam engine can transform all its fuel into energy available to do work because it loses energy to the environment in the form of heat. That heat, or the energy unavailable to do work, is entropy, a quantity that is always increasing in the environment. At each step in the Eltonian pyramid, heat is lost, and therefore the calories, or usable energy available for the next level, keep declining. For example, 10,000 calories at the level of producers might decline to 1,000 at the level of primary consumers, to 100 calories at the level of secondary consumers, 10 calories for tertiary consumers, and one calorie for the apex organism, such as the human being.

The economic approach to ecology was epitomized in 1968 by ecologist Kenneth Watt (1929–) of the University of California at Davis, who stated that the goal of ecology was to "optimize the harvest of useful tissue."[18] In Watt's *Ecology and Resource Management: A Quantitative Approach* (1968), the human being is seen as an economic animal, and economic ecology's goal is to maximize the productivity of each type of ecosystem and each level of that ecosystem for human benefit.

During the 1960s, ecologist Eugene Odum (1913–) at the University of Georgia argued that maximizing the productivity of ecosystems, as advocated by the economic approach, led to their degradation. Odum's concept of the balanced, homeostatic ecosystem integrated Clements' ideas about the evolution of the biotic community through successional stages, with thermodynamic approaches that employed food chains and energy transfers. In nature, there was a homeostatic balance within a mature ecosystem. The primary disturbance came from human beings. If humans disrupted the system by cutting forests or plowing prairies, a degraded ecosystem resulted. Expanding human populations, increasing demands for food, and growing industrial wastes were upsetting the balance of nature by polluting lakes, rivers, and wetlands, and depleting natural resources. But Odum argued that if humans cared about nature and were conservers of its processes, we could restore those degraded systems. Following a line of thinking similar to that of Ellen Swallow, Odum focused on the idea of the *oikos* as the human home and, like Aldo Leopold, argued that we need to be wise managers of the landscapes in which we live. We need to conserve our resources and use our capabilities as human managers to restore the lost balance of nature. Odum's synthesis of ideas from the organismic and economic approaches to ecology became the guideline for the environmental movement that emerged in the 1960s. Humanity might disrupt evolved ecosystems, but through scientific ecology and ethical principles, it could repair the damaged web of nature on which life depended for its continued existence.

The Influence of Chaos Theory

The idea of the balance of nature was challenged by a fourth approach to ecology, which began to take shape in the 1970s and 1980s with the development of chaos theory. The new science, according to Worster, represented "a radical shifting away from the thinking of Eugene Odum's generation, away from its assumptions of order and predictability, a shifting toward what we might call a new *ecology of chaos*."[19] Natural, rather than human disturbances, were viewed as playing a greater role in transforming nature than had previously been assumed by the balance of nature approach. Massachusetts biologists William Drury and Ian Nisbet, in a 1973 paper, "Succession," revived Henry Gleason's 1926 "individualistic" approach, arguing that succession leads neither to stability nor equilibrium. The following year, mathematical ecologist Robert May published "Biological Populations with Nonoverlapping Generations: Stable Points, Stable Cycles, and Chaos," demonstrating that older mathematical models in ecology could not predict aperiodic ecological events such as pest outbreaks.

Other ecologists who challenged both Clements' organismic approach and Odum's balance of nature included ecologists S.T.A. Pickett and P.S. White, whose work on *The Ecology of Natural Disturbance and Patch Dynamics* (1985) argued that ecosystems should be described in terms of patch dynamics. "Patch dynamics includes not only such coarse-scale, infrequent events as hurricanes, but also such fine-scale events as the shifting mosaic of badger mounds in a prairie." Ecosystems should be considered dynamic, fine-textured patches, rather than homogeneous stable systems. "Preservation of natural systems necessarily involves a paradox," they noted, "we seek to preserve systems that change. Success in a conservation effort thus requires an understanding of landscape patch structure and dynamics."[20]

During the 1980s and 1990s, the idea that nature itself was a major disturber of ecosystems became more widely incorporated into ecology through the work of population biologists. Chaos theory lent credence to the idea that natural disturbances, such as hurricanes, tornadoes, and earthquakes, could within a few minutes completely destroy an ecosystem that might have taken centuries to evolve. Unpredictable, chaotic disturbances began to play a much more important role in ecosystem studies.

Influenced by the new work in chaos theory, Eugene Odum, his brother Howard T. Odum, and his son William Odum wrote a paper entitled, "Nature's Pulsing Paradigm" (1995), in which they modified some of their earlier ideas about nature as a balanced ecosystem. A more realistic way to think about nature, they stated, is in terms of pulsing steady-state systems. A pulsing system is a repeating oscillation that is poised on the edge of chaos. Examples of pulsing ecosystems are tidal marshes and seasonally flooded wetlands. Ecosystem performance and species survival are enhanced when external and internal pulses are coupled together. They suggested that pulsing was a key to sustainability in ecosystems. In a similar vein, recent work in complexity theory characterizes a complex system as one that exists on the edge between order and chaos. Many ecosystems and human communities are examples of changing, perhaps pulsing, complex systems. Whereas an ethic based on the balance of nature grants humans the capacity and power to restore degraded systems, chaos and complexity theory challenge humanity to recognize nature as both predictable and unpredictable, orderly and disorderly. Both humans and nonhumans disrupt nature, and both can work in partnership to restore it.

Conclusion

The history of ecology can be seen as moving through several historical phases, each with differing assumptions about nature and the human ethical relationship toward it. Human ecology incorporates humans into nature and adapts to nature's limits; organismic ecology views humans as separate from nature, but as followers of its balanced, homeostatic processes; economic ecology asserts humans as scientific managers of a nature that can be controlled for human benefit; finally, chaotic ecology sees nature as having unpredictable characteristics, leaving humans as only partially able to manage its systems. Nature is thus far more complex than previously considered, and is best described as a disorderly order, rather than a harmonious balance.

Notes

1. Ernst Haeckel, *Generelle Morphologie der Organismen* (Berlin: Reimer, 1866), 2:286–87. English translation in Robert C. Stauffer, "Haeckel, Darwin, and Ecology," *Quarterly Review of Biology* (1957) 32: 138–44, at 140, 141.

2. Haeckel, *Generelle Morphologie*, 2:286–87.

3. Haeckel, "Ueber Entwickelungsgang und Aufgabe der Zoologie, in *Gesammelte populäre Vorträge aus dem Gebiete der Entwickelungslehre*" (January 1869), 17; English translation in Stauffer, "Haeckel, Darwin, and Ecology," 141.

4. Ernst Haeckel, *The History of Creation, or the Development of the Earth and Its Inhabitants by the Action of Natural Causes*, translation revised by E. Ray Lankester (London: Henry S. King, 1876 [1873]), 1:xiv; 2:354, italics in original.

5. Ernst Haeckel, *The Evolution of Man: A Popular Exposition of the Principal Points of Human Ontogeny and Phylogeny* (New York: D. Appleton, 1879), 1:114.

6. Robert Clarke, *Ellen Swallow: The Woman Who Founded Ecology* (Chicago: Follette, 1973), 39–40.

7. Robert P. McIntosh, *The Background of Ecology: Concept and Theory* (New York: Cambridge University Press, 1985), 20.

8. Ibid., p. 20.

9. Ellen Swallow Richards, *Sanitation in Daily Life* (Boston: Whitcomb and Barrows, 1910 [1907]), v.

10. Donald Worster, *Nature's Economy: A History of Ecological Ideas*, 2d ed. (New York: Cambridge University Press, 1994), 209.

11. Frederic Clements, *Plant Succession: An Analysis of the Development of Vegetation* (Washington, D.C.: Carnegie Institution of Washington, 1916), 124–25.

12. Ibid., 60–61.

13. Ronald Tobey, *Saving the Prairies: The Life Cycle of the Founding School of Plant Ecology, 1895–1955* (Berkeley: University of California Press, 1981), 207.

14. Gregg Mitman, *The State of Nature: Ecology, Community, and American Social Thought, 1900–1950*, 3; on T.H. Huxley's lecture "Evolution and Ethics" (1893), see 2.

15. Ibid., 7.

16. Worster, *Nature's Economy*, 311.

17. Henry Gleason, "The Individualistic Concept of the Plant Association," *Bulletin of the Torrey Botanical Club* 23 (1926): 26.

18. Kenneth Watt, *Ecology and Resource Management: A Quantitative Approach* (New York: McGraw-Hill, 1968), 54–56.

19. Donald Worster, "Ecology of Order and Chaos," *Environmental History Review* 14, no. 1–2 (Spring/Summer 1990): 4–16, at 8, emphasis in original.

20. S.T.A. Pickett and P.S. White, *The Ecology of Natural Disturbance and Patch Dynamics* (Orlando, Fla.: Academic, 1985), xiii, 5, 12.

10 The Era of Environmentalism, 1940–2000

During the latter half of the twentieth century, the resource conservation movement based on efficient use of natural resources changed to an environmental movement concerned with quality of life, species preservation, population growth, and the effects of humanity on the natural world. A multitude of government projects, policies, and laws, together with citizens' movements, increasingly regulated economic development and sought to preserve remaining wilderness areas. The rise of environmentalism is a core theme in environmental history, because it often influences the way contemporary historians look at the past and the topics they choose to investigate. This chapter examines the conservation movement of the mid-twentieth century and explores the emergence of the environmental movement of the late twentieth century, its regulatory framework, and its philosophy of nature.

From Conservation to Environmentalism

According to environmental historian Samuel P. Hays, the period between the conservation movement of the mid-twentieth century and the emergence of the environmental movement in the 1960s and 1970s displayed a shift in emphasis from resource efficiency to that of quality of life based on "beauty, health, and permanence." Hays notes: "We can observe a marked transition from the pre-World War II conservation themes of efficient management of physical resources, to the post-World War II environmental themes of environmental amenities, environmental protection, and human scale technology. Something new was happening in American society, arising out of the social changes and transformation in human values in the post-War years."[1]

Conservation policies in the first two decades of the century had focused on forests, rangelands, and water. During the 1930s and 1940s, resources such as grasses, soils, and wildlife came under the umbrella of conservation, leading to laws such as the Taylor Grazing Act (1934), the Soil Conservation Act (1935), and the Pittman-Robertson Wildlife Restoration Act (1937). But by the 1960s, quality of life issues, aesthetics, pollution, and environmental toxins began to take on greater importance. People became engaged with saving wild and scenic rivers for fishing and trout habitat, rather than damming rivers for massive irrigation projects. They enjoyed seeing wildlife and viewing birds, as opposed to treating them as game to be managed. They became concerned with the health effects of pesticides and polluted air and water as opposed to maximizing crop and timber yields. In general, the shift was away from production goals and toward consumption, health, and quality of life concerns.

New Deal Conservation

During the 1930s and 1940s, the federal government promoted conservation in ways that would benefit wage-workers and jobless and homeless people, as well as Native Americans. Such issues dovetailed with the creation of the welfare state. Franklin Delano Roosevelt's New Deal helped to repair the effects of the Great Depression that followed the stock market crash of 1929. The government provided relief for the unemployed during the Depression and took on the responsibility of promoting the general welfare. The conservation and preservation of natural resources became components of a federal policy that was tightly integrated with innovations in social welfare, such as the creation of the social security system. Wise management of natural resources became a hallmark of the New Deal era, along with the development of electric power from the construction of large dams and irrigation projects.

Large government-sponsored projects, such as the Tennessee Valley Authority (TVA) on the Tennessee River, Boulder (now Hoover) Dam on the Colorado, and the Grand Coulee and Bonneville Dams on the Columbia River, were developed during the 1930s as part of the program of promoting the national welfare by making hydroelectric power readily available to farmers, rural communities, and cities. TVA was created in 1933 to regulate water distribution and electrical power throughout the Southeast, in the states of Tennessee, North Carolina, Virginia, Kentucky, Alabama, and Mississippi. The goal was to provide cheap electric power for the country, along with flood control, recreation, and soil conservation.

Hoover Dam (originally called Boulder Dam), located just east of Las Vegas, was completed in 1935 and spans the Colorado River between the states of Nevada and Colorado. The structure reached 726 feet above the river, and created the 115-square-mile Lake Mead. The water captured and held behind the dam would cover the entire state of Connecticut up to a depth of 10 feet. The dam was created not only to provide flood control, but also to generate cheap electric power. Water and power were brought by electrical transmission lines, irrigation pumps, and aqueducts for distribution to the southwest, including Arizona, Nevada, and the Imperial Valley in southern California. The broader purpose of the dam was to enhance the health and comfort of the people of the southwest and to provide for their welfare.

Roosevelt's speech at the dam's 1935 inaugural exemplified the goals of the great era of dam building. He constructed a narrative of progress, a story of transformation of deserts into gardens. Before the dam was completed, the President noted, there had been only a desert and a steep, gloomy canyon. He described Boulder City, which was built to house the construction workers, as having been a cactus-covered waste, and the river as a raging torrent that, in times of drought, diminished to a mere trickle of water. Those below lived in constant fear of impending disaster. Each spring, they awaited with dread the coming flood that would destroy their crops. But in June of 1935, after completion of the dam, a great flood poured down the canyons of the Colorado, through the Grand Canyon, the Iceberg, and Boulder canyons, to be caught safely behind Boulder Dam. The water that was now trapped could be released in controlled flows and transported via aqueducts. Roosevelt celebrated the completion of the dam, proclaiming it the greatest in the world.

Another welfare state program was the Civilian Conservation Corps, or CCC, a federal program established in 1933 and operated through 1942, when it was abolished. Designed to relieve unemployment, it promoted the conservation of natural resources and administered educational and vocational training for 3 million young men ages 17 to 23. CCC workers developed parks, planted forests, and constructed fire towers. Their contributions enhanced many state parks and recreation areas throughout the country.

Other 1930s conservation measures were also aimed at improving the general welfare. In 1934, the Taylor Grazing Act limited grazing on the Great Plains and elsewhere to preserve soil and prevent erosion. Grazing was controlled, and the number of animals, measured in animal units, limited. The 1935 Soil Conservation Act established soil conservation districts and the federal Soil Conservation Service (SCS). Farmers in a given area joined together to develop their own programs of soil and water conservation, organizing legally as groups of landowners. They determined how to improve their land and water, in some cases leaving the land itself totally out of production, so that the soil would recover. By August of 1947, there were 1,670 districts initiated by farmers in the United States, with a total of 900 million acres and 4 million farms, with districts being created at the rate of 25 per month.

Attitudes toward wildlife also changed during the mid-twentieth century. The Pittman-Robertson Wildlife Restoration Act of 1937 provided federal aid for the restoration of wildlife and the acquisition of wildlife habitat by taxing sales of sportsman's guns and ammunition. The goal was to restore wildlife resources for economic, scientific, and recreational purposes. The dichotomy between predators (such as wolves, mountain lions, and coyotes) and game (such as deer, antelope, and elk) also began to give way to an appreciation of the wider roles of wildlife within natural systems. Poisoning programs and sports hunters were challenged by wildlife ecologists and conservationists, who argued that animal populations were valuable parts of functioning ecosystems, rather than varmints to be exterminated. By the 1960s, the newer attitudes would set the stage for citizens' campaigns against pesticides and for the preservation of endangered species.

The Rise of Environmentalism

In the 1960s, concerns over quality-of-life issues were the major drivers of a growing effort to expand environmental policy. During World War II, the country had focused on expanding production, but in its aftermath, people became engaged in rebuilding their lives. Women who worked in the war industries moved back into the home, and returning veterans took up the jobs women had filled. The problems of readjustment helped to fuel the rise of the women's movement of the 1960s. Additionally, the civil rights movement and anti-Vietnam War movements caused many people to question the social status quo. In this social milieu, issues of environmental quality came to the forefront of public concern.

Nuclear fallout was a major issue, resulting from atomic bomb tests conducted in the atmosphere in the 1950s, and the idea that radiation could appear in the food chain at vast distances from test sites became a public concern. Rachel Carson's *Silent Spring* (1962) polarized pesticide producers against society as a whole and brought to public attention concerns over DDT, chlorinated hydrocarbons, organophosphates, and other chemicals. Carson's language introduced new metaphors into environmental awareness. The landscape created by pesticides was filled with hazards for nature and nature's inhabitants. Toxic threats came from DDT, which, by accumulating in the food chain, caused birds to die and egg shells to break. Environmental and human health problems also stemmed from chlorinated hydrocarbons (such as aldrin, dieldrin, and chlorodane) and organophosphates (such as parathion and malathion). These pesticides had been used effectively to control mosquitoes, lice, and insect pests on crops in the post-World War II decades, but the side effects on human and ecosystem health were known mainly to scientists. Carson's compelling book brought them to the attention of the public.

Silent Spring not only mobilized Americans, but also mobilized entomologists, creating what environmental historian John Perkins, in *Insects, Experts, and the Insecticide Crisis* (1982), identified as a paradigm shift from the use of chemical controls to biological controls. Biological rather than chemical techniques were developed to enhance populations of natural enemies, such as birds and beneficial insects that reproduce along the borders of fields. Integrated pest management synthesized biological controls with limited uses of chemicals targeted for specific insects at limited times in the growing cycle.

Population and the Environment

During the late 1960s, Paul Ehrlich's *The Population Bomb* (1968) raised alarm over the rapidity of population growth, creating deep concerns that continue to the present. The book became a best-seller, alerting the world to the problem of increasing populations and the concomitant social and environmental consequences. The basis for the idea was Thomas Malthus' 1798 *Essay on Population*, which asserted that population was expanding at geometrical rates while food supplies were increasing at only arithmetical rates. Unless the human species initiated an era of birth control, the only curtailment of population growth would be through ecological and social collapse. Ehrlich looked at the exponential growth of global population from A.D. 1 to 2000. Eight thousand years ago, world population was about 5 million; in 1650 it had increased to 500 million; by 1850 it was about a billion. During the twentieth century, world population increased at rates of about 2.1 percent per year through the 1960s. The data showed that by the year 2000 the earth would contain more than six billion people, a prediction that came to pass on October 12, 1999.

Critics of *The Population Bomb* argued that if world population growth is looked at from a different perspective, rather than an alarming exponential growth curve, one can see a logistic curve, in which population growth rates begin to decline and level off during the twenty-first century. The world growth rate slowed from 2.1 percent in the 1960s to about 1.8 percent in 1990, and the doubling time for population increased from 33 years to almost 40 years. The length of time for the population to double is continuing to increase and is consistent with the decline of the growth rate.

Ehrlich later moderated his approach in *The Population Explosion* (1990), coauthored with his wife, Ann Ehrlich. The two scientists emphasized that one of the major means of dealing with population increase is through reduced fertility, obtained not only through birth control, but through better sanitation, education, healthcare, and especially equal rights for women. If women throughout the world were educated, especially in the developing countries, the number of children born to a woman would be reduced. Women would take better care of their families and, with higher standards of living, would tend to have fewer children. Men should also be educated, allowing them to obtain higher-income jobs than the low-paying jobs that have historically contributed to the necessity for larger numbers of children to assist in raising family income. The Ehrlichs put their emphasis both on birth control and the social structures that would help to achieve it.

Environmental Regulation

Also during the 1960s, new environmental policies emerged from interactions among the three branches of government. Congress played a major role in passing legislation, while the executive branch implemented the policy that Congress created, exhibiting a considerable range of discretion in how the policy was applied. Disputes were challenged in the courts and resolved, at least temporarily, by the judiciary. Then Congress responded by passing additional laws or, in some cases, modifying and responding to what the judiciary had decided.

Within the executive branch, charged with implementing policy and setting regulations, a number of agencies administer the public lands, set regulatory policy, and enforce environmental laws. The Department of the Interior contains the National Park Service (NPS), established in 1916 to administer the separately created national parks previously administered by the military. Also in the Interior, the Bureau of Land Management (BLM), created in 1946 out of a merger between the General Land Office and the Grazing Service, administers non-forested and desert lands, primarily in the arid West. The Fish and Wildlife Service (FWS), established in 1940 by merging the Bureau of Fisheries in the Department of Commerce with the Bureau of Biological Survey in the Department of Agriculture, is devoted to the study of wildlife and the conservation of species. The Bureau of Reclamation (BuRec), created in 1902, is responsible for the building and oversight of dams, especially those in the western states.

Within the Department of Agriculture, the U.S. Forest Service (USFS), created in 1886 as the Division of Forestry and established as the U.S. Forest Service in 1905, administers the cutting of timber on public lands, often providing low-cost timber sales to timber companies. The Soil Conservation Service (SCS), established in the Department of Agriculture in 1935, is dedicated to maintaining soil quality. It initiates efforts to prevent water and wind erosion and administers soil banks. The Department of Energy is concerned with nuclear energy and other appropriate energy technologies. Directly under the executive branch are also agencies established in the mid-twentieth century, such as the Tennessee Valley Authority (TVA), and those created in the 1960s and 1970s, such as the Environmental Protection Agency (EPA), the Water Resources Council (WRC), and the Nuclear Regulatory Commission (NRC). The Environmental Protection Agency (EPA) was established in 1970 through executive reorganization and given authority to regulate air and water quality, radiation and pesticide hazards, and solid-waste disposal. Whenever Congress passes an environmental law, the EPA is charged with issuing regulations to enforce it.

Congress, the second player in the environmental policy process, is responsible for passing environmental legislation. Attempts to legislate solutions to problems of pollution and depletion during the 1960s resulted in such laws as the Clean Air Act (1963) and the Water Quality Control Act (1965), both of which have been amended and updated several times. A major effort was mounted to preserve remaining wild areas by passing of the 1964 Wilderness Act, designating certain federal lands as wilderness areas in which commercial development was prohibited. The Wild and Scenic Rivers Act (1968) preserved free-flowing rivers for recreational and conservation purposes.

In 1969, the National Environmental Policy Act (or NEPA) was passed and signed into law on January 1, 1970 by Richard Nixon, creating a vast regulatory industry through the preparation of Environmental Impact Statements (EISs) at the federal level. NEPA began with the statement that Americans needed to recognize the profound impact on the environment of human activities, including urbanization, population growth, industrial pollution, resource exploitation, and technology. It asserted that humanity must promote the general welfare and bring "man and nature" into productive harmony with each other, in order to provide for the needs of future generations. Like the conservation movement of the early twentieth century, NEPA was concerned with long-term resource use and the quality of life. The law mandated that the U.S. government take "a systematic interdisciplinary approach to the integrated use of natural resources through the natural and social sciences." This was an integrated management approach that used both the social sciences and the natural sciences, drawing especially on the biological sciences. Scientists were to work with the Council on Environmental Quality (CEQ) to bring into the equation nonquantifiable aspects of natural resources. They would use not only cost-benefit analysis, placing monetary values on natural entities, but would take into consideration the aesthetic and scenic aspects of nature. NEPA mandated that "for every action that is going to affect the quality of the human environment, it is necessary to make a federal impact statement."[2]

The 1970s became known as the environmental decade, as well as the era of environmental regulation, a period in which increasing numbers of laws were passed to improve the environment. The Clean Air Amendments of 1970 were followed that same year by the Occupational Safety and Health Act (OSHA), which established standards for exposure to harmful substances in the workplace. The Environmental Pesticide Control Act of 1972 regulated pesticide use and controlled the sale of pesticides in interstate commerce. In 1973 the passage of the Endangered Species Act (ESA) established procedures for listing threatened and endangered species in critical habitats. The Safe Drinking Water Act (1974) authorized the EPA to set federal drinking water standards, while the Clean Water Act of 1977 required waters to be "fishable and swimmable" by 1983. Toxic and hazardous chemical wastes also came under scrutiny with the Resources Conservation and Recovery Act (RCRA) of 1976, which established standards for the disposal of hazardous industrial wastes, and the Toxic Substances Control Act (TOSCA) of the same year, which regulated public exposure to toxic materials.

The judiciary is the third player in the environmental policy process. Environmental litigation developed as an important tool for curtailing pollution and promoting conservation during the 1960s and 1970s. Under the guidance of Secretary of the Interior Stuart Udall and Senator Edmund

Muskie, and with the support and encouragement of President John F. Kennedy, a series of conferences was held with heads of major corporations who were polluting and extracting resources from the environment in ways that threatened the quality of life. The conferences made it apparent that discussion alone was inadequate. The litigation process developed by the Environmental Defense Fund (EDF, founded in 1967) and the National Resources Defense Council (NRDC, founded in 1970) advocated taking the polluters to court. In 1974, environmental lawyer Christopher Stone proposed legal rights for natural objects in his book *Do Trees Have Standing?* A number of prime legal cases occurred in the 1960s and 1970s, such as the Storm King Power Development proposed on the Hudson River and the Walt Disney Corporation's plan for a recreational complex in Mineral King Valley in Sequoia National Park, California. By the end of the 1980s there were twenty thousand environmental lawyers in the United States and the courts had reviewed more than three thousand environmental law cases.

The 1960s and 1970s also saw the rise of powerful citizens' movements, manifested in the successful fight against damming the Grand Canyon in 1966, led by the Sierra Club's David Brower, and the national outpouring of citizens on Earth Day, April 22, 1970, conceptualized by Wisconsin Senator Gaylord Nelson. Twenty million people, from grammar school children to college students, from ordinary citizens to politicians, participated in rallies, marches, teach-ins, and debates to save the environment. The League of Conservation Voters was created in 1970 to monitor the positions and voting patterns of legislators. Books such as Barry Commoner's *The Closing Circle* (1971) and *The Limits to Growth* by Dennis Meadows, et al (1972) contributed to sustained public interest in solving environmental problems. The 1973 oil crisis increased citizen awareness of energy conservation and drove home the need for alternatives to fossil fuels. All these developments, from laws and court cases to marches and publications, made the decade of the 1970s the environmental decade.

Reactions to Environmental Regulation

In the 1980s, a backlash against the 1970s regulatory era developed, in which environmental quality standards were relaxed. Under the administration of President Ronald Reagan (1981–89), policies evolved that allowed industries to play a much greater role in regulating themselves than was the case in the 1970s. The EPA and OSHA suffered severe budget cuts, and the Council on Environmental Quality (CEQ) became a minor player in comparison to its strength during the 1970s. In a controversial move, made shortly after Reagan took office in 1981, the new president appointed James Watt as Secretary of the Department of the Interior and Anne Gorsuch as director of the EPA, both of whom were associated with corporate interests that had lobbied for reduced regulations. Anti-environmentalist rhetoric developed by the Republicans and the party's Christian right wing pervaded the administration. Watt, first president of the conservative, anti-environmental Mountain State Legal Foundation (1977), funded by Joseph Coors, Sr. of the Coors Brewing Company, argued for the wise use of the environment. In 1981, he stated, "my responsibility is to follow the scriptures which call upon us to occupy the land until Jesus returns. I don't know how many generations we can count on before the Lord returns."[3] In the early 1990s, Republicans accused California democrat Barbara Boxer of holding the view that "spotted owls are higher on the food chain than we are."[4] Anti-environmental forces, therefore, had far greater influence on environmental policy under the Republican administrations of the 1980s and early 1990s than in the 1970s.

Industry also reacted to increased environmental regulation, because the additional costs reduced profits. Economists Daniel Faber and James O'Connor, in a 1989 article entitled "The Struggle for Nature," argued that "Environmental regulations added to the costs of capital but not to revenues.... [P]ollution abatement devices and clean-up technologies usually increase costs, hence, everything else being the same, reduce profits, or increase prices."[5] Large amounts of capital investment were required on the part of corporations to comply with the new regulations. The highest capital investments occurred in industries involved in oil, mining, electricity, and metallurgy, as well as in electrical plants operated by coal and nuclear power. In 1980, it was reported that the United States had spent $271 billion between 1972 and 1979 on pollution abatement. Much of that went into Clean Air Act enforcement. Corporations began to displace those costs onto the public, raising the prices of consumer commodities. Additionally, they displaced wastes onto landfills, inner city neighborhoods, and toxic waste dumps, and began exporting wastes to the Third World. The legislative victories of the environmental movement in the 1970s came back as a financial and ideological reaction, leading to failures in environmental regulation in the 1980s. As a result, they argue, pollution is now much worse and toxic waste sites more abundant in many areas, such as Indian reservations, inner cities, and throughout the Third World. There are also higher levels of dangerous chemicals such as nitrates, arsenic, and carcinogens in the environment than before.

Environmental Organizations

At the same time that regulations were being relaxed under the Reagan administration of the 1980s, citizens were responding to the policies of Watt and Gorsuch by increasing donations to mainstream environmental organizations. The "Group of Ten," or the ten major environmental organizations, experienced growth as cutbacks in environmental enforcement spearheaded private contributions to their organizations, increasing their budgets and memberships. Invigorated by financial gains and energized by the need to press for continued environmental regulation, they began to focus their main efforts in Washington as lobbyists and lawyers.

The National Wildlife Federation (NWF) constructed a $40 million building and added some 8,000 new members a month in the early 1980s, while the Sierra Club increased from 80,000 to 500,000 members. The National Resources Defense Council doubled its numbers in the 1980s. As the mainstream organizations received more donations from their new members, they also began to forge alliances with corporations.

By 1988, the National Wildlife Federation had a budget of $63 million, much of its funding coming from corporations such as Amoco, Arco, Coca-Cola, Dow, Dupont, Exxon, and Waste Management, and an executive from Waste Management was placed on the board of directors, moving the NWF closer to corporate interests. Other environmental organizations showed similar patterns. The National Audubon Society's 1988 budget was $38 million, with many donors coming from the same corporations. The Sierra Club, with a budget of $19 million, was receiving funding from Coca-Cola, Pepsi, petroleum companies, and major banks. Studies show that many members of environmental groups come from the corporate world and were wealthy persons who wanted the benefits of the Sierras for hiking; at the same time, ironically, they were involved in the management of companies whose activities destroyed those opportunities.[6]

One result of the "Group of Ten" environmental organizations' Washington focus was a tendency to neglect local issues. In the 1980s, therefore, in conjunction with the efforts of the "Group of Ten" in Washington, grassroots organizations sprang up throughout the country in response to local environmental problems.

The Antitoxics Movement

The antitoxics movement grew out of public attention to the pollution of Love Canal, a community near Niagara Falls, New York, in 1978. The main spokesperson was Lois Gibbs, whose son and daughter had both experienced major health problems as a result of living in an area formerly used as a waste site by Hooker Chemical Company. The company had sold the site to the city of Niagara Falls for one dollar, and a school was built on it in 1954. As the mothers of the school children in the late 1970s talked to each other, they found that many were having miscarriages, giving birth to babies with birth defects, and discovering that their children were dying or contracting cancer at higher rates than would normally be expected. Gibbs and others put pressure on the state of New York, and it eventually agreed to buy out the Love Canal homes. Gibbs went on to found a magazine called *Everyone's Backyard*, and to organize a major coalition, the Citizen's Clearinghouse for Hazardous Wastes. Operating out of Washington, D.C., the organization assisted local groups in moving beyond the Not in My Back Yard (NIMBY) phenomenon, to Not in Anyone's Backyard, in a major effort to reduce the health effects of pollution and toxic waste dumping.

The numerous local antitoxics groups merged together in a loose organization in the mid-1980s under the directorship of John O'Connor, to form the National Toxics Campaign. They worked closely with spokespersons, such as actress Meryl Streep of Mothers and Others, to reduce pesticides and toxics in the environment. Much of the focus of the antitoxics movement was on the differential health effects of hazardous wastes experienced by minority communities. A 1987 report by the United Church of Christ's Commission on Racial Justice, entitled, "Toxic Wastes and Race in the United States," spearheaded a nationwide environmental justice movement. The report, released under the sponsorship of director Ben Chavis, argued that communities with the greatest number of commercial hazardous waste facilities had the highest composition of residences occupied by racial and ethnic minorities. Fifty-eight percent of the country's blacks and 53 percent of its Hispanics lived in communities where hazardous waste dumping was uncontrolled, communities such as Emelle, Alabama, Houston, Texas, and Chicago's South Side.

The report also pinpointed the fifty metropolitan areas in the country with the greatest number of blacks living in communities with uncontrolled toxic waste sites. New York showed three cities — Buffalo, Rochester, and

New York City. In Indiana, both Gary and Indianapolis were listed. In Ohio, the cities included Cincinnati, Toledo, Columbus, and Cleveland. In California, locations included Oakland, Richmond, and Los Angeles. In Louisiana, the report listed Shreveport, Baton Rouge, and New Orleans. Virtually every major city in the country had hazardous waste sites located in areas defined by zip codes as being the addresses of people who were members of minority communities.

In 1993, for example, *Audubon* magazine published an article on the role that Dow Chemical Company played in polluting the African-American community of Morrisonville, Louisiana, near New Orleans, where Dow had a 1,400-acre plant on the edge of the Mississippi River. The plant opened in 1958 and employed some 2,000 people. By 1970, there were 100 chemical companies and refineries in the stretch between Baton Rouge and New Orleans. The region, called Chemical Corridor, experienced very high employment growth. By 1993, it was producing 16 billion pounds of chemicals and generating 3.3 million pounds of toxic wastes a year. Employees routinely complained of nausea, headaches, and vomiting, as they worked in plants where high incidences of cancer were reported, including lung, stomach, gall bladder, pancreas, liver, and colon cancer. The region was soon dubbed Cancer Alley, and in 1992, Louisiana ranked forty-ninth on the list of states with respect to its poor environmental health.

The federal "Community Right to Know" law of 1986 asserted that any community could obtain access to the types of chemicals used in local industries and the quantities of pollutants the industries released. The law gave community groups a basis for local action. According to studies done by Citizen's Clearinghouse for Hazardous Wastes, the movement for environmental justice has been dominated by women, working in coalitions at the grassroots level. Women comprised 80 to 85 percent of all local activists. Women of color, especially Native Americans, blacks, Hispanics, Pacific islanders, Asian Americans, and whites have worked together on problems of toxic wastes and their effects on human health.

An example of a women-organized protest occurred when, in the 1980s, the City of Los Angeles attempted to build a waste recovery plant, called the Los Angeles City Energy Recovery Project, known as LANCER. A network of three facilities, LANCER would incinerate 1,600 tons a day of garbage. The plant spawned an enormous protest. In 1988, the Mothers of East Los Angeles (MELA), a coalition of Latino women who had originally come together to oppose the construction of a prison, began to protest the plan. One thousand women and men demonstrated together to put pressure on the local air quality management district of the California Department of Health Services and on the Environmental Protection Agency to halt the project. In 1991, after lawsuits threatened to drive up the costs, the City of Los Angeles withdrew the project.

In 1990, West Harlem Environmental Action organized a protest over emissions from the North River Sewage Treatment Plant, which was causing Harlem residents numerous health problems. NYPIRG, the New York Public Interest Research Group, founded by Ralph Nader, helped to organize the action and worked with protesters in a campaign called "The Thousand Points of Blight," a phrase derived from President George Bush's "Thousand Points of Light" campaign slogan. In another example, in 1993, a group of New York City Hispanics staged a major protest over a proposed incinerator in the Brooklyn Navy Yard. Originally known as Toxics Avengers, they changed their name to El Puente Ojo Cafe, or Brown Eyed Bridge. The group was heavily dominated by high school students.

Native Americans were another minority adversely affected by hazardous waste problems. Much of the uranium mining for the nuclear power industry needed for atomic bombs and nuclear power plants was done on Indian reservations. The workers, especially Native American mine workers, experienced high rates of lung and other kinds of cancers. Children, some of whom played in the mine tailings, succumbed to leukemia. In 1977, Women of All Red Nations was formed, with the acronym WARN, to protest the declining quality of life on Indian reservations and problems of reproductive health. They said, "We call for the recognition of our responsibility to be stewards of the land and to treat with respect and love our Mother Earth who is the source of our physical nourishment and our spiritual strength." The White Earth Recovery Project in Minnesota exemplified another long-standing disagreement between native peoples and the United States government over access to land and its natural resources. Women on the reservation sought public support in a brochure that began: "Since our lands were stolen, we've been to the state, we've been to Congress, we've been to the Senate, we've been to the Supreme Court, and now we're coming to you."

The antitoxics movement and the 1986 federal Community Right to Know law began to have a positive effect by the late 1990s. In 1998, the Environmental Protection Agency released its Toxic Release Inventory of 644 chemicals put into the air and water by 21,000 companies for the year 1997. In just a decade the released chemicals had declined by 43 percent, but nevertheless totaled a staggering 2.58 billion pounds.

The Transformation of Consciousness

In conjunction with the emergence of grassroots movements, three social movements arose, dedicated to changing people's ecological consciousness and societal arrangements. The deep ecology movement began with a 1973 article by Norwegian philosopher Arne Naess, entitled "The Shallow and the Deep Long-Range Ecology Movement." Deep ecology was promoted in the United States by philosopher George Sessions and sociologist Bill Devall, who published *Deep Ecology: Living as if Nature Mattered* (1985), the goal of which was to change people's consciousness about the human relationship to nature. Deep ecology called for a new ecological worldview — a relational total-field image, as Naess put it — that pursues a philosophy of person-in-nature rather than a separation of people from nature. It proposed a new ontology, or science of being, to rethink the human manipulation of nature. It argued for a new psychology, in which individuals identified with nature as a Self-Writ-Large. When a redwood tree is cut down, for example, it is as if an individual has lost an arm or leg, because each self can identify with a larger cosmic Self. A new consciousness is critical, deep ecologists argue, to saving the planet.

Social ecology also developed as a movement during the 1970s and 1980s. Social philosopher Murray Bookchin founded the Institute for Social Ecology in Vermont in 1974, aimed at changing human ecological relations with nature, particularly at the local level. Social ecology is based on the idea that the evolution of dominance hierarchies in human societies led to the human domination of nature. To overcome domination and move toward an ecological society, Bookchin promoted communitarian democracy, modeled on New England town hall meetings and face-to-face decision-making. Bookchin and his followers saw an opportunity to enact these ideas through the emerging Green Party in the United States, identifying especially with the left Greens.

A third movement to emerge in the 1970s and 1980s was ecofeminism. The movement originated in Europe, where French feminist Françoise d'Eaubonne founded the Ecology-Feminism center in 1972 and published a chapter entitled "The Time for Ecofeminism" in her book *Feminism or Death* (1974). Ecofeminism examines women's historical and cultural connections to nature and draws on women's activism to resolve environmental problems. Numerous women throughout the world embraced and promoted ecofeminism through actions, articles, and books as a movement to liberate both women and nature. Among others, Ynestra King explicated the concept through a course at Vermont's Institute for Social Ecology, and helped to publicize it through a 1980 conference entitled "Women and Life on Earth: Ecofeminism in the '80s." Irene Diamond and Gloria Orenstein developed the movement further by organizing a major conference in southern California in 1987, celebrating the 25th anniversary of Rachel Carson's *Silent Spring*, and in 1990 published *Reweaving the World: The Emergence of Ecofeminism*.

The concerns of the environmental movement in the 1990s moved to the global level. Environmentalists warned of a global ecological crisis manifested by ozone depletion, global warming, population expansion, vanishing species, and declining biodiversity. Out of such global concerns, the Earth Summit was held in Rio de Janeiro in 1992. The Montreal Protocol over ozone depletion, signed by 24 countries in 1987, had successfully reduced chlorofluorocarbon (CFC) emissions worldwide by 1995, but emissions in developing countries continued to grow. Alarm over increasing signs of global warming led to the Kyoto Climate-Change Conference in 1997. The resulting Kyoto Protocol sought to establish limits on greenhouse gas emissions in developed countries that would reduce gases to 1990 levels by 2012, while exempting developing countries. Although the United States signed the agreement in 1998, the U.S. Senate had yet to ratify it at the century's end. The extinction of wildlife was another major concern. By 1999, the U.S. Fish and Wildlife Service reported 1,180 species as threatened or endangered under the provisions of the 1973 Endangered Species Act. Nine wild salmon species in the Pacific Northwest were threatened, resulting in mandatory economic and environmental restrictions. As the world's population passed the six billion mark in October 1999, the urgency of dealing with the environmental impacts of a potential doubling in population by 2040 heightened. Problems to be resolved included the encroachment of humanity on wilderness areas and wildlife from industry and housing expansion, energy development, air and water pollution, and increases in endangered species.

Over the last three decades of the twentieth century, environmental policies and laws that dealt effectively with global and national impacts achieved some success, but received opposition within the executive and legislative branches of government, illustrating the strengths and weaknesses of the environmental policy process and the influence of citizens' movements. Reversal of the environmental crisis, environmentalists maintain, will depend on reducing fossil fuel emissions and hazardous chemical wastes, on finding humane ways of reducing population growth rates, preserving biodiversity, and implementing sustainable forms of livelihood on the planet.

Conclusion

The conservation movement of the early to mid-twentieth century focused on the efficient use of natural resources and wildlife management, on
government-sponsored irrigation, dam, and power projects, and on conservation projects of the New Deal era. With the rise of the environmental
movement in the 1960s, the emphasis shifted to quality of life, aesthetic,
and health concerns. Fueled by popular issues that arose in the 1960s, such
as environmental pollution and population growth, an era of regulation
emerged in the 1970s in which Congress passed numerous environmental
laws to regulate resource use and environmental quality. The 1980s wit-
nessed a relaxation in regulation, the growth of mainstream environmental
organizations, and the rise of citizens' movements focused on local environmental issues, hazardous wastes, environmental justice, and green politics.
During the 1990s, environmental concerns became increasingly global,
stimulated by the 1992 Earth Summit, global warming, ozone depletion,
species extinctions, and population growth. All of these efforts at solving environmental problems have had numerous successes, but the problems facing humanity in the twenty-first century remain monumental in scope.

Notes

1. Samuel P. Hays, "From Conservation to Environment: Environmental Politics in the United States Since World War II," *Environmental Review* 6, no. 2 (fall
1982): 14–29, at 19.

2. National Environmental Policy Act of 1969, Public Law 91–190, 83 stat. 852
(1970).

3. James Watt, quoted in David Helvarg, *The War Against the Greens* (San Francisco: Sierra Club Books, 1994), 69.

4. Oliver North, quoted in Tom Hayden, *The Lost Gospel of the Earth: A Call for
Renewing Nature, Spirit, and Politics* (San Francisco: Sierra Club Books, 1996),
63–64.

5. Daniel Faber and James O'Connor, "The Struggle for Nature: Environmental
Crises and the Crisis of Environmentalism in the United States," *Capitalism, Nature, Socialism*, no. 2 (summer 1989): 12–39, at 20.

6. Brian Tokar, "Marketing the Environment," *Zeta Magazine* (February 1990):
16–17, compiled from 1988 annual reports of major environmental organizations.

Part II

American Environmental History A to Z: Agencies, Concepts, Laws, and People

Abbey, Edward (1927–89). An avid proponent of desert preservation through books and essays, Edward Abbey served as a National Park Service ranger and firefighter in the Southwest. His book *Desert Solitaire* (1968) opposed "industrial tourism" by automobiles and excessive development in the national parks as being both destructive to the parks and to those who visit them. *The Monkey Wrench Gang* (1975) and *Hayduke Lives!* (1990) made the case that the West was being destroyed by dams, irrigation systems, bulldozers, and logging trucks. His work inspired the movement Earth First! to advocate "monkeywrenching," or the practice of sabotaging the machines that were destroying the land by strip-mining, clear-cutting, and damming wild rivers.

Adams, Ansel (1902–84). Ansel Adams's powerful and austere black-and-white images inspired conservationists and the general public to set aside new natural areas, such as King's Canyon National Park, and to preserve existing ones. A careful technician, he took thousands of photographs in order to achieve the desired contrast and texture in the final prints. Recognizing the uplifting power of Adams's work, the Sierra Club made it a central feature of countless coffee-table books, calendars, and magazines.

Addams, Jane (1860–1935). Founder of Hull House, a settlement house in Chicago, Jane Addams worked to improve the lives of the urban poor by helping to upgrade their living and working environments. Over several decades, Addams and a number of other idealistic, university-educated, and financially secure women and men decided to engage with the problems and concerns of poor urban immigrants by living among them. In addition to establishing kindergartens, playgrounds, and cooperative housing for working women and their children, Hull House members were a pioneering force in bringing about environmental reforms. Noticing the piles of garbage left in streets and alleys as a result of municipal neglect, Addams and her colleagues investigated the city's system of collection and studied the implications of untended refuse on human health. Galvanizing political support in the community, they set up incinerators, collected tin cans, and arranged for the removal of dead animals. While respectful of the cultures and practices of immigrants, they opposed activities such as slaughtering sheep in basements, collecting rags from dumps, and baking bread in unhealthy spaces. Addams's autobiography, *Twenty Years at Hull House*, was published in 1910. Increasingly frustrated by the limitations of purely local action, she also helped to found the Women's International League for Peace and Freedom, served as its president from 1919 to 1935, and received the Nobel Peace Prize for her efforts in 1931.

Adirondack Forest Preserve Act (1885). Created in 1885 by an act of the New York State legislature, the Adirondack Forest Preserve was one of the nation's earliest efforts to preserve forests and watersheds and create recreational areas for the public. The movement to preserve the Adirondacks grew out of widespread concern for the loss of forests for lumber, paper, tanning, and mining, as well as the clogging of the Hudson River and Erie Canal with debris from such industries. The Adirondack Park was established in 1892 by the New York State legislature as an addition to the Preserve, and in 1894 the legislature mandated that the Preserve "be forever kept as wild forest lands." The Preserve now consists of a complex of 2.6 million acres of public and private lands, while the Park constitutes 6 million acres.

Agriculture. The science and art of cultivating the soil to produce fruits and vegetables for human and animal consumption.

Air Pollution Control Act (1955). In 1955, Congress passed the first of several laws regulating air pollution. The Air Pollution Control Act declared Congressional protection for the rights of the states and local governments to control air pollution, and appropriated funds to support technical research carried out in the public and private sectors.

Alaska National Interests Lands Conservation Act (1980). A product of much negotiation between environmentalists and development interests, the Alaska National Interests Lands Conservation Act of 1980 set aside 104 million acres of lands and waters in the state of Alaska having scenic, historic, and wilderness value. The act also declared the right of rural residents to continue their subsistence way of life. The coalition of environmentalists who backed the bill included 1,500 national and regional groups whose combined membership totaled about 10 million people. Development-oriented interests and the state of Alaska opposed the original plan for much higher acreages, resulting in the 1980 compromise signed into law by President Jimmy Carter.

Alaskan Native Claims Settlement Act (1971). In 1971, as a result of the Alaskan Native Claims Settlement Act, Alaskan Eskimos, Aleuts, and Indians received $462.5 million in federal grants, $500 million in federal and state mineral revenues, and 22 million acres of land in exchange for their agreement to settle long-standing land claims.

American Antiquities Act (1906). In recognition of the need for cultural conservation as well as for the conservation of natural resources, Congress passed a 1906 act authorizing the president to designate federal lands containing historic landmarks and prehistoric structures as national monuments. Reputable institutions were allowed to conduct archaeological research on the monuments and to gather objects for museums and educational institutions.

American Indian Movement (AIM). A movement organized to reclaim Indians' spiritual heritage and relationship to the land, the American Indian Movement (AIM) came alive in the mid-1960s. Its philosophy is that the land is "part of an integrated whole — of nature or the universe — which includes human beings and other living things.... Human beings ... must fit in with it and take care of it." The movement has worked to reassert its former treaty rights to land, water, fish, and wildlife resources, to reduce poverty, to reclaim health, expand education, create employment, and strengthen ties among Indian peoples.

Animism. The idea found in Native American and pagan philosophies that everything in nature, including what is now considered inanimate, is alive and has an inner spirit, soul, or organizing power.

Anthony, Carl (b. 1939). An architect and writer, described as combining the traditions of Martin Luther King and John Muir, Carl Anthony has worked for a more socially inclusive and culturally diverse environmental movement. Anthony, who is African American, has been especially active in San Francisco Bay Area initiatives such as the Urban Habitat Program, of which he is the founder and executive director, and the Earth Island Institute. His publications include a newsletter, *Race, Poverty and the Environment;* the *Natural Energy Design Handbook* (which he coauthored in 1973); and a number of articles urging African Americans to become environmentalists.

Anthropocentrism. The idea that the human species is the center of the natural world and that all other parts of that world exist for the sake of humankind.

Appalachian Mountain Club. A hiking and conservation club in the eastern United States, founded in 1876. The club's goal is to preserve the delicate balance of nature while maintaining environmentally safe ways for humans to enjoy the wilderness. It maintains huts and hiking trails throughout the Appalachians, organizes hikes, sponsors workshops for educational programs, and produces books, maps, newsletters, and souvenirs. Its "Conservation Action Network" updates members about conservation issues on which they can take action.

Audubon, John James (1785–1851). Famous for his paintings of the *Birds of America* (1827), Audubon was both an artist and writer who documented his impressions of American natural history, scenery, and landscape transformation in the East and Midwest. Audubon's paintings were made from freshly killed models, obtained by shooting and then quickly painting them in life-like poses. His *Delineations of American Scenery and Character,* published posthumously in 1926, chronicles his travels and observations during the years 1808–34, as Americans carved out homesteads on the frontiers, logged the forests, navigated the rivers, and transformed the land from swamps and thickets to cornfields and orchards. He both admired the industry of the settlers and lamented the rapid loss of wilderness and wildlife. Audubon's name is given to the National Audubon Society, founded in 1886 and reestablished in 1898, and whose mission is the protection of birds and their habitats.

Austin, Mary (1868–1934). In her numerous books on the western and southwestern deserts, novelist and essayist Mary Austin was a forceful advocate for the preservation of natural areas and Indian lifeways. The landscape and its natural forces are actors in her work, shaping a complex relationship between humans and the natural world. Having built an early career in California's interior Central Valley in the 1890s, she continued her development in the community of writers and artists on the ocean's edge in Carmel, and ultimately in Santa Fe, New Mexico. Austin's best known work includes her essays for the *Overland Monthly* in the 1890s, *Land of Little Rain* (1903), and *The Ford* (1917). The latter novel recounts the dramatic changes in the Central Valley landscape as a result of water diversion projects, oil drilling, corporate ranching, and cash crop agriculture, and suggests the alternative possibility of equitable rural productive partnerships.

Bailey, Florence Merriam (1863–1948). Ornithologist and nature writer, Florence Merriam Bailey was a distinguished contributor to ornithological studies in the Southwest. She was the author of *Handbook of Birds of the Western United* States (1902), which became a standard reference in the field, and *Birds of New Mexico* (1928), which was awarded the Brewster Medal by the American Ornithologists' Union, commemorating original scientific work. She collaborated with her husband Vernon Bailey, who was the chief field biologist in the Southwest for the United States Biological Survey, based in Washington and directed by her brother C. Hart Merriam. She also wrote several books that brought birding and conservation to public attention, such as *A-Birding on a Bronco* (1896) and *Birds of Village and Field* (1898). At a time when birds were being decimated for their plumage for decorating women's hats, she helped to raise public consciousness by teaching classes on birds, and in 1897 helped to found the Audubon Society of the District of Columbia.

Balance of nature. The tendency of earth and its ecosystems toward self-regulation, so as to return to states of equilibrium when disturbed.

Beckwourth, James (1798–1866). James Beckwourth was an important explorer and guide in the western mountains. Born in Virginia as the son of a slave and a white plantation owner, he traveled to the West and became a Crow war chief after gaining his freedom. While prospecting for gold in California in 1848, Beckwourth discovered the pass over the Sierra Nevada, near Reno, that now bears his name. His nearby ranch became a gateway for thousands of emigrants, whom he escorted over the pass and down the trail. In 1852, he dictated the details of his life to Thomas Bonner, who published the story as *The Life and Adventures of James P. Beckwourth* (1856, reprint 1972). Beckwourth moved to Colorado in 1864 and returned to the Crow Nation, among whom he lived for the remainder of his life. His name is commemorated in an annual festival in Marysville, California, and in the James P. Beckwourth Mountain Club.

Bennett, Hugh Hammond (1881–1960). Known as the father of soil conservation, Hugh Bennett joined the Bureau of Soils in the Department of Agriculture in 1903 as a soil surveyor and became head of the Soil Erosion Service in 1933. Owing to his efforts, the Soil Conservation Act was passed in 1935. The act created the Soil Conservation Service, to provide for the prevention and the control of soil erosion, and Bennett became its first head. Soil conservation had become a national priority with the coming of the Dust Bowl of the 1930s, during which loose topsoil was blown into devastating dust storms and washed downstream into rivers and oceans. Bennett's recommendations for better management of the soil gained added credibility when one such storm engulfed the United States Capitol during a 1934 congressional hearing on the subject. Bennett's books include *Soil Conservation* (1939) and *This Land We Defend* (1942).

Biocentrism. The idea that all plants and animals are centers of life, and as such have inherent worth and are worthy of moral consideration.

Biodiversity. The rich variety of animal and plant species in a given habitat.

Biogeochemistry/biogeochemical cycles. The transfer and flow of elements such as oxygen, carbon, nitrogen, and phosphorus through air, water, and soil.

Biological determinism. The priority of biological over cultural factors in deciding the outcome of events or development of personalities.

Biomass. The total weight of all living matter in a particular area.

Biome. A complex of animal and plant communities maintained under particular climatic conditions in a given area.

Bioregionalism. The idea that people, and other living and non-living things in a given region — usually a watershed — are interdependent, and should live as much as possible within the resources and ecological constraints of that place.

Biota/biotic community. The complex of interdependent living organisms in a given area. Biota comprise all living animals and plants.

Bird, Isabella (1831–1904). A noted nineteenth-century traveler and writer, Isabella Bird was among the earliest women to become active in mountaineering or to seek western travel adventures. Born and raised in England, Bird first traveled through the United States in her early twenties and wrote of her experiences in *The Englishwoman in America* (1856). Almost twenty years later, a voyage to the Hawaiian Islands, during which she climbed Kilauea volcano despite health problems, was the basis of her account in *The Hawaiian Archipelago* (1875). After crossing the Colorado Rockies on horseback, she depicted their peaks and canyons in perhaps her most acclaimed work, *A Ladies Life in the Rocky Mountains* (1879). Subsequent trips to Asia inspired her *Journeys in Asia and Persia* (1991), *Among the Tibetans* (1894), and *The Yangtze Valley and Beyond* (1899). Bird was the first woman to become a fellow of the Royal Geographic Society.

Broad Arrow Policy (1691). The Broad Arrow Policy, a plan by the British government to increase its share of the North American timber harvest, forced colonists to think about the preservation and distribution of natural resources. Included as part of the 1691 renewal of the charter of the Province of Massachusetts Bay, the policy reserved for the British Crown all trees not on private lands, 24 inches or more in diameter at 12 inches above the ground. It drew its name from the symbol carved into trees of a three legged "A" without the crossbar, designating property of the British Royal Navy. The law was extended to New Hampshire in 1708, to all of New England, New York, and New Jersey in 1711, and to Nova Scotia in 1721. It arose out of the increasing shortage of pines for ship masts in England and the North Sea countries, and the British Navy's quest for supremacy of the seas against its European challengers. The law and its enforcement in New England became increasingly controversial, and was part of the numerous complaints against British authority in the American colonies that led to the American Revolution.

Brower, David (1912–2000). David Brower was the first executive director of the Sierra Club, serving from 1952 to 1969. He was a defender of wild rivers and canyons and was opposed to the idea of damming any existing rivers, especially in places such as Dinosaur National Monument, Glen Canyon, and the Grand Canyon. In 1963, Brower and others reacted to the threat of dam construction in the Grand Canyon by commissioning a full-page advertisement in the *New York Times*, asking, "Should We Also Flood the Sistine Chapel So that Tourists Can Get Nearer to the Ceiling?" Another advertisement declared, "Now Only You Can Save the Grand Canyon From Being Flooded." In 1968, the Grand Canyon dams were deleted from the Pacific Southwest Water Plan. Under Brower's leadership, the club also won the fight to save Dinosaur, but lost on Glen Canyon, Brower's deepest defeat. He formed a global organization, Friends of the Earth (FOE), after political and fiscal policy differences led to his resignation from the Sierra Club in 1969. He left FOE in 1986 to become head of Earth Island Institute of San Francisco, an organization he founded in 1982. He received numerous awards for his work to save the environment.

Bullard, Robert (b. 1946). Robert Bullard is a major theoretician of environmental racism. In 1990 he wrote *Dumping in Dixie: Race, Class, and Environmental Quality*, and went on to edit a book entitled *Confronting Environmental Racism: Voices from the Grassroots* (1993). In these and later writings, he argued that minorities suffer disproportionately from the by-products of municipal landfills and incinerators, polluting industries, hazardous waste treatment facilities, and disposal facilities, and that institutional racism helped to shape the economic, political, and ecological landscape. He advocated a more effective incorporation of minority group concerns into plans for environmental protection.

Bureau of Indian Affairs (BIA). The Bureau of Indian Affairs is an agency created in 1824 in the Department of the Interior that is responsible for the administration of Indian lands and affairs. The BIA oversees the preservation of land and natural resources on Native American lands held in trust for the tribes by the federal government. It also operates offices, schools, and training programs on Indian reservations, and employs many Native American and Inuit peoples.

Bureau of Land Management (BLM). The Bureau of Land Management is an agency in the federal Department of the Interior, created in 1946 out of a merger of the General Land Office and the Grazing Service. It is responsible for the administration and management of public lands, including rangelands, deserts, and mineral resources. The agency manages some 272 million acres in the western states and Alaska, and reviews them for potential designation as wilderness. It issues grazing and mineral leases, and has offices and area managers in eleven Western states.

Bureau of Mines. The Bureau of Mines was created in 1910 with the goal of promoting mineral and mining technologies and mine safety. It deals with reduction of mine fires, operates safety training and rescue stations, and oversees remediation and restoration of mined lands.

Bureau of Reclamation (BuRec). An agency in the federal Department of the Interior, established by the passage of the Reclamation Act in 1902, the Bureau of Reclamation plans and administers dams and reclamation projects in the western states. It is charged with developing and promoting hydropower and irrigation delivery systems, managing recreational sites, and preserving ecosystems and parks associated with its projects.

Burroughs, John (1827–1921). A naturalist and lifelong resident of New York State, John Burroughs wrote books and essays on plants, mammals, birds, and the rural environment. He followed in the philosophical tradition of Henry David Thoreau and Ralph Waldo Emerson and hiked with John Muir and Theodore Roosevelt. Books such as *Wake Robin* (1871), *Birds and Poets* (1877), *The Ways of Nature* (1906), and *Bird and Bough* (1906) presented his nature observations in a literary style that captured public attention. He chronicled his western expeditions in a 1907 volume, *Camping and Tramping with Roosevelt.*

Capitalism. The economic system of private ownership of production conducted to gain a profit. Laissez-faire capitalism is an early form of United States capitalism, in which competition and the free market rather than government regulate production and distribution. Corporate capitalism is a later form, characterized by increased concentration of wealth and increased government regulation. Capitalism utilizes natural resources by extracting them from the environment and creating commodities for sale in the marketplace, often resulting in resource depletion and environmental pollution.

Carrying capacity. The ability of a given environment to support uses of its resources without permanent degradation.

Carson, Rachel (1907–64). A marine biologist with the U.S. Bureau of Fisheries and a well-known writer, Rachel Carson is credited with initiating the late twentieth-century environmental movement with her best-selling *Silent Spring* (1962). Carson's scientific knowledge, skillful use of language, and writing style — established in earlier works such as *Under the Sea Wind* (1941), *The Sea Around Us* (1951), and *The Edge of the Sea* (1956) — were major reasons why *Silent Spring* became popular. She described pesticides as new synthetic chemicals that entered into the most vital processes of animal bodies, changing them in sinister and deadly ways. Such new chemical conversions and transformations, striking at the heart of the living world, affected all of nature.

Excerpts from *Silent Spring* appeared in *The New Yorker* just prior to the publication of the book and these helped to build a ready audience for it. One of the most laudatory reviews appeared in the *Saturday Review*, which called her book a devastating, heavily documented, relentless attack on human carelessness, greed, and irresponsibility. The *Nation* said that civilized man was a disease of the biosphere who was destroying the natural system and himself in the process. On the other side came devastating reviews and responses from the chemical industry. Velsicol Chemical Company advised Carson's publisher, Houghton Mifflin, that it should reconsider its plan to publish the book. The company argued that by threatening the chemical industry, the book would indirectly act to degrade the United States' food supply. These tactics, intended to discredit her, instead aroused the nation, initiating the ecology movement, and causing a transformation in the human relationship to the natural world. Carson testified before Congress about the impacts of pesticides. She died of cancer in 1964.

Carver, George Washington (1864–1943). A scientist and pioneer in techniques for land conservation and renewable resources, George Washington Carver was raised by slave parents and privately tutored on a small farm in Missouri. After earning his bachelor's and master's degrees in agriculture at Iowa State College, he became the head of the Agricultural Department at Tuskegee Institute in Alabama under Booker T. Washington. Carver directed the only agricultural experiment station staffed entirely by African Americans, where he promoted ideals of public service and the interdependence of humans and nature, as well as practical methods for producing cost-effective and protein-rich foods such as peanuts. He received numerous awards for his scientific and humanitarian work. The George Washington Carver Foundation was established to promote further scientific research.

Cash crop. A crop produced for export profit rather than local consumption or subsistence.

Catlin, George (1796–1872). A noted nineteenth-century portrait painter, George Catlin traveled widely throughout the American West and Florida and painstakingly documented the vanishing lifeways and environments of Indians. During his childhood in northeastern Pennsylvania, he was influenced by stories of his mother, who had been a captive of Indians as a girl. Catlin abandoned an early career in law and moved to Philadelphia in 1823 to begin training as a portrait painter. Inspired by a visiting delegation of Indians from the West, he devoted himself to learning more about, and preserving, their cultures. Over the course of his lifetime, Catlin produced some six hundred portraits of Indians, as well as paintings of their villages, games, ceremonies, and clothing. He documented his travels and observations in several works, among them *Letters and Notes on the Manners, Customs, and Conditions of the North American Indians* (2 vols., 1841); *Life Among the Indians* (1867); and *Last Rambles Amongst the Indians of the Rocky Mountains and the Andes* (1867). Convinced that the Indian, buffalo, and the wilds of the West were doomed to extinction, Catlin proposed the setting aside of a "Nation's Park, containing man and beast, in all the wild[ness] and freshness of their nature's beauty."

Chaos theory. The mathematical theory that a small effect can lead to a large effect that cannot be predicted by linear (first-power) differential equations (those dealing with infinitesimal differences between variable quantities), and that such complex events may better be described by nonlinear equations.

Chief Joseph (c. 1840–1904). An eloquent and forceful orator and leader, Chief Joseph was a member of the Nez Perce who had declined to enter into a treaty with the U.S. Government in the 1850s and 1860s after whites began settling on their lands. From 1872 until 1876, Joseph sought non-violent ways of resolving the disputes, but finally led his band into war in 1877. The band spent thirteen days in Yellowstone National Park, where they captured and retained a group of nine tourists. U.S. troops pursued the Nez Perce to the Canadian border, where the Indians surrendered forty miles short of their intended destination and freedom. The incident ultimately led to the removal of Indians from Yellowstone and other national parks, and to governmental efforts to halt Indian hunting and gathering in the national parks.

Following their capture, the Nez Perce were sent to Indian Territory in Oklahoma, and in 1879 Joseph traveled to Washington to plead for justice for Indian peoples. In 1885 the non-treaty Nez Perce were relocated to the Colville Reservation in the state of Washington, whence Joseph continued to make speeches to colleges, governments, and the general public seeking redress for the injustices done to Indians.

Civilian Conservation Corps (CCC). The Civilian Conservation Corps, or CCC, was a federal program established in 1933 for the relief of unemployment, the promotion of the conservation of natural resources, and the administration of educational and vocational training. Created as the Office of Emergency Conservation Work of 1933, the CCC employed some three million unemployed young men, ages 17–23, to work on projects such as the development of parks, reforestation projects, and the construction of fire towers. The CCC was abolished in 1942, but the legacies of its work live on and are visible in many state parks and recreation areas.

Clean Air Act (1963, 1970, 1990). The first Clean Air Act of 1963 modified the 1955 Air Pollution Control Act by allocating permanent funding for the work of state pollution control agencies. The Clean Air Act of 1970 more actively established national ambient air quality standards for carbon monoxide, nitrogen dioxide, ozone, particulate matter, and sulfur dioxide, and — beginning in 1977 — lead. The 1970 Act charged the Environmental Protection Agency (EPA) with setting standards for the entire country, based on scientific research and affordable technologies, and with supervising state-level plans for their implementation. The 1970 Act also specified automobile emission standards for carbon monoxide, hydrocarbons, and nitrogen oxides. Amendments to the Act in 1977 aimed to prevent further deterioration of air quality by establishing classes, such as pristine quality in national parks and wilderness areas (Class I) to moderate air quality (Class II) to air quality that met the national secondary standards (Class III). A new version of the Act in 1990 addressed issues of acid rain, toxic air pollutants, ozone depletion, and nonattainment of standards. The amendments required a reduction in sulfur dioxide emissions, tighter automobile emission standards, and cleaner gasoline. The Clean Air Act and its amendments have helped to reduce particulate matter, lead, carbon monoxide, hydrocarbons, and nitrogen oxides from automobiles, but has been less successful in reducing ground level ozone, a component of smog.

Clean Water Act (1977). A federal act passed in 1977, the Clean Water Act changed the name of the federal Water Pollution Control Act of 1972 and required that waters be "fishable and swimmable" by 1983. The act gave states the authority to operate construction grants programs for water treatment facilities. In 1981 the types of projects were modified and applications streamlined.

Clements, Frederic (1874–1945). Botanist and ecologist Frederic Clements influenced the fields of ecology, forestry, and conservation with his theory of plant succession. A professor of botany at the University of Nebraska and the University of Minnesota, and a researcher at the Carnegie Institution in Washington, D.C., he wrote several pathbreaking books, including *Plant Succession* (1916); *Plant Ecology* (coauthored with John Weaver in 1929); and *Bio-Ecology* (coauthored with Victor Shelford in 1939). His theory of plant succession held that plants, as the natural units of classification, were dependent on climate, which included temperature, rainfall, and wind. The type of plant community developing in a particular locale depended on temperature and rainfall, with a stable equilibrium evolving over time. Much like a living organism, the vegetative community developed from an early youthful period into a mature adult, dying back as the individual organisms died off. In 1939, using the concept of the biome, Clements and Shelford integrated animals into the plant community. Plants mediated between the habitat and the animal population, providing a buffer against extremes of climate and supplying food for animals. In papers written in 1937, on "Environment and Life on the Great Plains," and "Succession and Human Problems," Clements identified farmers' reliance on monocultures, such as wheat and cotton, as causes of the Dust Bowl of the 1930s.

Coastal Zone Management Act (1972). A goal of the federal Coastal Zone Management Act of 1972 was to promote "the wise use of land and water resources for the coastal zone, giving full consideration to ecological, cultural, historic and aesthetic values as well as to needs for economic development." It stated that "there is a national interest in the effective management, beneficial use, protection, and development of the coastal zone." The law created a federal program to develop plans for the protection and development of coastal areas. The government provided funds to the states to implement management programs that included public recreational needs and environmental protection in the face of increased home building, dredging, filling, drilling, oil development, and industrial siting. The law was extended and amended in 1976, 1980, and 1990 to include natural resources protection, the redevelopment of urban waterfronts, and the issuance of grants to provide beach access and abatement of shoreline erosion.

Cole, Thomas (1801–48). Artist Thomas Cole was a premier painter of the nineteenth-century Hudson River school, which focused on natural landscapes as sites for human awe of sublime nature. Born in England, Cole moved to Philadelphia with his family in 1819 at the age of eighteen. In the mid-1820s, his early paintings of the New York State's Hudson River and Catskill Mountain areas brought him appreciation and fame among the eastern elite. Travels to England, Paris, Florence, and Rome in the 1830s resulted in a series of sketches of ancient temples and nature in the Italian hillsides that culminated, on his return, in the five-part series, "The Course of Empire," completed in 1836. His continued study of the New England and New York areas in the 1830s and 1840s was punctuated by another European trip in 1841–42. Cole painted many full-color illustrations of the Catskills, using as subjects rugged rocks, dark clouds, and wilderness. Nature filled with sublime light indicated the awe of God infused in the natural world. Dark forests, barren tree trunks, and hawks were emblematic of wildness at its highest moment.

Commodity. An article bought or sold. Nature, labor, and even the human body are sometimes regarded as commodities.

Commoner, Barry (b. 1917). Biologist and author, Barry Commoner is known for his writings and activism on behalf of the environment. His works include *Science and Survival* (1967), *The Closing Circle* (1971), *Energy and Human Welfare* (1975), *The Poverty of Power* (1976), *The Politics of Energy* (1979), and *Making Peace with the Planet* (1992). Commoner was a critic of the influx of nuclear power, chlorofluorocarbons, and petrochemicals into the environment and an advocate of alternatives based on the sun, organic agriculture, and the cogeneration of energy from waste heat. In *The Closing Circle*, he argued that, through pollution and degradation of the four elements of the ancients — earth (Illinois), air (Los Angeles), fire (nuclear power), and water (Lake Erie) — humanity had broken out of the cycle of life that has sustained the planet and added new, life-threatening, industrially produced chemicals to the biosphere. As a social activist, Commoner proposed the decentralization of the electric utility industry, and the forgiving of Third World debt. He argued that the "demographic transition" from higher to lower birth rates as countries industrialized would ultimately reduce population growth rates throughout the Third World. In 1980 he ran for President of the United States as the Citizen's Party candidate, with an environmental and socialist agenda.

Consciousness. The totality of an individual's thoughts, feelings, impressions, or a collective awareness of these by a group of individuals.

Conservation. The wise and frugal use of natural resources for the benefit of present and future generations.

Cooper, James Fenimore (1789–1851). A novelist and historian whose work presages both the environmental and Native American movements, James Fenimore Cooper is best known for his series of five Leatherstocking tales, featuring the growth and development of the hero Natty Bumppo (also known as Leatherstocking, Deerslayer, and Hawkeye) during the white settlement of New York State and the western prairies. The series consists of *The Pioneers* (1823), *The Last of the Mohicans* (1826), *The Prairie* (1827), *The Pathfinder* (1840), and *the Deerslayer* (1842). The first book in the series, *The Pioneers* (1823), set in the frontier village of Cooperstown, New York, highlighted the ecological destruction of massive numbers of passenger pigeons, and sensitively portrayed the option of living by hunting only those animals needed for human survival. Cooper also criticized the encroachment of white civilization on Native American lands and the decline of Indian lifeways.

Council on Environmental Quality (CEQ). An agency created by the National Environmental Policy Act of 1969 (NEPA) in the Executive Office of the President, to advise the President on matters of environmental quality, and to review agency compliance with NEPA. It existed until 1993, when President William Jefferson Clinton replaced it with the White House Office on Environmental Policy.

Culture. The concepts, habits, and institutions of a given people in a given period.

Dawes Act (1887). In 1887, Congress passed the General Allotment Act, known as the Dawes Act, after its author Congressman Henry Dawes of

Massachusetts. The act shifted the status of land on Indian reservations from communal to individual ownership by assigning 160-acre parcels to the heads of tribal families. The intent of the act was to bring Native Americans into greater conformity with the private property and farm ownership ideals of mainstream American settlement. The act broke up communally held lands on Indian reservations and stipulated that, after the individual plots had been allocated, the surplus land could be sold to non-Indians. The act not only resulted in Indian losses of "surplus" land, but additional losses accrued as speculators convinced many Indians to sell them their 160-acre allotments.

Declensionist. A narrative structure, or plot, that portrays environmental history as a downward spiral.

Deep ecology. The idea first proposed by Norwegian philosopher Arne Naess in his 1973 article in *Inquiry*, entitled, "The Shallow and the Deep, Long Range Ecology Movement: A Summary," that environmental problems cannot be resolved merely by legislation and regulation. They also require a deeper shift in basic philosophical assumptions about human relationships with the nonhuman world, based on principles such as diversity, symbiosis, biospherical egalitarianism, complexity, and local autonomy.

Desert Lands Act (1877). Passed in 1877, the Desert Lands Act provided for the sale of 640 acres of federal land unfit for cultivation in the western states, at 25 cents per acre and $1 per acre at final proof, to any settler who irrigated it within three years of filing. Stream water over and above that actually appropriated would remain free for use and appropriation by the public for irrigation, mining, and manufacturing purposes. Desert lands were defined as lands other than timber and mineral lands in the western states and territories that would not produce a crop without irrigation.

Diversity-stability hypothesis. The theory that complex biological and physical systems result in long-term maintenance of an ecosystem. Although accepted by many twentieth-century ecologists and conservationists as an argument for conserving biological diversity, the underlying concepts and validity of the hypothesis were questioned in the last quarter of the century.

Douglas, Marjory Stoneman (1890–1998). Florida writer and environmental activist, Mary Stoneman Douglas is best known for her efforts to create the Everglades National Park in 1947. She worked as a reporter and editor for her father's newspaper, the *Miami Herald*, and for the American Red Cross until 1920. Her book *The Everglades: River of Grass* (1947) characterized the Everglades as an interconnected system of water, grass, wildlife, and land threatened by drainage, dumping, and fill. In 1969, she helped to found Friends of the Everglades, to preserve the delicate ecosystem. She detailed her lifelong struggle to save the Everglades in an autobiography, *Marjorie Stoneman Douglas: Voice of the River* (1987). She died at the age of 108.

Downing, Andrew Jackson (1815–52). A founder of landscape architecture, Andrew Jackson Downing began his career as a horticulturist in his father's nursery in Newburgh, New York, and became its sole owner about 1838. He designed country cottages for the average rural American and country estates for the well-to-do along the Hudson River. He also planned the grounds for the United States Capitol, the White House, and the Smithsonian Institution in Washington, D.C. He influenced the work of New York City's Central Park designer Frederick Law Olmsted and worked with Olmsted's future partner, Calvert Vaux. Downing's books, which became classics in the field, include, *Treatise on the Theory and Practice of Landscape Gardening, Adapted to North America* (1841); *Cottage Residences* (1842); *The Fruits and Fruit Trees of North America* (1845), coauthored with his brother Charles; and *Architecture for Country Homes* (1850).

Dust Bowl. The term used to describe the effects of wind erosion on the soils of the Great Plains during the drought years of the 1930s.

Earth First! A movement founded in 1979 by Dave Foreman and other preservationists, in the belief that mainstream environmentalism had become too compromised by corporate interests, Earth First! vowed to make "no compromise in defense of Mother Earth." The group had no official membership list, and used "monkeywrenching" techniques to protest logging, dam building, and billboarding that degraded wilderness and endangered biota. Its methods included tree-spiking, sabotaging the gas tanks of bulldozers, and the use of guerrilla theater to dramatize the loss of the wild. It also promoted biodiversity projects, lower consumption and birth rates, and the restoration of native species such as grizzly bears. It encouraged respect for wilderness, living simply on the land, and acts of civil disobedience to preserve wild nature.

Earth Island Institute (EII). The goal of San Francisco's Earth Island Institute, founded in 1982, is conservation, preservation, and restoration, or as founder David Brower put it, to administer CPR (a play on cardiopulmonary resuscitation) to the environment. EII's international mission has included such projects as holding several conferences on the Fate and Hope of the Earth, sponsoring an environmental program on Central America, promoting dolphin-safe tuna catches, protecting old-growth and tropical rainforests, restoring urban habitats, and reducing inner city poverty. Within the United States, it helped citizens to develop legal strategies for enforcing environmental statutes, to promote the Green political movement, and to support environmental justice.

Ecocentrism. The idea that the natural world is an integrated ecological whole and, as such, is worthy of moral consideration.

Ecofeminism. The idea that an historical and cultural association exists between women and nature, which have both been dominated in Western culture, and that women can liberate themselves and nature through environmental activism.

Ecological imperialism. The idea proposed by environmental historian Alfred Crosby, in a 1978 article in *The Texas Quarterly*, that biota (such as people, animals, pathogens, and weeds) are all part of colonization, intentionally or not.

Ecological revolution. A concept developed by environmental historian Carolyn Merchant to characterize major transformations in relationships between humans and nonhuman nature. These include the colonial ecological revolution that characterized the transition from Native American ways of life to colonial settlement, and the capitalist ecological revolution that marked the transition from the colonial to the market economy of the nineteenth century.

Ecological succession. The gradual replacement, in a given environment, of one community of organisms by another.

Ecology. A branch of science, named by Ernst Haeckel in 1866, that explores the relationships between and among organisms and their abiotic surroundings, or environment.

Ecosystem. A self-regulating and self-sustaining community of organisms that relate to each other and to the larger environment.

Ecotage. The idea, promoted by the environmental group Earth First! during the 1980s, that nature can be saved by sabotaging the machines and equipment (but not people) that destroy it.

Ehrlich, Paul (b. 1932). Biologist and conservationist Paul Ehrlich became famous for his book *The Population Bomb* (1968), in which he outlined the negative environmental effects of uncontrolled population growth. Adopting the approach developed by Thomas Malthus in his 1798 *Essay on Population*, Ehrlich argued that world population is increasing exponentially while resources such as food are increasing at a far slower rate. Ten thousand years ago, the world population was about 5 million, but by 1650 it had increased to 500 million, and by 1850 it was about a billion. If population continued to grow at rates between 1.7 and 2.1 percent per year, Ehrlich predicted that, by the year 2000, the earth would have upwards of 6 billion people, a prediction that came to pass on October 12, 1999. Unless the human species initiated an era of birth control, population growth would lead to ecological and social collapse. To help achieve population control, Ehrlich founded Zero Population Growth in 1968 to push for a balance between population and the environment. In 1990, Ehrlich and his wife, Ann Ehrlich, moderated their approach in a book entitled *The Population Explosion* (1990), arguing that if women, especially those in developing countries, were educated, they would achieve higher standards of living and tend to have fewer children. Two years later, the Ehrlichs expanded their approach to the problem in *Healing the Planet: Strategies for Solving the Environmental Crisis* (1992).

Emerson, Ralph Waldo (1803–82). An American essayist and philosopher of the Transcendentalist school of thought, Emerson spent most of his life in Concord, Massachusetts, where he wrote and lectured. His essays, which were published in several collections, dealt with topics ranging from nature and the beauties of his surroundings to wealth and self-reliance. He graduated from Harvard College in 1821, and in 1825 began divinity school in Cambridge to prepare for the Unitarian ministry, but did not remain in the profession. His philosophy was influenced by European writers Samuel Taylor Coleridge (1732–1834), William Wordsworth (1770–1850), and Thomas Carlyle (1795–1881). He believed that the human soul and the

universal Oversoul corresponded, and that humans could look within themselves in order to find larger truths. Each person was contained within the
unity of the Oversoul, and the eternal One existed within every individual.
In his 1836 essay "Nature," he wrote that the natural world was emblematic
of the universal spirit. Emerson influenced the work of Henry David
Thoreau (1817–62), who placed greater emphasis on the material world as
a source of truth, and John Muir (1838–1914), who held that nature was
like a mirror that reflected the Creator. Emerson visited Muir in Yosemite
Valley in 1872, but, to Muir's disappointment, did not join him for a month
of immersion in the high Sierra mountains. Of the three, Emerson, in addition to his worship of nature, extolled the virtues of the commercial spirit,
while Thoreau eschewed them for the middle ground of Walden Pond, and
Muir escaped them for the mountain wilderness.

Endangered American Wilderness Act (1978). Passed in 1978 to augment
the Wilderness Act of 1964, the Endangered American Wilderness Act set
aside many additional undeveloped national forest lands. Designated lands
in the Western states were added to the National Wilderness Preservation
System for their value in areas such as watershed preservation, wildlife habitat protection, scenic and historic preservation, scientific research and education use, primitive recreation, solitude, and physical and mental challenge.

Endangered Species Act (ESA) (1973). The Endangered Species Act
(ESA) is a federal measure, passed in 1973, that established procedures for
identifying and protecting endangered plants and animals in critical habitats, and for prohibiting the taking, harvesting, hunting, and harming of
listed species. The ESA, applying stricter standards than an earlier act in
1969, defined two levels of enforcement: an "endangered" species, as one
"in danger of extinction throughout all or a significant portion of its range,"
excluding those species considered to be pests or risks to humanity; and a
"threatened" species, as one "likely to become an endangered species
within the foreseeable future." ESA prohibited the importation and exporta-
tion of endangered species and the "taking" of such species within the
United States or its possessions, but it allowed limited taking of threatened
species. The law also prohibited federal agencies from destroying the critical
habitat of an endangered species, a provision that met with controversy
when the snail darter's only known habitat was threatened by the completion of the Tellico Dam in Tennessee. Provisions added to ESA in 1978 de-

fined a process for public participation in the listing of species, a five-year review of all species on the list to determine if any should be removed, a requirement that the critical habitat of a species be specified at the time of listing, and a recovery plan for listed species. Further amendments in 1988
provided more details on the recovery plans and reauthorized ESA for four
years. By 1992, when the list included 389 animals and 373 plants native to
the United States, further reauthorization encountered political opposition
due to the perceived confiscatory power of the "takings" provision. Instead,
a compromise emerged that called for development mitigations through
creating habitat conservation plans (HCPs). Developers and landowners
agreed to leave portions of forests and wetlands untouched and to preserve
wildlife corridors in exchange for development permits. Although many
HCPs were subsequently approved, critics argued that they represented
shortsighted compromises that favored development at the expense of
wildlife.

Endangered Species Conservation Act (1969). The goal of the Endangered Species Conservation Act, passed in 1969, was to prevent the importation of endangered species threatened with worldwide extinction into the
United States, and to prevent the shipment within the United States of rep-
tiles, amphibians, and other wildlife that were protected by state laws. The
best scientific and commercial data available were to be used by the Secretary of the Interior to establish lists of endangered species. A species could
be endangered owing to the destruction of its habitat, its overutilization for
commercial purposes, its degradation from disease or predation, or other
natural or man-made factors that threatened its continued existence. The
act was superseded by the Endangered Species Act (ESA) in 1973.

Environment. The surroundings or aggregate of external conditions that influence the lives of individuals, populations, communities, and societies.

Environmental Defense Fund (EDF). A citizens' litigating organization,
the Environmental Defense Fund (EDF) was founded in 1967. It links
science, economics, and law in an effort to create cost-effective solutions to
environmental problems. From its earliest strategy of suing the polluters
over issues such as the spraying of DDT to its recent concerns over acid rain
and corporate emissions trading, it brings legal expertise to bear on issues of
environmental quality. EDF has worked with the fast-food industry to reduce its solid waste stream and to substitute paper for polystyrene contain-

ers, implicated in upper-atmospheric ozone depletion. It has also worked with utilities to promote the use of renewable energy sources and economically efficient energy conservation programs.

Environmental ethics. That branch of philosophy that addresses questions of what one *ought* to do to save the environment.

Environmental Impact Statement (EIS). Mandated by the passage of the National Environmental Policy Act of 1969, the Environmental Impact Statement (EIS) is a requirement that every federal agency prepare and circulate statements "on proposals for legislation and other major federal actions significantly affecting the quality of the human environment."

Environmental Justice Movement. The idea that minority communities have been disproportionately affected by the negative aspects of environmental pollution and resource depletion and that steps must be taken to reverse those effects. The movement arose during the 1980s when African and Native Americans in Warren County, North Carolina, protested a proposed disposal site for PCBs (polychlorinated biphenyls) in 1982, and a United Church of Christ report, "Toxic Wastes and Race in the United States" (1987), showed that communities with the greatest number of commercial hazardous waste facilities had the highest composition of racial and ethnic residents. The movement against inequities and for environmental justice has grown to include numerous ethnic and racial communities in both urban and rural environments throughout the country and the world.

Environmental Protection Agency (EPA). Created by executive reorganization in 1970, the Environmental Protection Agency (or EPA) is an agency in the executive branch of the federal government, with authority to regulate air and water quality, radiation and pesticide hazards, and solid waste disposal. In the late 1980s it also began to address issues related to global climate change and its impact on the environment. The EPA, from its main office in Washington, D.C., and ten regional offices throughout the country, works with the states to conduct research, oversee environmental programs, and evaluate federally mandated Environmental Impact Statements (EIS). The EPA also has the authority to develop regulations that enforce congressional legislation concerning environmental quality. It maintains field laboratories that collect data for monitoring and enforcing its regulations.

Environmentalism. Beliefs and actions taken to preserve the environment, encompassing a range of impulses from preservationism, or the intention to save undisturbed nature for its own sake, to conservation, or the maintenance of nature for human use by present and future generations.

Equilibrium. The tendency for a system to restore itself to a particular condition.

Exploit. (1) to make use of; (2) to use for one's own advantage, profit, or selfish gain.

Federal-Aid Highway Act (1956). Culminating several years of study and planning, Congress passed the Federal-Aid Highway Act in 1956, creating the Dwight D. Eisenhower System of Interstate and Defense Highways. The system has been lauded as increasing economic productivity, competitiveness, safety, and freedom, and criticized for creating automobile dependent urban sprawl, air and water pollution, and for slowing construction of mass transit systems. The federal government's cost share was 90 percent, to be paid for by an increase in gasoline taxes, with the remainder being paid by the states. The system required interstates to be divided highways with a minimum of two lanes in each direction, moderate grades, high-speed curves, and conveniently spaced rest areas, with no intersections or traffic signals. It incorporated existing turnpikes and highways that met the standards, and by the 1990s consisted of over 40,000 miles of highway, representing about one percent of the nation's roads.

Federal Land Policy and Management Act (FLPMA) (1976). With the passage of the Federal Land Policy and Management Act (FLPMA), Congress granted the Bureau of Land Management (BLM) the authority to manage lands under its jurisdiction more effectively. The act retained public lands in public ownership, effectively closing the public domain to private entry; established guidelines for the sale of public lands; and gave the BLM the authority to set grazing, preservation, and mining policy. According to the act's provisions, Congress would have final approval over the reauthorization of ten-year grazing permits, as well as the state-by-state designation of BLM lands as wilderness areas.

Fernow, Bernhard (1851–1923). One of several influential forest conservationists at the turn of the nineteenth century, Bernhard Fernow was born in

Prussia and educated in forestry at the Hanover-Mueden Academy before coming to the United States in 1876. He managed a private forest in Pennsylvania, became secretary of the American Forestry Association (1883–95), and served as chief of the Division of Forestry in the executive branch of the U.S. government from 1886–98. His books on forestry, including *Economics of Forestry* (1902), *Care of* Trees (1910), and A *Brief History of Forestry* (1913), helped to educate the public on the decline of forests and the need for conservation. He fought for the passage of the Forest Reserve Act of 1891, which authorized the president to create reserves from forested lands in the public domain. A critic of laissez-faire individualism, a philosophy that supported private control of the forests, Fernow advocated a policy of *faire marcher* (literally, "to make it work") with the central government representing the communal and future interests of the people.

Fish and Wildlife Service (FWS). The Fish and Wildlife Service (FWS) is a unit in the Department of the Interior of the executive branch of the federal government that conserves, protects, and enhances fish and wildlife, as well as their habitats, for the benefit of the American people. It was created in 1940 by merging the Bureau of Fisheries in the Department of Commerce with the Bureau of Biological Survey in the Department of Agriculture. It oversees national wildlife refuges and fish hatcheries, develops recovery plans for endangered species, restores fisheries, monitors and enhances wetlands, and enforces wildlife laws.

Food web. Interlocking systems of organisms and food chains in which energy, in the form of food, is transferred from one trophic (nutritional) level to another.

Foreman, Dave (b. 1947). Controversial founder of the radical environmental organization, Earth First!, Dave Foreman began his career in Washington, D.C., with the Wilderness Society in the 1970s. He was angered by the extreme profit motive of the resource extraction industry and became disillusioned with the ineffectual strategies of both government regulatory agencies (such as the Forest Service, the Bureau of Land Management, and the Department of Agriculture) and conservation groups (such as the Sierra Club, Friends of the Earth, and the Wilderness Society). Influenced by Edward Abbey's *The Monkey Wrench Gang* (1975), in 1979 Foreman founded Earth First!, a group of anonymous activists who advocated attacking the machines that were destroying the last remaining wilderness. Following its motto, "No compromise in defense of Mother Earth," the group filled the gas tanks of bulldozers with sugar, spiked trees to break chain saws, cut down high-voltage power lines, and rolled black plastic strips down the outside of dams to simulate cracks. Foreman's book *Ecodefense: A Field Guide to Monkeywrenching* (1987) and his autobiography *Confessions of an Eco-Warrior* (1991) promoted the destruction of property, but not people, in defending wilderness. In the wake of an FBI enforcement campaign and Foreman's arrest and plea bargain in 1989, Earth First! split into two factions. Foreman went on to found the Wildlands Project in Tucson, Arizona, and to publish a guide to books on the "Big Outside."

Forest Management (Organic) Act (1897). The Forest Management Act stated that the purpose of forest reservations was "to improve and protect the forest within the reservation or for the purpose of securing favorable conditions of water flows, and to furnish a continuous supply of timber for the use and necessities of citizens of the United States." Included in an 1897 appropriations act about "Surveying the Public Lands," the act provided that the Secretary of the Interior Department (the department that administered the forests at the time), could permit timber harvesting, mining, and water use on the forest reservations by settlers, miners, residents, and prospectors. Private owners could exchange lands within the reservations for equivalent lands outside them, in some cases leading to abuse when companies and railroads cut the forests before exchanging them.

Forest Reserve Act (1891). The Forest Reserve Act was the last section of an act that repealed the Timber Culture Act of 1873 and authorized the president to "set apart and reserve in any State or Territory having public land bearing forests, in any part … covered with timber or undergrowth, whether of commercial value or not, as public reservations." The act arose out of concern to create watershed protection to prevent flooding and soil erosion, to preserve the nation's forests and wildlife, and to reserve the forests for democratic development. Division of Forestry head Gifford Pinchot later stated that he considered the act to be the most important legislation in the history of American forestry. The act was used extensively by President Theodore Roosevelt in the first decade of the twentieth century, under Pinchot's guidance, to create the national forest system.

Forest Service (USFS). The Forest Service is a federal agency in the Department of Agriculture that was created in 1886 as the Division of Forestry and renamed when the forest reserves were transferred from the Department of the Interior to Agriculture in 1905. The forest reserves, created through the advocacy of forest division chiefs Bernhard Fernow and Gifford Pinchot and the powers of President Theodore Roosevelt, were declared national forests in 1907. The Forest Service, responsible for the administration of the nation's forests, has a mandate to provide multiple uses of the forest resource, including timber, wildlife, recreation, range, and water. The USFS, and some of its sustained yield policies, have evoked controversy: among them old-growth and clear-cutting practices, the issuance of grazing permits on forested lands, and the failure to establish adequate protection for particular threatened species such as the northern spotted owl.

Forestry. The science of cultivating forests and managing the production of timber.

Free Timber Act (1878). The Free Timber Act was a federal act, passed in 1878, that gave residents of nine western states the privilege of cutting timber on public mineral lands for building, agriculture, mining, or other domestic purposes.

Friends of the Earth (FOE). Friends of the Earth (FOE) is an international environmental organization, founded in 1969, dedicated to protecting the planet from environmental disaster and to preserving biological, cultural, and ethnic diversity. It was founded by David Brower and other environmentalists, who promoted an activist approach to dealing with urgent environmental degradation using boycotts, demonstrations, sit-ins, and marches. It later adopted a more mainstream approach of endorsing green products and promoting legislation. Through its worldwide organization, it works at all levels of society to create networks that can influence legislation and arouse public awareness of environmental problems.

General Land Office (GLO). Established in 1812, the General Land Office was created by the federal government to conduct land surveys and to process and record sales, entries, withdrawals, reservations, and leases on the public lands. It had primary jurisdiction over the entire public domain until 1946, when it was merged with the Grazing Service into the Bureau of Land Management (BLM).

General (Land) Ordinance of 1785. Passed by the Continental Congress, the General Land Ordinance established the rectangular land survey system of townships, 6 miles square divided into 36 sections of 640 acres each, that could be sold at auction for not less than $1 per acre. Section 16 of each township was reserved for schools and four sections for government disposal. The ordinance fostered both widespread land speculation and squatting and was the basis of the Homestead Act's provision of 160 acres, or one-fourth of a section, for individual land claims.

Gibbs, Lois (b. 1951). An activist for healthy, toxic-free environments, in 1978 Lois Gibbs organized neighbors of a toxic chemical dump underlying Love Canal, an area of Niagara Falls where her children attended school, to seek relocation and retribution from the state. The chemical manufacturer initially denied allegations that the high rates of cancer, birth defects, and miscarriages experienced by the surrounding population were caused by leaking chemicals, but in August of 1978, state officials closed the school and 239 residents were moved out of the area. In 1980, an emergency declaration by President Jimmy Carter moved nine hundred families from Love Canal. In 1981, Gibbs founded the Citizen's Clearing House for Hazardous Wastes, and the newsletter "Everyone's Backyard." The goals were to coordinate grassroots efforts nationwide to oppose and remove hazardous waste sites and to promote environmental justice. She received numerous awards for her work, was the subject of a prime-time movie — "Lois Gibbs: the Love Canal Story" — and is the author of *Love Canal, the Story Continues* (1998).

Global warming. The idea that the earth's surface temperature is gradually rising, owing to industrial processes that raise gas levels, trapping heat.

Hamilton, Alice (1869–1970). Public heath activist, physician, and founder of occupational medicine, Alice Hamilton was a pioneer in the field of industrial health and safety and the remediation of hazardous effects of the workplace. After a medical education at the University of Michigan, from which she received a degree in 1893, and in Germany, she became a bacteriologist at the Memorial Institute for Infectious Diseases in Chicago, beginning in 1902. During her time in that city, she lived and worked at Jane Addams's Hull House, publicized the problem of the spread of typhoid fever from flies on sewage during the epidemic of 1902, and pressured the city to correct the unhealthy working conditions of immigrant laborers. Heading an Occupational Disease Commission created by the Governor of Illinois in 1910, Hamilton succeeded in pressing for legislation for job-related health and injury compensation. As a founding member of the Women's International League for Peace and Freedom, which promoted peace and an end to World War I, she explored wider concerns about social welfare. Hamilton's later career centered on Harvard University, where she served as an assistant professor of industrial medicine in the School of Medicine between 1919 and 1925 and at the School of Public Health until 1935. She subsequently worked as a consultant for the Division of Labor Standards in the U.S. Labor Department and was awarded many honorary degrees and awards for her work. Her books include *Industrial Poisons in the United States* (1925), *Industrial Toxicology* (1934), and an autobiography, *Exploring the Dangerous Trades* (1943). She died in 1970 at the age of 100.

Hatch Act (1887). The Hatch Act is a federal act, passed in 1887, for the purpose of establishing agricultural experiment stations. The stations, associated with land grant universities, undertake research on the production of food, fiber, and nutrition. The results are disseminated to the public through the Cooperative Extension Service, whose agents work with growers to improve production techniques and crop yields and with the public to promote better nutritional standards.

Homeostatic. Maintenance of a constant value or steady state within a narrow range, such as the human body temperature.

Homestead Act (1862). A federal act, passed in 1862, which aimed to encourage rapid settlement of the western lands of the United States and to distribute lands equitably to all settlers. The act specifically authorized any person who was the head of a family or over twenty-one years of age, and who was a citizen of the United States or had declared an intention to become one, to enter upon no more than 160 acres (one-fourth of a section) of unappropriated land subject to preemption and sale at a minimum price of $1.25 per acre, or not more than 80 acres subject to sale at a minimum price of $2.50 per acre. The act was consistent with the Jeffersonian ideal that every person should be able to own enough land for farming. It provided opportunities to those who were unable to purchase land outright, to single, divorced, and widowed women, and to European immigrants. Owing to grants to railroads and to fraud by speculators, only about one in six to nine acres actually were acquired by homesteaders by the end of the nineteenth century. Repercussions of the act included extensive railroad development, land acquisition by speculators and absentee landlords, tenant farming, decimation of the Plains buffalo and Indians, and degradation of the prairies and arid lands due to extensive cultivation.

Homocentrism. The idea that humans are the center of all existence and deserve moral consideration above all other parts of the natural world.

Hudson River school. A group of nineteenth-century American artists, including Thomas Doughty, Asher B. Durand, Thomas Cole, Thomas Moran, and Albert Bierstadt, who explored the relationship between the land and human society. Their paintings of the Hudson River and a number of other areas, done primarily between 1820 and 1880, made landscapes the main subject and contrasted wild and forested nature with human effects on the land. The paintings were both realistic depictions of nature and romantic renderings of a "wilderness" that once covered most of the American continent, but by the mid-nineteenth century was seen as vanishing before the advancement of civilization.

Human ecology. The study of human communities and their changes over time with respect to their interactions with other communities, species, and the natural environment.

Ickes, Harold (1874–1952). Harold Ickes, as the Secretary of the Interior under Franklin Delano Roosevelt, beginning in 1933, and continuing under Harry Truman until 1946, did much to modernize the administration of federal lands. Arguing that conservation was a major responsibility of the federal government, he expanded the National Park System to include Olympic National Park in Washington, Shenandoah National Park in Virginia, and King's Canyon National Park in California, and took steps toward creating the Grand Teton National Park in Wyoming. In 1939, under his auspices, the National Fish and Wildlife Service created 160 wildlife refuges. Progressive in his policies toward minority groups, he supported John Collier of the Bureau of Indian Affairs in passing the Indian Reorganization Act of 1934 and worked to elevate the status of African Americans in his department. He was responsible, moreover, for outlawing the killing of predators, such as wolves and coyotes, in the national parks.

Imperialism. The view that the role of human beings and governments is to establish power over nature and over as great an area and as many peoples as possible.

Indian Civil Rights Act (1968). A federal act, passed in 1968, denying states the authority to extend their civil and criminal jurisdiction over Indian lands without tribal consent, and imposing certain restraints on tribal governments.

Irrigation. The means of causing water to flow onto land for cultivating plants.

Jefferson, Thomas (1743–1826). As the nation's third president, serving from 1801–1808, Thomas Jefferson contributed to the growth of diplomacy, democracy, agriculture, science, and natural history. His *Notes on the State of Virginia* (1787) set out the natural history of Virginia, expounded on its physical features, biota, agriculture, and commercial products and discussed Indians, slavery, religion, and society. Jefferson extolled the nation's farmers as the "chosen people of God," and advocated keeping manufacturing in Europe to prevent degrading human labor systems from taking hold in America. Jefferson's legacy was the agrarian ideal that every individual should be able to own a plot of land. His finalization of the Louisiana purchase in 1803 more than doubled the size of the country and was followed by the explorations of Meriwether Lewis and William Clark, whose mission was to map the new territory and to collect and record species of plants, animals, and minerals. As a farmer who yearned to give up politics to return to his fields, Jefferson experimented with agricultural improvement, employing methods of restoring soil fertility and improving crop yields.

Jewett, Sarah Orne (1849–1909). Short-story and nature writer, Sarah Orne Jewett is best known for her acclaimed literary work *The Country of the Pointed Firs* (1896) and for her impassioned plea for plume-bird conservation in her short story "A White Heron" (1886). In the latter story, a vagrant snowy egret being hunted by a gun-bearing ornithologist looking for a specimen is saved by a young girl who intimately knows the marshes where the bird nests, but decides not to reveal its location to the scientist. The story, written at a time when public consciousness about bird endangerment was only beginning, was a critique not only of the milliner's trade in bird plumage, but also of collecting scientific specimens as a hobby and for museums. Beyond that, the work reveals cultural as well as gendered tensions between civilization, represented by the male urban ornithologist, and untouched nature, symbolized by the heron and the small girl.

Kemble, Frances Anne (1809–93). As an actress and author, Frances Kemble was a passionate critic of the environmental degradation and brutal labor systems associated with cotton production in the nineteenth-century American South. Born in England, Kemble and her father came to the United States in 1832 to join the Park Theatre Company in New York. After marrying Pierce Butler, she left the stage, and in 1838–39 accompanied him to his plantation on the Sea Islands of Georgia. Horrified by slavery, but delighted by the natural surroundings, Kemble penned a caustic, critical treatise, entitled *Journal of a Residence on a Georgian Plantation in 1838–1839* (first published in 1863), which criticized the treatment of slaves and the waste of the soil under careless cotton cultivation, while extolling the beautiful flowers, birds, and marshes of the islands. She subsequently left her husband and used her acting talents to give well-attended public readings of Shakespeare, settling in the Berkshire Mountains in Lenox, Massachusetts. Her other writings included a two-volume work, *Journal of a Residence in America*, published in 1835, the critical tone of which met with mixed reviews in the United States.

Land and Water Conservation Fund Act (1965). The Land and Water Conservation Fund Act of 1965 has been amended numerous times to preserve, develop, and assure accessibility to outdoor recreation resources and to develop land and water areas and facilities. It assesses user fees at recreation areas, such as national parks, monuments, and scenic areas, in order to subsidize state and federal acquisition of lands and waters for recreational and conservation purposes. These acquisitions include additions to the national parks, national forests, and national wildlife refuges.

Land ethic. A term used by twentieth-century ecologist Aldo Leopold in his 1949 *A Sand County Almanac* to refer to the extension of the ethics of the human community to include "soils, waters, plants, and animals, or collectively: the land."

Land stewardship. The responsible use and care of the environment and its natural resources.

Leopold, Aldo (1887–1948). Aldo Leopold, a pioneer in such fields as wildlife and restoration ecology, strongly influenced emerging ideas about environmental ethics and decision-making. After completing his degree at the Yale Forestry School in 1909, he worked for the U.S. Forest Service in Arizona and New Mexico. In 1925, Leopold moved to Madison, Wisconsin, where he was the associate director of the United States Forest Products Laboratory, and in 1933 became professor, holding the new Chair of Game Management, at the University of Wisconsin at Madison. Early in his career, he thought of wild animals as a crop to be harvested for human use. In

his book on *Game Management* (1933), Leopold considered deer to be a resource to be managed by controlling the environmental factors pertaining
to the seed stock.

Two decades later, Leopold looked back to an Arizona experience that
he identified as a transformative moment in his thinking. In his book *A
Sand County Almanac*, published posthumously in 1949, he included a
chapter entitled "Thinking Like a Mountain," stating that his conviction
arose from the day he saw a wolf die. A wolf came out of the river and "in a
second we were pumping lead into the pack.... I thought that because
fewer wolves meant more deer, that no wolves would be hunters' paradise.
But after seeing the fierce green fire dying in her eyes, I sensed that neither
the wolf nor the mountain agreed with such a view."

A prolific writer, Leopold founded what became the science of restoration ecology when he bought an old farm in the sand counties of Wisconsin
and began to replant it. His observations about nature and ecology published in *A Sand County Almanac* became a bible for environmentalists and
environmental ethicists. The "land ethic," he wrote, "simply enlarges the
boundaries of the community to include soils, waters, plants, and animals,
or collectively: the land." In implementing the ethic, Leopold stated, "A
thing is right when it tends to preserve the integrity, stability, and beauty of
the biotic community. It is wrong when it tends otherwise."

MacKaye, Benton (1879–1975). A regional planner with a master's degree
in Forestry from Harvard University (1905), Benton MacKaye was a co-
founder of the Wilderness Society in 1935 and a strong advocate for the
preservation of forests and wildlands for hiking and outdoor recreation. He
conceptualized the Appalachian Trail in a 1921 article entitled, "An Appalachian Trail: A Project in Regional Planning," and worked for its implementation and completion. The goal was to create an interlinked chain
along the Appalachian crest, from northern Maine to the southern tip of the
Appalachians, that would provide an escape for overworked eastern urbanites. Work on the trail commenced in the 1920s and was completed by the
Civilian Conservation Corps in 1937. MacKaye published *The New Exploration: A Philosophy of Regional Planning* in 1928, served on the regional
planning commission of the Tennessee Valley Authority (TVA) from 1934 to
1936, and on the staff of the Rural Electrification Administration from 1942
to 1945.

Marsh, George Perkins (1801–82). George Perkins Marsh was one of the
earliest advocates for the wise use of land. Born in Woodstock, Vermont, he

became fish commissioner from that state. As the Vermont fish commissioner, he could see that cutting the forests was polluting rivers, causing them to silt up and preventing fish from spawning. There was thus a direct relationship between forest use and other resources, such as fisheries. He later turned his attention to national and international matters as a U.S. congressman and, ultimately, a twenty-one year career in the diplomatic corps. In the latter post, he drew explicit connections between the historical destruction of Mediterranean lands — how the forests had been devastated, the soil eroded, and pastures over-grazed — and what he regarded as similar changes in America.

His book *Man and Nature*, published in 1864, sold 100,000 copies in a few months, testimony to the increasing numbers of people becoming concerned about the waste of resources and desiring to take action. "Man is everywhere a disturbing agent," he wrote. "Wherever he plants his foot, the harmonies of nature are turned to discords." Marsh's book put forward the basis for an ethic of human cooperation with the land. He said that man should become "a co-worker with nature in the reconstruction of the damaged fabric," and he must "reclothe the mountains as the valleys are deforested." Humans ought to use natural resources, but work with the land in order to restore it. Ten years later, he republished the book under the title *The Earth as Modified by Human Action*. Today Marsh is looked upon as America's first prophet of ecology.

Marshall, Robert (1901–39). Robert Marshall was an influential and controversial activist on behalf of wilderness conservation. He attended Syracuse University's School of Forestry, which had been endowed by his father. After serving as a silviculturist in Montana between 1925 and 1928, he spent two years studying tree growth, climate, and society in the Alaskan village of Wiseman. From 1932 to 1933 he worked for the U.S. Forest Service, drafting a National Plan for American Forestry. In 1933, as chief of the Division of Recreation and Lands of the Bureau of Indian Affairs (BIA), he proposed a plan for indigenous management of Indian reservation forests. Marshall sharpened his appreciation of the delicate balance of forest life, and the way it was being disrupted by lumber companies, when he headed the Outdoors and Recreation Office in the National Forest Service in the late 1930s.

In his most important book, *The People's Forests* (1933), Marshall criticized the Forest Service and the timber companies for their emphasis on profits and advocated full governmental control over all forests. In Marshall's view, timber companies, which artificially inflated consumers' need for wood products, had not proven to be able caretakers of the land; rather, he proposed, the public should acquire at least 562 million acres out of the 670 million existing acres of forest, for their social as well as natural value.

Marshall's activism continued throughout his life. Working with Olaus Murie, Aldo Leopold, and Benton MacKaye to establish the Wilderness Society in 1935, he proposed that the society be dedicated to creating and protecting wilderness areas and to preserving public lands, including forests, sea shores, wildlife refuges, and lands administered by the Bureau of Land Management. Wilderness areas, in contrast to national parks, were to be places free of cars or established guest facilities, in which people could walk, backpack, or travel with pack trains and camp out. Marshall's insistence that wilderness areas be set aside near large concentrations of factory workers and underprivileged populations put him at odds with many of his colleagues, who feared that popular use of the wilderness would destroy its special virtues. His advocacy for causes associated with socialism opened him to harassment by the House Un-American Activities Committee in the late 1930s. After his death in 1939, his $1.5 million estate was distributed to trust funds benefitting social advocacy, civil liberties, and the preservation of wilderness areas. In order to commemorate his passion, the Wilderness Society created the Bob Marshall Wilderness in northwestern Montana.

Maya, Esperanza (b. 1942). A co-founder of *El Pueblo Para Aire y Agua Limpio* (People for Clean Air and Water) in 1989, as well as California Communities Against Toxics, Maya helped to devise tactics by which communities can fight environmental hazards. Responding to a plan by Chemical Waste Management in 1989 to add a toxic waste incinerator to a landfill site above an area of predominately Hispanic farm workers in Kettleman City, California, she and other members of the community began to intervene in company-dominated planning meetings, and to demand that there be simultaneous translation in Spanish so that residents could hear what was actually proposed.

Working with black activist Jesse Jackson and with Lois Gibbs of Citizens Clearinghouse for Hazardous Waste, *El Pueblo Para Aire y Agua Limpio* staged marches and coordinated with other communities throughout the country who were mounting similar protests. The group filed and in 1992 won a lawsuit protesting the adequacy of the Environmental Impact Report (EIR). Chemical Waste Management withdrew the incinerator proposal, and the people in Kettleman City won. The group went on to become part of a network of green activists throughout the country.

McHarg, Ian (1920–2001). Landscape architect and urban planner, Ian McHarg is noted for his innovative designs integrating ecology and landscape with human communities. Born in Scotland, McHarg received master's degrees in Landscape Architecture and City Planning from Harvard University before moving to the University of Pennsylvania, where he became a leader in the emerging field of ecological city planning. His book *Design with Nature* (1969) became a bible for city planners seeking to integrate ecological factors, such as climate, geology, soil, vegetation, wildlife, and historical landmarks, into urban and suburban design. He developed integrative plans for cities such as Denver and Minneapolis and designed communities in Woodlands, Texas and Davis, California. In addition to *Design with Nature* (1969), his books include a collection of his writings, *To Heal the Earth* (1999), and his autobiography, *A Quest for Life* (1996).

Mechanism. A philosophy that arose in the scientific revolution of the seventeenth century that likens the world to a vast machine made up of inert parts that obey the laws of physics and chemistry.

Miller, Olive Thorne (1831–1918). A writer of numerous books and short stories for both children and adults, Olive Thorne Miller (Harriet Mann Miller) was an early advocate of birdwatching and bird preservation. She popularized birds and their behavior in books such as *Bird Ways* (1885), *In Nesting Time* (1888), *A Bird-Lover in the West* (1894) and *With the Birds in Maine* (1904). She contributed articles to the journal of the Audubon Society, *Bird Lore*, and her work helped to strengthen the movement to protect birds from plume hunters for the millinery trade and to promote nature as a refuge and source of spiritual renewal.

Mining Law (1872). The General Mining Law of 1872, signed by President Ulysses S. Grant to encourage western settlement and development of the country's mineral resources, has come under criticism for its failure to extract royalties for the government, and numerous attempts have been made to modify it. It allowed prospecting by individuals and corporations on the nation's public lands and the staking of claims for development of a deposit. The law covered placer deposits of gold and other minerals in streams and lode deposits of gold, silver, lead, zinc, copper, tungsten, uranium, and all other minerals except coal. Modifications to the law require claimants to pay an annual holding fee of $100 per claim and to purchase surface and mineral rights of $2.50 per acre for placer claims and $5 per acre for lode claims. There is no limit on the number of claims a person can hold and no royalties for the minerals extracted paid to the federal government.

Monoculture. The raising of only one crop in a field or on a given piece of land.

Morrill Act (1862). A federal act, passed in 1862, granting each state a minimum of 30,000 acres of public land for sale, for the purpose of creating a fund to establish agricultural and technical colleges of agriculture.

Moses, Robert (1888–1981). Parks Commissioner of New York City from 1933 to 1959, Robert Moses transformed the urban landscape of the city and its surroundings through a series of parks, playgrounds, tunnels, bridges, and parkways. He led the city's efforts to construct public housing to replace and improve tenements. He also held a variety of public and state offices, including chair of New York States's Council of Parks in 1924 and Secretary of State for New York (1927–28), ran for governor of the state in 1934, and wrote *Working for People* (1956).

Muir, John (1838–1914). One of the country's foremost advocates of wilderness preservation, John Muir was born in Scotland but grew up on a farm near Portage, Wisconsin, to which his family emigrated in 1849. In *The Story of My Boyhood and Youth* (1913), Muir described the severity of the work extracted from him by a Protestant father, his self-education, and mechanical inventions, experiences that later led him to leave established religion to seek God in the richness and beauty of nature. After studying chemistry, geology, and botany at the University of Wisconsin and surviving an eye injury at an Indiana factory where he worked, he determined to "study the inventions of God," and in 1867 set out on a journey described in *A Thousand Mile Walk to the Gulf* (1916). Muir traveled to the West in 1868, where he explored the Sierras and the Yosemite Valley, interspersed with explorations in the Northwest and Alaska, before marrying in 1880 and settling on his father-in-law's farm in Martinez, California.

With Robert Underwood Johnson, Muir mounted a campaign to save western forests from logging and grazing, which was influential in establishing Yosemite National Park in 1890. He founded the Sierra Club in 1892 in order to preserve the forests and valleys of the Sierra Nevada. Muir described his experiences in the Sierras in *The Mountains of California*, which appeared in 1894. His impassioned writing, including such articles as "Forest Reservations and National Parks" (*Harper's Weekly*, June 5, 1897) and "The American Forests" (*Atlantic Monthly*, August 1897), as well as a book, *Our National Parks* (1901), proved vital for raising consciousness about the need to save the national forests. His growing reputation enabled Muir to camp with Theodore Roosevelt in Yosemite in 1903 and convey to him the urgency of setting aside even more land as national forests and parks. Victory, however, did not come within Muir's lifetime, and his greatest defeat occurred when Congress passed the Raker Act in 1913, authorizing a dam for water and power for San Francisco in Yosemite National Park's Hetch Hetchy Valley. Dispirited by his failed campaign, Muir died in 1914, two years before Congress passed the National Parks Service Act to administer the country's thirteen then existing and future national parks.

National Audubon Society. Initially organized in 1886 and reestablished in 1905, the Audubon Society, named after ornithologist John James Audubon (1785–1851), promotes the study and protection of bird life and publishes *Audubon Magazine* (formerly entitled *Bird Lore*) for its members. Its goal is to preserve birds, their habitats, and migratory routes, to conserve and restore ecosystems, and to educate the public on the importance of wildlife conservation. The society originated in New York City in the 1880s in the attempt to protect birds whose plumes were used in the millinery trade to decorate women's hats, bonnets, and garments. In 1886, it succeeded in having a "model law" passed in New York State to protect the state's birds, and by 1905 twenty-eight states had passed the law. From its early work in protecting wild birds, the society expanded toward promoting appreciation of avifauna through birdwatching, compiling an annual census of birds, and producing educational materials on birds, their habits, and habitats. It is now one of the oldest and largest conservation groups in the world.

National Environmental Policy Act (NEPA) (1969). The National Environmental Policy Act, passed in 1969 and signed into law on January 1, 1970 by President Richard Nixon, requires every federal agency to prepare and circulate Environmental Impact Statements (EIS) on proposed legislation and actions affecting the quality of the human environment. The law created the Council on Environmental Quality (CEQ) in the Executive Office of the President, which was replaced by the White House Office on Environmental Policy in 1993. NEPA's broad mandate is to "create and maintain conditions under which man and nature can exist in productive harmony, and fulfill the social, economic, and other requirements of present and future generations of Americans."

National Park Service Act (1916). In 1916, Congress passed the National Park Service Act to transfer administrative power over national parks from the United States Cavalry to the Department of the Interior. Its mission was to "conserve the scenery and the natural and historic objects and the wild life therein and to provide for the enjoyment of the same in such a manner and by such means as will leave them unimpaired for the enjoyment of future generations." The law allowed for limited timber cutting to control insect damage or to conserve scenery, the removal of detrimental animals and plants, the leasing of land for guest accommodations, and limited livestock grazing. From thirteen parks in 1916, the National Park System had grown by the late twentieth century to include approximately 355 sites that covered over 125,000 square miles of natural, cultural, historical, and recreational lands.

National Wildlife Federation (NWF). Founded in 1936, the National Wildlife Federation is dedicated to the conservation of fish, wildlife, and other natural resources. The organization lobbies for legislation to conserve wildlife, maintains an institute for wildlife research, sponsors programs and camps for children, issues wildlife stamps, and publishes several magazines and directories. In contrast to absolute preservation of all biota, the NWF promotes hunting and trapping as a means for managing wildlife populations.

National Wildlife Refuge System. A collection of federally owned properties, the National Wildlife Refuge System is designed to preserve threatened wildlife habitats. The earliest federal wildlife preservation efforts, including a northern seal reserve on Alaska's Pribiloff Islands declared by President Ulysses S. Grant in 1868, were accomplished by executive fiat. President Benjamin Harrison, twenty-five years later, used the newly authorized presidential power to set aside timber reserves (1891) as an opportunity to create the first wildlife refuge on Alaska's Afognak Island. With a proclamation that stressed the common good, Harrison barred fishing and settlement on the island, and protected the resident salmon, sea lions, and sea otters. President Theodore Roosevelt, beginning in 1903, began aggressively to build a system of over fifty refuges. In his first such action, Roosevelt asked if any law prevented him from declaring Florida's Pelican Island, whose birds were being decimated by plume hunters, as "a Federal Bird Reservation." When told that no such law prevented it, he said, "Very well, then I so declare it." The National Wildlife Refuges are currently administered by the U.S. Fish and Wildlife Service, and included 520 units, covering 93 million acres, by the late twentieth century.

Natural resources. Objects of use or sources of wealth found in or arising from nature, as opposed to being produced by human labor. Renewable resources consist of living entities that can reproduce themselves after use, such as trees, grass, or fish. Nonrenewable resources are entities that cannot replace themselves once used or extracted, such as minerals and fossil fuels.

Natural Resources Defense Council (NRDC). Founded in 1970, thehe Natural Resources Defense Council (NRDC) is a citizen's litigating organization working on behalf of the environment. It focuses on drafting environmental laws, lobbying for their passage, and litigating for their implementation. Among issues it has taken up are the phase-out of lead from gasoline, electrical utility conservation and efficiency, old-growth forest preservation, and the protection of the environment from oil drilling, pollution, nuclear weapons proliferation, and pesticides. NRDC, which publishes the journal *Amicus*, advocates the creation of a sustainable society in which people can live in a harmonious relationship with the environment.

Nature. *Nature* is one of the most complex words in the English language and has several interrelated meanings that have evolved over time. On the one hand, it is the physical universe, including all its physical features, processes, organisms, and their interactions. Although it technically includes human beings, it is often used to refer to the world as distinct from or prior to human beings and their social and cultural institutions. Another meaning is that of an inherent force or impulse to act and sustain action; hence to go against nature is to go against this inherent impulse. A third meaning is the fundamental character or quality of an individual entity.

Historically, Nature has often been personified as a mother, virgin, or goddess who acted in accordance with her own dictates, or capricious impulses, or who carried out the mandates of a superior deity. The latter meaning was elaborated as the unchanging natural laws of the universe which could be discovered, rationally understood, classified, or used for prediction; the idea of the state of nature was used as the basis for an ideal society obeying natural laws. Natural selection referred to an active evolutionary process through which species competed and survived or went extinct. Ideas of nature as unchanging and competitive or changing and cooperative were projected onto society and used to justify political and social goals and programs.

Nature Conservancy. Founded in 1951, the Nature Conservancy is a citizens' environmental organization dedicated to protecting the habitats of plants, animals, and natural communities that represent the diversity of life on earth. The Conservancy works with individuals and communities to set aside habitats as nature preserves and has acquired more than 1,500 preserves, encompassing over 9 million acres in North America. It has pioneered debt-for-nature swaps by which developing countries can reduce their debt to the United States by setting aside nature preserves in their own countries. The organization oversees over 90 million acres in Latin America and the Caribbean, and maintains a global database of endangered species.

New Ecology. A school of ecology, prominent in the 1930s through the 1960s, that emphasized the quantitative study of energy, energy exchanges, and ecological efficiency. Major figures included Arthur Tansley, Raymond Lindeman, and Eugene Odum.

Newell, Frederick (1862–1932). A hydraulic engineer and hydrographer with the U.S. Geological Survey beginning in 1888, Frederick Newell made irrigation surveys in the West. After passage of the Reclamation Act of 1902, he became director of the Reclamation Service (now the Bureau of Reclamation), serving in that post from 1907 to 1914. Newell was instrumental in organizing President Theodore Roosevelt's White House Conference of Governors in 1908, which promoted the conservation of the nation's natural resources. He wrote several books on irrigation, including *Hydrography of Arid Regions* (1891); *Agriculture by Irrigation* (1894); *The Public Lands of the United States and their Water Supply* (1895); *Irrigation in the United States* (1902); and *Water Resources, Present and Future Uses* (1919).

Odum, Eugene (b. 1913). An ecologist and environmental philosopher, Eugene Odum developed a systems approach to interactions between humans and the environment based on the concept of balanced, homeostatic ecosystems. Influential in the 1960s and 1970s, his work integrated Frederic Clements' ideas of ecological succession with thermodynamic approaches focused on energy transfers within the system. While nature (conceptualized as the *oikos*, or human home) tended toward a homeostatic balance within particular ecosystems, disturbances came from both human and natural causes. Nevertheless, humanity had the possibility of restoring degraded ecosystems through science, ethics, and the wise management of landscapes. Odum's *Fundamentals of Ecology* (1953), a definitive work for a generation of ecologists and environmentalists, set out the basic concepts and principles of ecology and applied them to aquatic and terrestrial ecosystems and the natural resources they contained. Odum won several prizes, including the Tyler Prize, for his achievements in environmentalism (1977) and served as the president of the Ecological Society of America.

Olmsted, Frederick Law (1822–1903). Noted both as a writer and landscape architect. Frederick Law Olmsted was an early advocate for, and designer of, public parks in the United States. Before taking up a career in design, he wrote several accounts of his travels, including A *Journey in the Seaboard Slave States* (1856), and a report on "The Yosemite Valley and the Mariposa Big Trees" (1864) that led to the preservation of Yosemite Valley. In collaboration with architect Calvert Vaux, Olmsted won the commission to design New York's Central Park in 1866. Although reconfiguring and managing every aspect of the landscape, Olmsted wanted the park to appear as though it had simply been preserved from development. Park workers created a lake, planted a forest, and filled a former dump with topsoil to restore native meadows. In addition to Central Park, he designed the Boston Fens, participated in campaigns to preserve Niagara Falls and the Adirondacks of New York State, and designed several university campuses, city parks, cemeteries, and hospital grounds. A guiding philosophy of his work was the necessity for accessibility to nature by all walks of society for a democratic culture to thrive.

Organicism. A philosophy that the world is like a living organism, such as the human body. Each part of nature, including animals, plants, and minerals, is part of a larger whole and participates in it.

Pinchot, Gifford (1865–1946). Gifford Pinchot is considered the founder of modern forestry in the United States. He went to Yale University and studied forestry in Germany and France. When he returned to the United States, he experimented with different forestry techniques while managing the Biltmore Estate in North Carolina. Succeeding Bernhard Fernow, he served as chief of the Division of Forestry (now the Forest Service) in the U.S. government between 1898 and 1910, during which time he strove to create forest reserves and to train a core of young foresters in the latest management techniques. Pinchot's policies included transferring the forest reserves out of the Department of the Interior and into the Department of Agriculture in 1905, on the grounds that forests ought to be managed as a crop, instituting fire suppression programs in the national forests, and devising a sustained yield program in which any trees cut would be replanted. Some of his decisions, such as allowing grazing permits in the national forests and supporting the damming of Hetch Hetchy Valley in Yosemite National Park, put him at odds with the preservationist movement. After leaving government office in 1910, he pressed for the passage of the Weeks Act in 1911, which authorized the purchase of additional forest reserves, and was a founder and president of the Society of American Foresters between 1900 and 1911. Pinchot served as Governor of Pennsylvania from 1922 to 1926. He wrote about conservation in *The Fight for Conservation* (1910), and described his life's work in *Breaking New Ground* (1947).

Pittman-Robertson Wildlife Restoration Act (Federal Aid in Wildlife Restoration Act) (1937). The Pittman-Robertson Wildlife Restoration Act was first passed in 1937, and has been amended numerous times since then. It provides federal aid to the states to restore and manage wildlife and to acquire wildlife habitat, funded through a tax on sporting guns and ammunition. The states submit to the Secretary of the Interior detailed five-year plans for fish and wildlife management or for wildlife restoration that ensure the perpetuation of wildlife resources for economic, scientific, and recreational purposes.

Plant associations. Groups of plants usually found together in areas having similar ecological conditions.

Polyculture. The raising of several crops together in one field or piece of land.

Population. The total number of people or organisms living in a given area.

Population biology. The study of those factors that determine the size and distribution of organisms in a given area and the ways they change over time.

Portmanteau biota. The term used by environmental historian Alfred Crosby in his book *Ecological Imperialism* (1986) to characterize associated animals, plants, weeds, pests, and pathogens accompanying people on ships when establishing New World colonies.

Positivism. The philosophical theory that only mathematical statements and empirically verifiable statements are true and lead to positive knowledge of the external world.

Possibilism. The philosophical theory that allows for a multitude of choices, influences, or explanations of an event or outcome.

Powell, John Wesley (1834–1902). John Wesley Powell was an ethnologist and geologist who strongly influenced Americans' view of the western landscape. Born in Mount Morris, New York, Powell fought in the Civil War on the side of the Union, becoming a major, after which he took a position as professor of geology at Illinois Wesleyan College and subsequently became curator of the museum of Illinois Normal University. He conducted a number of exploratory surveys of the West, including a hazardous journey down the Colorado River in 1869 (described in *Explorations of the Colorado River*, 1875), a federally funded scientific mission in 1871, and a survey of the Rocky Mountain region in 1875–79. His efforts helped establish the U.S. Geological Survey, which he led from 1881 to 1894 and at which he instigated the compilation of topographic maps, geological, soil, and groundwater surveys, and proposals for flood control and irrigation. He was also the first director of the Bureau of Ethnology, a position he held from 1789 until his death in 1902.

Powell's most important legacy is his 1878 *Report on the Lands of the Arid Region of the United States*, which identified a fundamental difference between the lands of the western and eastern United States and proposed strategies for constructive human settlement in the West. The meager rainfall in most areas west of the 98th meridian, averaging less than twenty inches annually, divided the landscape into distinctive zones: whereas fertile areas in river valleys and adjacent irrigated lowlands could support farms as small as 80 acres, non-irrigated lands could only be used for grazing, and required tracts upwards of 2,560 acres. A third classification of valuable timberland, he asserted, should be set aside and reserved for commercial use. These findings cast into doubt the idea of the standard 160-acre allotment that had been prescribed by the Homestead Act of 1860.

The most radical position Powell took was that people should continue the medieval and New England ideal of the commons on western lands. He argued that cooperative irrigation districts should be created, so that a group of people could irrigate a district together, grazing their animals on the commons. The federal government should supply funds to aid people in cooperative irrigation projects by setting aside money to help them build reservoirs to store the water, from which the irrigation ditches could be channeled. Scientific surveys would be conducted to identify soil characteristics suitable for farming and pasturage. Although the more radical features of his work were not accepted, Powell's *Report* led ultimately to the passage of the Reclamation Act of 1902 and the creation of the Bureau of Reclamation.

Preservation. The belief that natural areas should be left undisturbed by human beings.

Public domain. The lands owned and managed by the United States government.

Railroad Land Grants. A federal policy of granting even-numbered sections (640 acres) of land on either side of a right of way to western railroads. The land grants, meant to spur the rapid construction of transcontinental routes for the carrying of mail and troops as well as the general settlement of western lands, resulted in a great deal of land speculation.

Raker Act (1913). The passage of the Raker Act in 1913 was the culmination of the famous struggle during the first decade of the twentieth century over whether or not to dam Hetch Hetchy Valley in Yosemite National Park. The controversy polarized the followers of John Muir, who likened Hetch Hetchy to a cathedral, and the followers of Gifford Pinchot, who saw a greater public good in damming the valley as a source of water for the city of San Francisco. The act, which provided for land appropriations within, and rights of way through, federal lands in and around Yosemite, stipulated that the resulting water and power resources must be kept in public hands, a requirement that was afterwards ignored.

RARE I and II. Two studies, Roadless Areas Review and Evaluation I (1971) and II (1977), that were undertaken by the U.S. Forest Service, with massive public involvement, to identify wilderness areas for protection and open non-wilderness roadless areas to multiple-use management.

Reclamation. The process of making the land productive for agriculture through irrigation. Also, the process of restoring disturbed land to an ecologically healthy state.

Reclamation Act (1902). The Reclamation Act, otherwise known as the Newlands Act, was passed in 1902 to aid in the settlement of the arid West. It established a fund, generated from the sale of public lands, for the construction of dams and irrigation systems. Water from federal projects was restricted to land parcels of 160 acres or less, or 320 acres for married couples, and to bona fide occupiers of the land. Speculators thwarted the act's provisions for the democratic distribution of land and water, as had been the case with the earlier Homestead Act.

Recycle. To reuse materials over again in new products, as opposed to throwing them away as waste.

Resource. Money, property, or natural entities that can be used to the advantage of an individual or a country.

Resource Conservation and Recovery Act (RCRA) (1976). Congress passed the Resource Conservation and Recovery Act in 1976 in an effort to manage the increasing volume of solid waste with better methods. The goals of RCRA, which amended the Solid Waste Disposal Act of 1965, are to protect human health and the environment, to reduce waste, and to conserve energy and natural resources. RCRA's Hazardous and Solid Waste Amendments were added in 1984 to address the problems of production and reduction of hazardous waste. RCRA created a program to handle hazardous and medical wastes from "cradle to grave," and to develop standards for underground petroleum storage tanks to prevent leakage into groundwater. The act delegated to the states the development and implementation of solid waste management plans in accord with Environmental Protection Agency (EPA) standards for landfills, and is implemented with EPA technical and financial assistance. Waste recovery policies developed by the EPA follow the hierarchy of first reduce, then reuse, recycle, incinerate with energy recovery, and finally, only as a last step, to resort to landfills.

Richards, Ellen Swallow (1842–1911). Ellen Swallow Richards, a chemist and educator, conceptualized ecology as the study of the human home, or community. After undergraduate work at Vassar, from which she received a bachelor's degree in 1870, Richards continued her research in chemistry as the first woman to attend the Massachusetts Institute of Technology (MIT). At MIT, where she remained throughout her career, she incorporated into her framework the ideas of Frank Storer, a professor studying the atmosphere, and concepts of the earth derived from the work of her future husband, Robert Richards.

The concept of oekology as the science of living, which Richards introduced in 1892, led to a later definition of "human ecology" as "the study of the surroundings of human beings in the effects they produce on the lives of men" in her book *Sanitation in Daily Life* (1907). The features of the environment, according to this theory, came both from the climate, which was natural, and from human activity, which was artificial, or superimposed on it. "Human ecology" laid the groundwork for the sanitation and home economics movements, and raised public consciousness of problems caused by humans living in industrializing cities, such as noise, smoke, and water pollution.

Richards was a pioneering educator as well as researcher and theorist. She used funds from the Women's Education Association to establish a Woman's Laboratory in 1876, where she and growing ranks of scientists devised protocols for testing food, water, and other materials. In 1881, she extended this initiative to include a summer laboratory in Massachusetts, which later moved to Woods Hole and eventually became the Marine Biological Laboratory. Her books include *The Chemistry of Cooking and Cleaning* (1882), *Home Sanitation* (1887), *Air, Water, and Food* (1890), *Euthenics, the Science of Controllable Environment* (1910), and *Sanitation in Daily Life* (1907; 2d ed., 1910).

Right of Way Act (1901). The Right of Way Act was passed in 1901 to allow the Secretary of the Interior to permit rights of way through public lands, national forests, and national parks. Specifying Yosemite, Sequoia, and General Grant National Parks in California, the act allowed the construction of electrical and water transmission reservoirs, conduits, and equipment for mining, domestic, and other beneficial uses on rights of way. Although the act technically would have allowed the construction of Hetch Hetchy Dam and the transmission of its water and power through Yosemite National Park, permission for right of way for the development was denied until the Raker Act in 1913.

Roosevelt, Theodore (1858–1919). Conservationist, outdoors adventurer, politician, and President of the United States from 1901 to 1908, Theodore Roosevelt made major contributions to the conservation movement at the turn of the nineteenth century. A bird-watcher, hunter, and rancher, Roosevelt pursued a vigorous outdoor life in the western United States in his youth and through trips after his presidency to Africa and South America. After serving as Governor of New York State from 1898 to 1900, where he became an advocate for forest and bird preservation, Roosevelt ran for the vice-presidency and, after the assassination of William McKinley in 1901, became the nation's twenty-sixth president. While in office, he frequently collaborated with advocates such as Gifford Pinchot to promote efficiency in natural resources use and forest conservation and John Muir to preserve national parks. To these ends, he worked with Congress to pass the Reclamation Act of 1902, appointed commissions to study the use of public lands (1903) and inland waterways (1907), and held a White House Conference of Governors on the conservation of natural resources (1908). He increased the forest reserves from 50 million to 150 million acres and created fifty wildlife refuges. Roosevelt's legacy in conservation endures through his efforts to preserve the resources of his day for future generations.

Sears, Paul (1891–1990). A botanist, ecologist, and conservationist, Paul Sears studied the link between human disruption of nature and the desertification of landscapes. He argued that planners and government officials at every level should consult with ecologists to understand how both settlement and agricultural and extraction schemes relate to the degradation, conservation, and the health and stability of nature. As chair of the Conservation Program at Yale University, he created a graduate program in Conservation of Natural Resources and publicized the dependency of humanity on other biota, natural resources, and habitats. He called ecology a "subversive science" for its capacity to question the assumption, central to the development of modern Western culture, that human society progresses through the transformation of nature. His books include *Deserts on the March* (1935), *This Is Our World* (1937), *Life and the Environment* (1939), *The Living Landscape* (1964), and *Lands Beyond the Forest* (1969).

Sierra Club. The Sierra Club was founded in 1892, under the leadership of John Muir (1838–1914), to promote the enjoyment and protection of the Sierra Nevada Mountains. From an initial focus on creating and preserving parks in the western mountains, the organization has undertaken wider efforts to preserve wilderness areas, such as old-growth redwoods, national seashores, and the Grand Canyon, and to lobby for clean air and other environmental protection legislation. The club has more than 58 chapters throughout the United States, which work on issues from the local to the national and beyond.

Social ecology. The study of human social and economic institutions and communities with respect to their interactions with other humans, species, and the natural environment.

Soil Conservation Act (1935). Enacted by Congress in 1935 out of concern over soil erosion and the Dust Bowl, the Soil Conservation Act established a permanent Soil Conservation Service in the United States Department of Agriculture.

Soil Conservation Service (SCS) (1935). An agency in the Department of Agriculture in the executive branch of the federal government, established in 1935 by passage of the Soil Conservation Act, to provide for the control and prevention of soil erosion. It makes soil surveys throughout the country, reports on salinization, wetlands, and water quality, and combats soil erosion by wind and water.

Stability. Relative constancy over time of the numbers of individuals in a species or of different species in a given area.

Standing Bear, Luther (1829?–1908). Luther Standing Bear was a leader of the Ponca Indian tribe of northern Nebraska. In addition to his successful legal campaign to preserve his tribe's ancestral lands, he is noted for his statement that he and his people did not consider their beautiful lands as wild. "Only to the white man was nature a 'wilderness' and only to him was the land 'infested' with 'wild' animals and 'savage' people. To us it was tame. Earth was bountiful…" The statement is seen by some as challenging the idea that the North American continent was a pristine wilderness at the time of European settlement and therefore that Indians both managed and were stewards of the landscape. Standing Bear's memoirs were published posthumously in *Land of the Spotted Eagle* (1933).

Staple. (1) A chief or leading crop or commodity regularly grown or sold in a particular place, such as sugar, wheat, or cotton. (2) The fiber of cotton, wool, or flax, with reference to its length and fineness.

Sustainable development. The idea that each generation can satisfy current needs for natural resources while respecting those of future generations.

Steady state. A dynamic balance between inputs and outputs of a system, such as an organism or an ecoystem.

Subsistence. The means of providing sustenance, support, or livelihood for the continued existence of a human being.

Tansley, Arthur G. (1871–1955). British plant ecologist Arthur Tansley introduced the concept of the ecosystem to the field of ecology. A member of an international botanical expedition to the United States in 1913, Tansley toured the Great Plains, the Sierras, and the Southwest and prepared the final report of the expedition, noting that, despite the inevitability of economic development, "tracts of original untouched nature can and should be preserved for the enjoyment and use of our successors." As Professor of Botany at Oxford University from 1927 to 1937, he published a foundational essay in the journal *Ecology*, "The Use and Abuse of Vegetative Concepts and Terms" (1935), defining the ecosystem as "the biome [animals and plants] considered together with all the effective inorganic factors of its environment.... In an ecosystem the organisms and the inorganic factors alike are components which are in relatively stable dynamic equilibrium." In 1939, he published a classic book on plant ecology entitled *British Islands and Their Vegetation* (1939), and in 1946 put out a revision of his 1923 textbook entitled *An Introduction to Plant Ecology*. Tansley's later years were devoted to the cause of conservation in Great Britain, where he served as the first head of the Nature Conservancy.

Taylor Grazing Act (1934). First passed in 1934, the act retained all unreserved public domain lands and federally regulated grazing on public lands through establishing grazing districts and managing livestock grazing in those districts. The Secretary of the Interior is charged with providing for the preservation of the land and for issuing permits to graze livestock to settlers, residents, and livestock owners for periods of not more than ten years, and may suspend permits in times of range depletion due to drought or during epidemics. Other resources such as stone, timber, and coal within the districts must remain accessible. Hearings are held in the states affected before grazing districts are created and adjacent landowners may apply for rights-of-way through the districts.

Tennessee Valley Authority (TVA). The Tennessee Valley Authority (TVA) is an agency created in 1933, for unified development of the resources of the Tennessee River Valley to achieve flood control, navigation, electric power generation, reforestation, and the economic well-being of the people of the valley. A product of Franklin Delano Roosevelt's "New Deal," it was designed to bring electricity to rural households and to provide employment during the Depression era and beyond. Originally based on hydroelectric power, the project later turned to coal and nuclear sources to maintain its electrical distribution systems. TVA has had impacts on the environment in the form of air pollution, habitat degradation, and species depletions in the region.

Thoreau, Henry David (1817–62). Nature writer and philosopher, Henry David Thoreau was an early advocate for an environmental consciousness. After attending Harvard College from 1833 to 1837, he moved to Concord, Massachusetts, and became engaged in the intellectual currents there, including intense interest in the natural world and the philosophy of transcendentalism, or the idea of an overarching unity in all things. At Concord, he became a follower and protégé of Ralph Waldo Emerson, with whom he also worked on abolitionism during the year 1844. This passion led to his essay "Civil Disobedience" (1847) and to the later use of his home as a station on the Underground Railroad. In 1845, he moved to Walden Pond to undertake an experiment in solitary living for two-and-one-half years, from which emerged his most famous book, *Walden* (1854). *Walden* is a landmark in American nature writing, natural history observation, and environmental history. It reflects on the qualities of life that can be found in and from nature and the ambiguities associated with so-called advancements such as the railroad and commerce. Thoreau's essay "The Succession of Forest Trees" (1860) is heralded as a major early contribution to the ecological theory of plant succession. Journals kept throughout his life, contain much of his best writing and nature observations. His collected works were published in 1906 as *The Writings of Henry David Thoreau* (20 volumes). He died of tuberculosis in Concord in 1862.

Timber and Stone Act (1878). A federal act, passed in 1878, that provided for the sale in Washington, Oregon, California, and Nevada of 160 acres of land valuable for timber and stone, but unfit for cultivation, for not less than $2.50 per acre.

Timber Culture Act (1873). A federal act, passed in 1873, to donate 160 acres of public land to any person who would plant 40 acres with trees and keep them healthy for a period of ten years. Due to the practical difficulties of planting trees on the prairies and plains, as well as problems of fraud and speculation, the act's provisions were repealed by the Forest Reserve Act of 1891.

Toxic Substances Control Act (TSCA) (1976). Congress, in response to nationwide concerns over asbestos, chlorofluorocarbons (CFCs), mercury, polychlorinated biphenyls (PCBs), and vinyl chloride, passed the Toxic Substances Control Act in 1976. The act mandated that manufacturers and processors keep records about the production and transportation process, as well as health and environmental effects, of all chemicals regulated by the law. The Environmental Protection Agency (EPA), was charged with determining whether new chemicals can be manufactured, a task it has had difficulty fulfilling.

Tragedy of the commons. The theory that a group of self-interested people collectively using resources will degrade and deplete those resources. Examples include the overgrazing of open rangelands, the overfishing of the oceans, and the pollution of air and water.

Transcendentalism. The idealist, Platonist philosophy held by a group of individuals in nineteenth-century United States, most notably Ralph Waldo Emerson. Reality was ideal and the material world only a constantly changing appearance. Symbols and emblems in the material world provided clues to ideal truths through visionary insight and spiritual intuition, enabling the human soul to transcend physical limitations and gain insight into the Oversoul, or divine One.

Trophic. Food and nutritional processes entailing the transfer of energy from one organism to another.

Tucker, Cora (1940–1997). Cora Tucker founded Citizens for a Better America, an early spearhead in the movement for environmental justice. An African-American grandmother, activist, and self-described homemaker, Tucker began to fight uranium mining projects and nuclear waste dump sites while living in rural Virginia's Halifax County. In 1986, she organized the successful efforts of a coalition of residents of Halifax County to protest a disposal site for a high-level nuclear dump, using a federal law requiring that siting decisions take minorities and Indians into consideration. She is also credited with turning the derogatory phrase, "hysterical housewives," into a term of empowerment for organizing. In 1991, Tucker received a Bannerman Fellowship for Long-Time Activists of Color. She was the chair of the Board of Grassroots Leadership in Charlotte, North Carolina and the first President of Virginia Action.

Turner, Frederick Jackson (1861–1932). A pioneering historian of the development of western lands, Frederick Jackson Turner is famous for his "frontier thesis." Born in Portage, Wisconsin, Turner taught at the University of Wisconsin from 1885 until 1910 and Harvard University until his retirement in 1924. His essay entitled "The Significance of the Frontier in American History," first presented at the Columbian Exhibition at Chicago in July 1893, is one of the most influential and hotly debated explanatory principles in the interpretation of American history. Turner stated: "The existence of an area of free land, its continuous recession, and the advance of American settlement westward, explains American development." Continuous waves of trappers, ranchers, prospectors, and farmers, drawn to rich natural resources in the West, were stripped of the trappings of European culture and transformed into a new, and uniquely democratic, type of American. In addition to his frontier thesis, Turner is noted for *The Significance of Sections in American History* (1932), delineating the differentiation of the United States into economic sections in the early nineteenth century and those sections' formative roles in American historical development and conflict.

Udall, Stewart (b. 1920). Noted for his book *The Quiet Crisis* (1963), Steward Udall served as Secretary of the Interior under Presidents John F. Kennedy and Lyndon B. Johnson from 1961 to 1969. Appearing the year after Rachel Carson's *Silent Spring* had alerted Americans to the threat to human and biotic health from the use of pesticides, Udall's book was noted for its warning about the creeping effects of pollution on the landscape and the need for conservation education. While science and technology had opened up new sources of energy and possibilities for the control of nature, he wrote, they had also contributed to the increase in garbage, roadside litter, contaminated lakes and rivers, polluted air, and declining wildlife. Udall called for new educational initiatives, legislation, and technologies to respond to the crisis and restore the balance of nature. In addition to his service in the Cabinet, Udall served as a congressman from Arizona from 1955–60, and was a member of the Joint Committee on Navajo-Hopi Indian Administration. He contributed to *National Parks of America* (1966), *The National Parks* (1974), and, beginning in 1970, wrote "Udall on the Environment," a column syndicated in many newspapers.

United States Department of Agriculture (USDA). The Department of Agriculture is an agency within the executive branch of the United States government that oversees agricultural production and crop protection. From a division within the Patent Office in charge of seed distribution, statistics, and research started in 1840, Agriculture became a department in 1862 and rose to cabinet rank in 1889. The Department oversees farm production, sets health standards for farm products, develops overseas markets, provides assistance to rural residents, and is charged with protecting soil, forests, and other resources. The Department's Natural Resources and Environment Division contains the Forest Service and the Soil Conservation Service.

United States Department of the Interior. The U.S. Department of the Interior was founded in 1849 to manage the western lands and their allocation to settlers, railroads, colleges, and universities. In the twentieth century, as public lands became increasingly closed to settlement, the Department shifted its energies towards caring for public lands and natural resources. It administers wildlife refuges, wetlands, national parks and monuments, wild and scenic rivers, and Indian reservations. It oversees the National Park Service, the Bureau of Land Management, the Bureau of Reclamation, the Fish and Wildlife Service, the Geological Survey, the Bureau of Mines, the Office of Surface Mining, Reclamation, and Enforcement, and the Bureau of Indian Affairs. While criticized by some for the construction of large dams and reclamation systems that result in ecological degradation, the Department also is a primary conserver of ecosystems and wildlands.

United States Geological Survey (USGS). The U.S. Geological Survey is a federal agency in the Department of the Interior, created in 1879 to survey mineral resources and rivers in the American West. The agency's first two directors included geologist Clarence King (1879–81), who conducted a survey of the fortieth parallel, and explorer John Wesley Powell (1881–94), who conducted the first survey expeditions down the Colorado River in 1869 and 1871. The USGS maps topography, catalogs natural resources such as minerals, fuels, and water, and conducts research on fossil and glacial history, hydrology, oceanography, geophysics, and geochemistry. It compiles information on volcanic action, landslides, floods, and droughts, and monitors underground aquifers, streamflows, and water quality.

Washington, Booker T. (1856–1915). A scientist and educator, Booker T. Washington effectively pioneered improved farming methods and worked to diffuse practical knowledge among African Americans in the South. Born

a slave in the Virginia Piedmont, Washington grew up on a 207-acre tobacco farm now a national monument. As a young man he worked in the
West Virginia coal mines and was educated at the Hampton Normal and
Agricultural Institute in Norfolk, Virginia, graduating in 1875. In 1881, he
moved to the Tuskegee Institute in Tuskegee, Alabama, and built it from a
school for prospective schoolteachers into a regional center of agricultural
and industrial training, research into better farming methods, and outreach.
To better equip blacks to be successful farmers and owners of land, he developed the Tuskegee Negro Conference for distributing agricultural techniques through methods such as county fairs, short courses, and leafleting.
During his career, Washington delivered numerous speeches and wrote
books on the education and potential of African Americans, including his
Autobiography (1901).

Water Pollution Control Act (1948, 1972). The Water Pollution Control
Act, passed in 1948, was the first federal law that officially dealt with water
pollution. It provided funding for state and local governments to identify
and improve polluted waters. The Water Pollution Control Amendments
passed in 1972 shifted the major responsibility for directing water pollution
control programs to the federal government. As revised and expanded, the
act mandated that discharges of polluting and toxic materials be halted by
1985 and that water quality improve to a level that would protect fish and
wildlife. Funds for the construction of waste treatment plants were authorized and the states were charged with introducing technologies to halt pollution and to control polluting forms of runoff. The Environmental Protection Agency (EPA) was given the responsibility of issuing guidelines to the
states and polluting industries for the treatment of wastewater using the best
practicable technologies to control organic wastes, sediment, acid, bacteria,
viruses, nutrients, oil and grease, and temperature levels. The EPA was also
directed to publish a list of toxic pollutants and standards for releasing each
substance into the environment.

Water Quality Act (1987). The Water Quality Act passed by Congress in
1987 ended federal funding for treatment of wastewater and instead emphasized control of toxic substances in sewage and storm water. The states were
directed to list toxic hot spots unable to be improved by control technologies. The act left many states with unmet wastewater treatment needs and
waters that did not meet federal water quality standards. Runoff and toxic
substances pollution continued to present problems as well.

Water Quality Control Act (1965). In 1965, Congress passed the Water Quality Control Act, giving the Federal Water Pollution Control Administration the power to establish water quality standards.

Wheatley, Phillis (1753–84). Born in Gambia, West Africa, and brought to Boston as a slave in 1761, Phillis Wheatley used nature imagery in her poetry, drawing on classical and biblical references in the style of the eighteenth-century neoclassical Enlightenment. She was bought by Susanna Wheatley, who had several black slaves as servants in her household, and schooled by Wheatley's youngest daughter, Mary, who taught Phillis English and the alphabet. At twelve, Phillis began to study Latin, and after four years began translating portions of Ovid, gaining a reputation in America and abroad. Under the influence of the Latin classicists, she began to write her own poetry, published in 1773 as *Poems on Various Subjects, Religious and Moral*, which brought her to the attention of Benjamin Franklin, George Washington, Thomas Jefferson, and others. Although much of her poetry commemorates individuals, those which extol nature use classical allusions such as Aurora as dawn, gentle zephyr as wind, the sun as king, and birds as the feathered race in appreciation of God's power in the natural world.

White, Laura Lyon (1839–1916). Conservationist Laura (Mrs. Lovell) White, founder and president of the California Club, was instrumental in the preservation of California's redwoods. From 1900–1909, she led a nationwide campaign to save the Calaveras Big Trees (*Sequoia gigantea*) in the Stanislaus River basin, the largest known redwoods in existence, measuring over 12 feet in diameter. Taking office in 1903 as president of the Sempervirens Club (which later became the Save the Redwoods League), she worked with others to save the coastal redwood (*Sequoia sempervirens*). An advocate of scientifically based forestry, she chaired the Forestry Committee of the General Federation of Women's Clubs from 1910–12, which coordinated efforts nationwide to preserve forest reserves and watersheds. She gained a national reputation for her work on behalf of forest preservation through her executive abilities and tireless advocacy in the cause of conservation.

Wild and Scenic Rivers Act (1968). The Wild and Scenic Rivers Act, passed in 1968 in the wake of the controversial damming of western rivers, began the process of designating and protecting wild rivers. The act designated free-flowing rivers that were free of dams and had outstanding scenic, geologic, historic, recreational, or wildlife features. The National Rivers Inventory, published in 1982, listed 1,524 portions of rivers eligible for the designation. When so listed, the river — whether in a remote, rural, or urban area — would be ineligible for hydroelectric or other water projects or for mineral extraction. By the mid-1990s, 152 rivers with a total of 10,516 miles of river segments had been designated as wild and scenic under the act. The rivers are managed by the Department of Agriculture's Forest Service, the Department of the Interior's National Parks Service, or its Bureau of Land Management.

Wilderness. The word *wilderness* derives from medieval terms such as *wildern* (wild savage land), *wilddeor* (wild deer), and *wilddren* (wild man). Wild places were uncultivated, uninhabited forests, wastes, and deserts. Deep forests were dark, unknown places in which one might become bewildered and lost. Wastes were open, unused lands with little vegetation. *Wilde* and *wylde* pertained to untamed animals living in a state of nature and to uncultivated, undomesticated plants. Although in colonial America wilderness denoted an unknown forested place, it later took on positive characteristics of awe, beauty, and the sublime, and ultimately of untrammeled lands worthy of preservation for scientific or recreational purposes.

Wilderness Act (1964). The Wilderness Act, passed in 1964, designated certain federal lands as wilderness areas, preventing their development, and defining wilderness as an area "where the earth and its community of life are untrammeled by man, where man himself is a visitor who does not remain." It established the National Wilderness Preservation system, which grew to comprise over 95 million acres of land in its "natural condition." In 1975, Congress created sixteen wilderness areas in the eastern states with the Eastern Wilderness Act. The Roadless Area Review and Evaluation (RARE I) and its successor, RARE II, played a key role in the designation process.

Wilderness Society (TWS). The Wilderness Society is an environmental organization, founded in 1935, dedicated to the creation and protection of wilderness areas in particular and to the preservation and management of the nation's public lands in general. These lands include national parks, national seashores, national forests, national wildlife refuges, and other federal lands. The Society was co-founded by forester Robert Marshall, wildlife biologist Aldo Leopold, Appalachian Trail founder Benton MacKaye, and others. It was instrumental in the passage of the Wilderness Act (1964), as well as other conservation laws such as the Wild and Scenic Rivers Act (1968), the National Forest Management Act (1976), and the Alaska National Interest Conservation Land Act (1980). Its goals are to save old-growth forests, to prevent below-cost timber sales in the national forests, to prohibit oil drilling in the Arctic National Wildlife Refuge, to expand the number of wild and scenic rivers, and to double the size of the National Wilderness System.

Winters decision. A decision by the United States Supreme Court, in January 1908, explicitly affirming water rights claims by Indian tribes in the western states.

Works Progress Administration (WPA). The Works Progress Administration was an agency founded by the Roosevelt administration between 1935 and 1943 to create employment during the depression in areas such as conservation, education, the arts, writing, and theater production. The workers constructed parks, bridges, and public buildings.

World Wildlife Fund (WWF). Founded in 1961, the World Wildlife Fund is an international conservation organization dedicated to worldwide conservation and the maintenance of habitats, endangered species, and essential ecological processes. It monitors international trade in wildlife and endangered species and works with other national and international conservation organizations to conduct scientific research that will preserve biodiversity and protect wildlife, such as the African elephant and the giant panda, which is its logo.

Wright, Mabel Osgood (1859–1934). Mabel Osgood Wright, president of the Connecticut Audubon Society in Fairfield, Connecticut from 1898 to 1925, helped to build a national network of activists on behalf of wildlife. She was an editor of *Bird Lore*, which became *Audubon* magazine, and the director of the National Association of Audubon Societies (which became the Audubon Society) from 1905–28. She encouraged the secretaries of the other state Audubon societies to send in news items to the magazine in order to publicize the plight of song and shore birds, whose plumes were being used for women's hats. This work was influential in the passage of a "model law," restricting the harvesting of birds for their feathers, a Tariff Act in 1913, outlawing the importation of wild bird feathers, and the success of the campaign to convince women to change their hat styles. Devoted to the observation and teaching of birds in their habitats, Wright taught bird classes to children and wrote several children's books on nature, including *The Friendship of Nature* (1894), *Birdcraft* (1895), *Birds of Village and Field* (1898), and *Citizen Bird* (1897).

Yellowstone Park Act (1872). The Yellowstone Park Act of 1872, which set aside "a tract of land lying near the headwaters of the Yellowstone River as a public park," created the nation's first national park. It was "set apart as a public park or pleasuring-ground for the benefit and enjoyment of the people," including "all timber, mineral deposits, natural curiosities, or wonders" so that they might be retained in their "natural condition." The act allowed the construction of buildings to accommodate visitors and forbade the destruction of fish and game within the park boundaries. The area had been explored in 1871 by the Hayden Expedition, which brought back photographs taken by William Henry Jackson and paintings by Thomas Moran. The legislation was promoted by the Northern Pacific Railroad, which saw the potential for tourism in the American West.

Zahniser, Howard (1906–64). A major influence in the writing and passage of the Wilderness Act of 1964, Howard Zahniser served the federal government and the Wilderness Society in numerous capacities between 1930 and his death in 1964. He was a writer and researcher with the Bureau of Biological Survey (later to become the United States Fish and Wildlife Service) from 1931 until 1942, and for the U.S. Department of Agriculture's Division of Information until he joined the staff of the Wilderness Society in 1945. He edited the society's magazine *The Living Wilderness* and wrote "Parks and Wilderness" in *America's Natural Resources* (1957) and "Wilderness Forever," in *Wilderness: America's Living Heritage* (1961).

Part III

Chronology:
An Environmental History Timeline

13,000 B.P.–A.D. 1500: Settlement of North America
The Bering Land Bridge allowed people to cross from Eurasia to North America. Humans arrived in present-day Alaska at least by 13,000 B.P., and in the present-day lower United States by 11,500 B.P.

1492–1580: European exploration of North America
In 1492, following Viking voyages, Columbus began the exploration of the New World. He was followed by John and Sebastian Cabot's voyage from England to the present-day Canadian coast in 1497, the French explorer Jacques Cartier in 1534–45, and by Giovanni da Verrazano's 1524 exploration of the Atlantic coast of the present-day United States. Between 1540 and 1542, Francisco Vasquez de Coronado explored the American Southwest, and from 1539 to 1542 Hernando DeSoto explored the Southeast. Sir Francis Drake sailed up the West Coast of North America between 1577 and 1580.

1580: Spanish settlement of Southwest
Between 1598 and 1605, Don Juan de Oñate explored New Mexico, with the desire to Christianize and "civilize" the Pueblo Indians. The Mission system and its Franciscan priests/friars attempted to establish Christianity as the Native Americans' new religion; they substituted the Virgin Mary for the Pueblos' Corn Mother.

1585–90: John White's depictions of Roanoke/Indians
John White was a member of the 1585–86 failed attempt to settle Roanoke

Island (off the coast of North Carolina). He served as leader of the 1587 "Lost Colony" of Roanoke. He went to England to get more supplies for the colonists, only to find they had deserted the settlement, saying they had gone to "Croatoan" (area on south end of island), but the colonists were never found. White's drawings of Native American life were published in 1590.

1607: English settlement of Chesapeake Bay area

In 1607, Jamestown, Virginia, was settled by John Smith and others. By 1614, tobacco was found to grow well in Maryland and Virginia, giving the settlers an important crop to trade with Europe.

1620–30: English settlement of Massachusetts Bay area

The Pilgrims came to the New World on the *Mayflower*, founding Plymouth Colony for religious and economic reasons. In 1630, the Puritans under John Winthrop settled the Massachusetts Bay Colony.

1691–1729: Broad Arrow Policy

This English policy reserved the white pines of the New England forest that were more than two feet in diameter (at one foot off the ground) for the King (for masts for the Royal Navy).

1705: Robert Beverley's *History and Present State of Virginia*

Robert Beverley cataloged the natural resources and peoples of Virginia, describing the tobacco South as a garden that could make the colonists lazy and cause them to fall into a naïve harmony with nature.

1760–1820: Soil conservation experiments in Virginia

Gentleman farmers and agricultural improvers in Virginia — such as Washington, Jefferson, Madison, John Taylor, and others — experimented with soil improvement. They used contour and deep plowing, soil rotations, animal and vegetable manures, legumes and grasses, and chemicals such as lime, marl, and gypsum to neutralize acid soils.

1773: Phillis Wheatley's poems about nature

Phillis Wheatley was born in 1753 on the west coast of Africa, enslaved, and educated in the Wheatley's Boston home by the mother and daughter. She was certified as an authentic poet, well versed in classical and biblical literature, by members of Boston's upper class. Her book was the first published

book of poetry by an African American, and the first published book of literature by a female African American. Her poems indicated to the culture of the
times that African Americans could be educated and think creatively.

1775–83: American Revolution

The American Revolution began with Paul Revere's ride to Lexington to
warn Samuel Adams and John Hancock of an imminent attack by the
British and ended with the surrender of Lord Cornwallis at Yorktown, Virginia in October 1781. The Treaty of Paris that officially ended the war in
1783 created boundaries for the new republic that extended from the Atlantic coast westward as far as the Mississippi River.

1776: Declaration of Independence

The Declaration of Independence drew on the concepts of nature and reason that prevailed during the eighteenth-century Enlightenment, asserting
that it had become "necessary" in the "Course of human events" for the
American people "to assume among the Powers of the earth, the separate
and equal station to which the Laws of Nature and of Nature's God entitle
them." The Declaration stated: "We hold these truths to be self-evident, that
all men are created equal, that they are endowed by their Creator with certain unalienable Rights, that among these are Life, Liberty, and the pursuit
of Happiness."

1785: General (Land) Ordinance of 1785

The land ordinance that went into effect two years after the American Revolution established the rectangular survey system of 6-mile square townships
divided into 36 sections of 640 acres each. The system became the basis for
settlement and land entry patterns such as the 160-acre, or quarter section
homestead.

1787: Thomas Jefferson's Notes of the State of Virginia; agrarian ideal

Jefferson's Notes on the State of Virginia described nature and farming practices in Virginia, arguing that wheat was better for the soil than tobacco and
that independent farmers were the chosen people of God.

1803: Louisiana Purchase

The Louisiana Territory, which had been transferred from Spain to France
in 1800, was purchased by the United States in 1803, doubling the size of
the nation and extending its boundaries westward of the Mississippi along

the concourses of the Arkansas, Platte, and Missouri Rivers to the Continental Divide.

1804–6: Lewis and Clark Expedition

Under Thomas Jefferson's urging, Congress appropriated funds in 1803 for an expedition to explore the upper reaches of the Missouri River and the Oregon Territory. Merriwether Lewis and William Clark led the expedition to study the flora, fauna, and minerals of the region and to map its latitude and longitude. The expedition left St. Louis in the spring of 1804, spent the winter of 1805–1806 at the mouth of the Columbia River on the Pacific Ocean, and returned to St. Louis in September of 1806.

1808–34: John James Audubon's depictions of birds and life in the United States

Audubon's *Delineations of American Scenery and Character* (published 1926) depicted nature and life in America, the depletion of forests, the transformation of the land through settlement, and America's vanishing wildlife. In an age before binoculars, Audubon shot birds in order to paint them. His paintings often portrayed the bird as the subject, in a stylized background.

1812–15: War of 1812

The War of 1812 with the British over free maritime trade ended in 1815 with the Treaty of Ghent, with U.S. territory intact. Trade resumed between the United States and Europe, opening an extended period of economic expansion called the market revolution. New lands for settlement opened up westward of the Mississippi, and the concurrent transportation revolution facilitated the development of a dynamic internal market economy.

1836: Ralph Waldo Emerson's *Nature*

In 1836, Emerson published *Nature* and helped to found the Transcendental Club, both of which promoted the philosophy of Transcendentalism, which looked to nature for manifestations of ideal, spiritual truths from which to gain insights into the divine One.

1836–49: Hudson River school of painters

The Hudson River painters made nature the subject of their paintings (as opposed to human portraits). The paintings often portrayed the dualism of light versus dark, and wild versus civilized. Humans were small in size and often depicted as intruders in nature, rather than as the main subjects of the paintings.

1844: George Catlin's paintings of American Indians and nature in the Great Plains

George Catlin's work draws a parallel between the disappearing American wilderness and the disappearing Native Americans. He showed European Americans the everyday lives of the "noble savages." Catlin proposed a national park for both Indians and nature on the Great Plains.

1846: Thomas Hart Benton's "Manifest Destiny"

Missouri Senator Benton urged settlement of the newly acquired Oregon, based on what he perceived as the superiority of the "white" race over the "red, yellow, black, and brown" races.

1849: California Gold Rush

The mining that ensued after James Marshall's 1848 discovery was detrimental to California's environment: mining camps polluted air and water, hydraulic mining clogged rivers with debris (affecting farmers downstream and the aquatic life in the rivers); mercury used in the mining process killed wildlife in the areas surrounding the infected streams.

1854: Henry David Thoreau's *Walden*

Thoreau lived in a cabin at Walden Pond, near Concord and Boston, Massachusetts between the years 1845 and 1848. He went to Walden to try to understand the world and to live simply on the land. He believed that nature was alive, and said that "In Wildness is the preservation of the World." He was one of the most famous members of the Transcendentalist school of thought.

1861–65: Civil War

The Civil War over the freedom of African Americans from bodily bondage was fueled by the northern states' abolitionism, political power, population growth, economic expansion, and their failure to enforce the Fugitive Slave Law resulting from the Compromise of 1850, over freedom versus slavery, in the new western territories ceded by Mexico. After the Civil War, the South entered a period of reconstruction, in which plantation owners resorted to sharecropping contracts with free blacks to obtain labor for recovering cotton production.

1862: Homestead Act

This act authorized any head of a household (male or female) who was a citizen (or declared their intention to become one) to gain title to 160 acres of unappropriated land after five years of settlement. It was often abused by speculators illegally in order to gain title to large amounts of land.

1862: Morrill Land Grant Act

This act granted each state a minimum of 30,000 acres of public land to sell, to create a fund for the establishment of a college of agriculture and mechanic arts. Its purpose was to increase technology for developing resources.

1863: Frances Anne Kemble's *Journal of a Residence on a Georgian Plantation in 1838–1839.*

Kemble was an English actress before marrying an American plantation heir. Her journal and letters to friends contrast her disgust with the slave system and her love of the natural beauty of the American South.

1872: Yellowstone Park Act

The Yellowstone Park Act set aside "a tract of land lying near the headwaters of the Yellowstone River as a public park," creating the nation's first national park.

1873: Isabella Bird's *A Lady's Life in the Rockies*

Bird's letters describing her appreciation for female-personified nature contrast with more modern ideas of nature as a commodity for resources and profit. Women, like Bird, were especially important in the preservationist movement.

1873: Timber Culture Act

In a effort to create more wooded vegetation and moisture on the Great Plains, the federal government authorized the donation of 160 acres of land to individuals who would plant 40 acres of it in trees and maintain them over ten years.

1876: Appalachian Mountain Club founded

The club was founded as a hiking and conservation club in order to preserve the balance of nature while maintaining ways for people to enjoy the wilderness.

1878: Free Timber Act

This act promoted settlement by giving residents of nine western states the privilege of cutting timber on public mineral lands.

1878: Timber and Stone Act

In an effort to encourage settlement of the western United States, the government allowed the sale of 160 acres of land in Washington, Oregon, California, and Nevada unfit for agriculture, but useful for tree harvesting and stone extraction.

1878: John Wesley Powell's Report on the Arid Lands of the United States

Powell served as head of the Bureau of Ethnology, and also as the director of the U.S. Geological Survey. He declared that the land west of the 100th meridian — the 20-inch rainfall line, where rainfall was consistently not more than 20 inches per year — was arid. He believed that 2,560 acres were necessary for western farms, if the land was not to be irrigated, whereas 80 acres was sufficient for irrigated farms.

1885: Adirondack Forest Preservation Act

The Adirondack Forest Preserve, created by the New York State legislature, was one of the nation's earliest efforts to preserve forests and watersheds and create recreational areas for the public.

1886: Creation of the Division of Forestry

The Division of Forestry was created in 1886 to administer the country's forest reserves. Under pressure from forester Gifford Pinchot, the reserves were transferred into the Department of Agriculture in 1905 on the grounds that trees were crops that could be planted and harvested.

1887: Dawes Act

Passed as the General Allotment Act, the Dawes Act was an effort to induce Indians to become farmers by allotting communally held tribal lands in 160-acre parcels to tribal members.

1887: Desert Lands Act

To encourage the settlement of arid lands in the western United States, the federal government authorized the sale of 640 acres (one section) of land to anyone who irrigated it within three years.

1887: Hatch Act

The Hatch Act established agricultural experiment stations to do research on the production of food, fiber, and nutrition and disseminate the results to the public.

1892: Founding of the Sierra Club

Under the leadership of John Muir, the Sierra Club promoted protection of the environment and facilitated outdoor enjoyment of the mountains of California and the West.

1892: Ellen Swallow introduces the term *oekology* to America

Swallow's term *oekology* soon became *ecology*. She intended it to be the study of the "earth's household," the environment in its role as the home of all people.

1893: Frederick Jackson Turner's "Significance of the Frontier in American History"

The 1890 census declared the American frontier non-existent. Turner argued that the frontier produced rugged individualism and democracy, and thus this environment determined the character of American people and institutions.

1897: Forest Management (Organic) Act

To promote development of the West, the act authorized the Secretary of the Interior (the department that administered the forests at the time) to permit timber harvesting, mining, and water use on the forest reservations by settlers, miners, residents, and prospectors.

1897: Louis Hughes's *Thirty Years a Slave*

In his 1897 autobiography, ex-slave Louis Hughes described life on a southern cotton plantation, including the techniques of cotton planting, pest control, harvesting, and ginning.

1900: Lacey Act

In response to the decimation of birds used for their feathers in the millinery trade, Congress passed the Lacey Act, authored by Iowa Congressman John Fletcher Lacey, prohibiting the interstate and international trade in illegally taken wildlife.

1900–13: Progressive Conservation Movement

"Conservation" was defined as the wise use of natural resources to benefit the greatest number of people for the longest time. The term conservation was introduced in 1907 by Gifford Pinchot and WJ McGee to embrace the collective use and preservation of forests, waters, soils, and minerals.

1901: Right of Way Act

In 1901, Congress allowed the Secretary of the Interior to permit rights of way through public lands, national forests, and national parks.

1901: Booker T. Washington's *Up from Slavery*

Washington's autobiography describes his life as a slave in Virginia and his rise to the head of Tuskegee Institute in Alabama, where he promoted the use of agricultural improvement through courses, conferences, and county fairs.

1902: Reclamation Act

The Reclamation Act financed federal irrigation projects in the arid West by selling public lands. It also established the Bureau of Reclamation to oversee the administration of western water and irrigation works.

1902: Bernhard Fernow's *Economics of Forestry*

Fernow served as first chief of the Division of Forestry. He viewed the individualism established on the frontier as detrimental to the conservation of natural resources, because it assaulted the rights of the many in favor of the rights of the few. He believed that forests should be managed to save them for future generations.

1906: American Antiquities Act

In 1906, Congress passed an act that authorized the President to create national monuments by preserving federal lands that contained historic landmarks and prehistoric structures.

1908: Winters Doctrine

In 1908, in response to complaints by western Indian tribes, the United States Supreme Court affirmed the water rights of Indian tribes in the western states.

1908: Theodore Roosevelt's Conference of Governors
Roosevelt's speech at the Conference of Governors launched the conservationist movement as a national cause.

1909–13: National Conservation Congresses
Five National Conservation Congresses were held between the years of 1909 and 1913, in order to bring public attention to the need to conserve natural resources. Many groups participated, including women's organizations. The resulting contributions to the conservationist cause were immensely significant.

1910: Jane Addams's *Twenty Years at Hull-House*
As founder of the Hull House refuge for the poor, Addams encouraged women in their efforts to control garbage and other pollution in the tenement-house districts of Chicago.

1913: Hetch Hetchy Dam (Raker Act)
Passage of the Raker Act — authorizing construction of a dam in Hetch Hetchy Valley in Yosemite National Park, to supply water and power to the city of San Francisco — was the culmination of several years of national controversy between preservationist John Muir and forester Gifford Pinchot over the damming of a river in a national park.

1914: Extinction of the passenger pigeon
On September 1, 1914, after decades of shooting for sport and market profits, Martha, the last passenger pigeon, died in the Cincinnati Zoo.

1914–18: World War I
During World War I, the United States eventually joined the Western Allies in Europe in the fight against the Central Powers of Germany, Austria-Hungary, Turkey, and Bulgaria. During the war and its aftermath, natural resource conservation was superceded by resource and agricultural development, which led ultimately to environmental decline, economic depression, and the disastrous Dust Bowl of the 1930s.

1916: National Park Service Act
Congress passed the National Parks Act, creating the National Park Service to administer the thirteen existing national parks and future parks and monuments set aside for their scenery and aesthetic beauty.

1916: Frederic Clements's *Plant Succession*

Clements believed that plant communities' development is similar to that of a complex organism (a human, for example) and that the plant community grows in such as a way as to reach a climax state.

1918: Migratory Bird Treaty Act

The act protected migratory birds from excessive hunting, selling, and shipping between the United States, Canada, and, later, Mexico.

1929–38: The Great Depression and the New Deal

Franklin Delano Roosevelt's New Deal to overcome economic depression is closely linked with conservation and preservation. The New Deal established and promoted numerous agencies involved with the environment.

1933: Tennessee Valley Authority (TVA) created

The Tennessee Valley Authority (TVA) was created to develop the resources of the Tennessee River Valley to achieve flood control, navigation, electrical power generation, and reforestation; and to improve the economic well-being of the people of the valley bringing electricity to rural households and providing employment during the Depression.

1933: Robert Marshall's *The People's Forests*

Under the conservationism of the New Deal, Marshall's ideas for a program to turn over private forests to government were well received. Although he came from a wealthy family, he promoted access to the outdoors for people of all classes.

1933–45: New Deal Conservation

New Deal conservationist programs included TVA, or the Tennessee Valley Authority, which was chartered to "develop" resources in the valley; the construction of many dam projects in the western U.S.; the CCC, or Civilian Conservation Corps, which enlisted young men to develop parks and wilderness areas; and the SCS, or Soil Conservation Service, which advocated the creation of soil banks and the prevention of soil erosion.

1933: Founding of the Civilian Conservation Corps (CCC)

The Civilian Conservation Corps (CCC) was a New Deal program to relieve unemployment by allowing young, unmarried men to receive food, housing, and a small salary in exchange for working on federal projects to conserve soils and other natural resources. It was abolished in 1942.

1934: Taylor Grazing Act

The Taylor Grazing Act was passed to establish grazing districts, manage livestock grazing, and to retain unreserved public domain lands.

1935: Soil Conservation Act

In response to the devastations of the Dust Bowl during the 1930s, Congress passed the Soil Conservation Act, establishing a permanent Soil Conservation Service in the United States Department of Agriculture.

1935: Founding of the Wilderness Society

Founded by Bob Marshall and others in 1935. The Wilderness Society proposed the Bob Marshall Wilderness (one million acres in northwest Montana) in 1940 to commemorate Marshall's 1939 death.

1936: National Wildlife Federation founded

The National Wildlife Federation was founded to conserve fish, wildlife, and other natural resources and to lobby for legislation to conserve wildlife.

1937: Pittman-Robertson Wildlife Restoration Act

In 1937, Congress passed the Pittman-Robertson Act in order to provide federal aid to the states to restore and manage wildlife and acquire wildlife habitat.

1941–45: World War II

Environmentally, World War II served as the dividing line between New Deal responses to the environment and the modern environmental movement. Technological advances invented or brought into wide use during the war (atomic bomb/energy, DDT) posed threats to the safety of the environment and humans. Before the war, efficient management of resources dominated environmental thought; after the war, people began to emphasis environmental quality and human and ecological health.

1946: Creation of Bureau of Land Management (BLM)

As part of the Department of the Interior, the BLM, created by merging the General Land Office and Grazing Service, is responsible for the administration and management of federal lands, deserts, and minerals.

1948: Water Pollution Control Act

As the first federal law to deal officially with water pollution, the Water Pollution Control Act authorized funding for state and local governments to identify and improve polluted waters.

1949: Aldo Leopold's *A Sand County Almanac*
In his book *A Sand County Almanac*, Leopold put forward the land ethic: "A thing is right when it tends to preserve the integrity, stability, and beauty of the biotic community. It is wrong when it tends otherwise." He also proposed the "A-B Cleavage": land as a resource/commodity to be used versus land as biota, a place where organisms should be left to flourish. The book provided the basis for the modern environmental revolution.

1951: Nature Conservancy founded
The Nature Conservancy was founded as a citizen's environmental organization dedicated to purchasing and protecting the habitats of plants, animals, and natural communities that represent the diversity of life on earth.

1962: Rachel Carson's *Silent Spring*
Carson was a well-respected author and scientist, and her book was both an environmental and a popular success. The book identified and discussed the environmental impact of pesticides (especially DDT), including their concentration as they move up the food chain and insects' development of genetic immunity (requiring still stronger pesticides).

1963: Clean Air Act
The Clean Air Act of 1963 modified the 1955 Air Pollution Control Act by allocating permanent funding for the work of state pollution control agencies.

1964: Wilderness Act
Congress designated certain federal lands as wilderness areas, "where the earth and its community of life are untrammeled by man, and where man himself is a visitor who does not remain."

1965: Land and Water Conservation Fund Act
The Land and Water Conservation Fund Act was passed in order to preserve, develop, and assure accessibility to outdoor recreation resources.

1965 Water Quality Control Act
The Water Quality Control Act gave the Federal Water Pollution Control Administration the power to establish water quality standards.

1968: Wild and Scenic Rivers Act

The Wild and Scenic Rivers Act protected free flowing, undammed rivers that had outstanding scenic, geologic, historic, recreational, or wildlife features.

1968: Indian Civil Rights Act

The states' increasing control over American Indian lands was reversed by the Indian Civil Rights act, which mandated tribal consent in civil and criminal juridical matters concerning Indian lands.

1968: Paul Ehrlich's *The Population Bomb*

The *Population Bomb* discussed the negative impacts that the population explosion of the late twentieth century would have on food and other resources. It drew on Malthus's idea that the rich control their population, but the poor multiply.

1969: Santa Barbara oil spill

On January 28, 1969, an oil well off the coast of Santa Barbara California erupted, spewing oil for ten days into the ocean, polluting the coastline of California, and triggering safeguards and reforms in the energy industry.

1969: Friends of the Earth founded

The international organization Friends of the Earth (FOE) was founded in 1969 to protect the planet from environmental disaster and to preserve biological, cultural, and ethnic diversity.

1969: The National Environmental Policy Act (NEPA)

NEPA required every federal agency to prepare an Environmental Impact Statement (EIS) for any legislation or project that would affect the quality of the human environment. It also created the Council on Environmental Quality (CEQ), which advises the president on environmental quality and compliance with NEPA.

1970: Founding of the Environmental Protection Agency (EPA)

The Environmental Protection Agency was created in the executive branch of the federal government for the purpose of regulating air and water quality, radiation and pesticide hazards, and solid-waste disposal, amalgamating earlier separate federal programs.

1970: Earth Day / Environmental Movement

Earth Day was promoted by Wisconsin Senator Gaylord Nelson, who called on students to fight for environmental causes and to oppose environmental degradation (such as the 1969 Santa Barbara oil spill) with the same vigor they used for fighting the Vietnam War.

1970: Clean Air Act

The Clean Air Act regulated air emissions and authorized the U.S. Environmental Protection Agency to establish National Ambient Air Quality Standards (NAAQS) to protect public health and the environment.

1970: Natural Resources Defense Council founded

The Natural Resources Defense Council (NRDC) was founded as a citizen's organization to draft environmental laws, lobby for their passage, and litigate for their implementation.

1971: Alaskan Native Claims Settlement Act

Alaskan Eskimos, Aleuts, and Indians received federal grants, federal and state mineral revenues, and land in exchange for their agreement to settle long-standing land claims.

1972: Federal Water Pollution Control Act

The Federal Water Pollution Control Act set the basic structure for regulating discharges of pollutants to waters of the United States and shifted responsibility for control programs from the states to the federal government.

1972: Coastal Zone Management Act

The law created a federal program to develop plans for the protection and development of coastal areas for beneficial use of the coastal zone.

1973: Endangered Species Act (ESA)

The law gave authority to list threatened and endangered species and to protect their vital habitat. The subject of intense controversy and conflict over resource development, the law has nevertheless been repeatedly extended.

1974: Institute for Social Ecology founded

The Institute for Social Ecology was founded in 1974 in Plainfield, Vermont, to study social ecology, an interdisciplinary field drawing on philosophy, political and social theory, anthropology, history, economics, the natural sciences, and feminism.

1976: Toxic Substances Control Act (TOSCA)
In response to public pressure over pesticides and environmental toxins, Congress passed a law regulating public exposure to toxic materials.

1976: Federal Land Policy and Management Act
The act retained public lands in public ownership, established guidelines for the sale of public lands, and gave the Bureau of Land Management (BLM) the statutory authority to set grazing, preservation, and mining policy, and to undertake long term planning based on multiple use.

1976: Resource Conservation and Recovery Act (RCRA)
Congress passed the Resource Conservation and Recovery Act in 1976 to protect human health and the environment, to reduce waste, and to conserve energy and natural resources.

1978: Endangered American Wilderness Act
Designated lands in the Western states were added to the National Wilderness Preservation System to increase watershed, wildlife-habitat, and scenic and historic preservation.

1980: Comprehensive Environmental Response, Compensation, and Liability Act (CERCLA; Superfund)
Allowed the Environmental Protection Agency to designate and respond to abandoned hazardous waste dumps and to sue to recover cleanup costs. Set up a five-year, $1.6-billion trust fund financed by a tax on industrial feedstock chemicals.

1980: Alaska National Interest Lands Conservation Act (ANILCA)
This act was designed to preserve Alaskan lands and waters that have "scenic, historic, or wilderness values." It provided rural Alaskan residents with the right to continue a subsistence way of life.

1980: "Women and Life on Earth" conference
In 1980 the conference on "Women and Life on Earth: Ecofeminism in the '80s" in Amherst, Massachusetts, marked the beginning of ecofeminism as a movement in the United States and undertook to explore and act on the cultural connections between women and nature.

1982: Environmental Justice Movement

In 1982 a group of African Americans protested the designation of a landfill site in Warren County, North Carolina, for disposal of toxic PCBs (polychlorinated biphenyls), initiating the environmental justice movement.

1985: Publication of Deep Ecology: Living as if Nature Mattered

In 1985, philosopher George Sessions and sociologist Bill Devall published *Deep Ecology: Living as if Nature Mattered*, drawing on philosopher Arne Naess's 1973 term, "deep ecology" and promoting the concept as a new consciousness about the connection between humanity and nature.

1986: Emergency Planning and Community Right-to-Know Act

Industries were required to report toxic releases and communities were asked to plan emergency responses.

1987: Water Quality Act

In 1987, Congress ended federal funding for treatment of wastewater and directed the states to list toxic hot spots.

1987: Montreal Protocol

An agreement was signed in 1987 by twenty-four countries, including the United States, to phase out, by 1999, production of chlorofluorocarbons (CFCs) that break down the earth's protective ozone layer.

1989: North American Wetlands Conservation Act

The act authorized funding for the conservation of wetland habitats in Canada, the United States, and Mexico to implement the American Waterfowl Management Plan developed in 1986.

1989: *Exxon Valdez* Oil Spill

In March 1989, the *Exxon Valdez* oil tanker ran aground in Alaska's Prince William Sound, spilling 11 million gallons of crude oil, killing thousands of sea otters, shorebirds, salmon, and herring, and jeopardizing coastal habitats and the state's economy.

1990: Clean Air Act amended

Amendments to the Clean Air Act tightened restrictions on air pollution emissions that cause smog, acid rain, airborne toxins, and chlorofluorocarbons.

1990: Pollution Prevention Act
A law was passed in 1990 that gave corporations incentives to reduce pollutants by implementing changes in production, operation, and use of raw materials.

1990: Spotted Owl listed as a threatened species
In June 1990, the U.S. Fish and Wildlife Service listed the spotted owl as a threatened species under the Endangered Species Act, initiating steps to protect its habitat in the forests of the Pacific Northwest and setting off numerous protests by timber industry advocates.

1991: First National People of Color Environmental Leadership Summit
Held in Washington, D.C., in October of 1991, the Leadership Summit gathered together peoples of color to form a movement to promote environmental justice as a basis for public policy, calling for the cessation of production of and protection from hazardous wastes and toxic substances.

1992: Earth Summit
In 1992, 172 countries sent representatives to the Earth Summit (in Rio de Janeiro, Brazil) to discuss and create a plan to combat environmental, economic, and social problems facing the international community.

1994: California Desert Protection Act
A law was passed to set aside and protect as wilderness millions of acres of land in the California deserts.

1996: Grand Staircase-Escalante National Monument
The Grand Staircase-Escalante National Monument in southern Utah was created by President Clinton in an executive action that set aside 1.7 acres of Utah as wilderness.

1996: Safe Drinking Water Act Renewed
Congress reauthorized and amended the Safe Drinking Water Act of 1974 to emphasize sound science, community-empowered source water assessment and protection, public right-to-know, and water system infrastructure assistance.

1997: Kyoto Protocol
An agreement was reached in December 1997 by the Climate-Change

Conference, held in Kyoto, Japan, that would set legally binding limits on greenhouse gas emissions and require industrialized nations to reduce their emissions below 1990 levels by 2012. It required ratification by 55 nations responsible for 55 percent of greenhouse gas emissions.

2000: World Population surpasses six billion people

In October 1999, the world population reached six billion people and population continued to swell with the possibility of another doubling by 2040.

Part IV

Resource Guide

Visual Resources: Films and Videos

Video purchase information accompanies each entry; prices may vary.

1. The American Environment and Native-European Encounters

The Columbian Exchange, video, 60 min. $40.00 purchase. (From *Columbus and the Age of Discovery* series. Available on interlibrary loan from The Valley Library, Oregon State University, Corvallis, Ore.)

http://osulibrary.orst.edu/video/hist69.htm (E-mail: valley.circ@orst.edu)

Interchange of horses, cattle, corn, potatoes, and sugar cane between the Old and New Worlds and its impact on people of both worlds. Illustrates Alfred Crosby's concepts in his 1972 book *The Columbian Exchange*. (1991)

"Confronting the Wilderness," video, 58 min. (From *The Land of the Eagle* series Time-Life, No. 2 in series of 8 videos. Available on interlibrary loan from The Valley Library, Oregon State University, Corvallis, Ore.)

http://osulibrary.orst.edu/video/bio62.htm (E-mail: valley.circ@orst.edu)

The Huron, Ottawa, and Cree Indians and the French in the Northeast. (1991)

"The Environment: A Historical Perspective," video, 52 min. $89.95 purchase. (Films for the Humanities and Sciences, Princeton, N.J. 1-800-257-5126; http://www.films.com) http://www.films.com/item.cfm?bin=10462

A history of environmental transformation from the first agricultural settlements to the Industrial Revolution to agribusiness and widespread urbanization. The environmental impact of humans on the planet and perspectives on ecological stewardship for the twenty-first century. (1999)

The Environmental Revolution, video, 60 min. $39.95 purchase; $169 series. (From *Race to Save the Planet* series, No. 1, 1-800-LEARNER, The Annenberg/CPB Project, c/o Intellimation, P.O. Box 1922, Santa Barbara, Calif. 93116-1922)

http://www.learner.org/catalog/ordering/series66.html

http://www.pbs.org/wgbh/shop/ipsaveplanet.html

Hosted by Meryl Streep and narrated by Roy Scheider. Broad history of environmental transformation from agriculture in the Middle East to the Industrial Revolution and Rachel Carson. (1990)

"Hopi: Songs of the Fourth World," video, 58 min. $199 purchase, $75 rental. (New Day Films, 1-888-367-9154)

A compelling study of the Hopi that captures their deep spirituality and reveals their integration of art and daily life. (n.d.)

"Living on the Edge," video, 58 min. (From *The Land of the Eagle* series, Time-Life, No. 6 in series of 8 videos. Available on interlibrary loan from The Valley Library, Oregon State University, Corvallis, Ore.)

http://osulibrary.orst.edu/video/bio66.htm (E-mail: valley.circ@orst.edu)

Missionaries in the Southwest and their impact on Papago, Pima, and Apache peoples. (1991)

"The Story of Hawaii: Children of the Long Canoes," video, 56 min. $19.95. (Island Heritage, 1-800-468-2800)

The history of the Hawaiian Islands from their discovery by Polynesians who sailed across uncharted waters in their "long canoes" to the Hawaii of the 1930s. (n.d.)

"The Sun Dagger," video, 58 min. $30 purchase (Bullfrog Films, P.O. Box 149, Oley, Pa. 19547; 1-610-779-8226)

http://www.rock-art.com/books/sofaer1.htm

Explores the complex culture of the Anasazi Indians, who constructed the calendar and thrived both spiritually and materially in the harsh environment of Chaco Canyon a thousand years ago. Narrated by Robert Redford. (1983)

The Sword and the Cross, video, 60 min. (From *Columbus and the Age of Discovery* series. Available on interlibrary loan from The Valley Library, Oregon State University, Corvallis, Ore.)

http://osulibrary.orst.edu/video/hist74.htm (E-mail: valley.circ@orst.edu)

The Conquistadors and the Church and their effect on indigenous peoples. (1991)

2. The New England Wilderness Transformed

"New Found Land," film, 52 min. (*America* series, narrated by Alistair Cooke) (Available from Documentary-Video, 28 west 44th St., Suite 2100, New York, N.Y. 10036; 1-800-526-4663)

http://www.documentary-video.com/displayitem.cfm?vid=626

Columbus's 1492 voyage, routes of Spanish Conquistadors such as Coronado, French-Canadian hunters and trappers, and La Salle's journey to the mouth of the Mississippi. (1973)

Plymouth Plantation, video, $29.95 purchase. (Teacher's Video Co., 1-800-262-8837)

Life in seventeenth-century Plymouth Plantation as reconstructed and reenacted. (1989)

The Shape of the Land, film, 60 min. $37 rental (*America by Design* series) (#37644 Center for Media and Independent Learning, University of California, Berkeley, 1-510-642-0460)

History and consequences of American efforts to control, shape, and alter the land and its rivers, marshes, and forests. Engineering feats of mining, dam, and waterway construction and efforts at preservation of wilderness and national parks. (1987)

3. The Tobacco and Cotton South

"Booker T. Washington: Life and Legacy," video, 32 min. $19.95. (Harper's Ferry Historical Association, P.O. Box 197, Harper's Ferry, W.Va. 25425, 1-800-821-5206)

Details the life of Booker T. Washington and the founding of Tuskegee Institute. (1982)

"Colonial Naturalist: The Life and Work of Mark Catesby," video, 55 min. $24.95 purchase. (Colonial Williamsburg, 1-804-220-7645)

http://www.buyindies.com/listings/6/8/FCTS-6813.html#

In the 1700s, Mark Catsby led the way to documenting plant and animal life in America by observing the natural environment, making drawing, and sending plant specimens back to London. (1964)

"Conquering the Swamps," video, 58 min. (From *The Land of the Eagle* series, No. 3 in series of 8 videos. Available on interlibrary loan from The Valley Library, Oregon State University, Corvallis, Ore.)

http://osulibrary.orst.edu/video/bio63.htm (E-mail: valley.circ@orst.edu)

The Timuca, Seminole, and Calusa peoples meet Spanish in the Caribbean and Florida. (1991)

"Cotton Production," video, 35 min. $79 purchase. (Visual Education Productions, 1-800-235-4146; fax: 1-800-243-6398)

http://www.cev-inc.com

Field trip to view West Texas cotton production, cotton gin, burr and seed removal, cleaning of cotton lint, baling of cotton, weaving and dying of cotton denim. (n.d.)

Dark Passages, video, 60 min. $69.95 purchase. (PBS videos)

The slave trade conducted in West Africa, narrated through interviews and dramatization. (1990)

"Family Across the Sea," video, 56 min. $75 rental, $195 purchase (5 or more videos from California Newsreel are $99 each) (California Newsreel, 1-415-621-6196)

Uses the research of historians to demonstrate the efforts of African-Americans to maintain their ties with their African heritage. (1991)

"The Great Encounter," video, 58 min. (From *The Land of the Eagle* series, Time-Life, No. 1 in series of 8 videos. Available on interlibrary loan from The Valley Library, Oregon State University, Corvallis, Ore.)

http://osulibrary.orst.edu/video/bio61.htm (E-mail: valley.circ@orst.edu)

English in the Chesapeake Bay area and the Powhatan and Cherokee tribes. (1991)

"Homecoming," video, 56 min., $75 rental, $195 purchase, 5 or more videos from California Newsreel are $99 each. (California Newsreel, 1-415-621-6196)

Explores the relationship between African-Americans and the land they have worked for centuries. Brings the story to the present. (1998)

"Longing to Learn," video, 14 min. $13.95 purchase. (America's National Parks, 470 Maryland Dr. Suite 2, Ft. Washington, Pa. 19034, 1-877-Nat-Park).

http://www.eparks.com/eparks/find.asp

Based on Booker T. Washington's autobiography, *Up from Slavery* (1901). Washington's own words are used to describe his boyhood years, spent as a slave on the Burroughs Plantation in Franklin County, Virginia. (1990)

"Lumber Production," video, 35 min. $85 purchase. (Visual Education Productions, 1-800-235-4146; Fax: 1-800-243-6398)

http://www.cev-inc.com

Field trip to logging production in West Texas. Lumberjacks, shearers, sawhands, skidders, and loaders at Spurlock Logging. Debarking, soaking, peeling, and cutting logs and plywood production at Champion International Corporation in Camden Texas. (n.d.)

Roots of Resistance — A Story of the Underground Railroad, video, 60 min., $59.95 purchase. (From *The American Experience* series. PBS videos, Alexandria, Va., 1-800-344-3337)

The secret passage traveled by black men and women out of American slavery. (1989)

"A Son of Africa: The Slave Narrative of Olaudah Equiano," video, 28 min. $75 rental, $195 purchase (5 or more videos from California Newsreel are $99 each) (California Newsreel, 1-415-621-6196)

The powerful story of a black man's capture into slavery and his struggle to free himself and fight the institution of slavery. (1996)

Where America Began, video, 60 min. $29.95 purchase. (Teacher's Video Co., 1-800-262-8837)

http://www.teachersvideo.com

Colonial Williamsburg, Jamestown, and Yorktown. Historic sites and events that shaped life and times in the colonial Chesapeake. (1988)

4. Nature and the Market Economy

Hands to Work, Hearts to God, video, 60 min. $34.95 purchase. (Hancock Shaker Village, Hancock, Mass. 1-413-443-0188; 1-800-817-1137, x246)

Ken Burns portrays the life, work, and worship of the Shakers at the restored nineteenth-century Hancock Shaker Village in Massachusetts. (1985)

"Land," video. 25 min. $90 purchase (Visual Education Productions, 1-800-235-4146) (Part 2 of *America the Bountiful* Collection — set of 6, $395.)

Land-development concerns of Thomas Jefferson and William Penn, religious attitudes toward agricultural experiments, rise of cotton and tobacco and the institutionalization of slavery. (n.d.)

Old Sturbridge Village, video, $19.95 purchase. (Old Sturbridge Village, Mass., 1-508-347-3362

http://www.osv.org/Gifts/Index.html

Life and work in nineteenth-century Sturbridge Village as reconstructed and reenacted. (1989)

"Swords and Plowshares," video, 26 min. $90 purchase (Part 3 of *America the Bountiful* Collection — set of 6, $395) (Visual Education Productions, 1-800-235-4146)

How strains of Indian corn made possible the formation of the midwestern corn belt. Work of agricultural experimenters Seaman A. Knapp, Perry G. Holden, and Henry Wallace. (n.d.)

"Talking with Thoreau," video, 29 min. $79 purchase. (*Humanities* series, Encyclopedia Britannica Educational Corporation, 1-800-554-9862)

Imagined conversations with Thoreau at Walden Pond as visited by David Brower, B. F. Skinner, Rosa Parks, and Elliott Richardson. (1975)

Thomas Jefferson, a biography by Ken Burns. 180 min. $29.98 purchase. (PBS videos, 1-800-344-3332)

http://www.shop.pbs.org

The life and accomplishments of one of the country's founding fathers. (1997)

"Torchlight," video. 24 min. $90 purchase. (Part 4 of *America the Bountiful* Collection — set of 6, $395) (Visual Education Productions, 1-800-235-4146)

Growth of the transportation infrastructure, rise of commerce, and the agrarian revolt. Agricultural chaos following the Civil War and the work of George Washington Carver. (n.d.)

5. Western Frontiers: The Settlement of California and the Great Plains

"Across the Sea of Grass," video, 58 min. (From: *The Land of the Eagle* series, Time-Life, No. 4 in series of 8 videos.) Available on interlibrary loan from The Valley Library, Oregon State University, Corvallis, Ore.

http://osulibrary.orst.edu/video/bio64.htm (E-mail: valley.circ@orst.edu)

The Mandan, Sioux, Pawnee, and Kiowa of the Great Plains. (1991)

Alaska: A History in Five Parts, video, 90 min. $29.95 purchase. (Alaska Natural History Association, 750 West Second Ave, Suite 100, Anchorage, Alaska 99501-2167. 1-907-274-8440)

http://www.alaskanha.org

(Or, Portage Glacier Recreation Area, Boggs Visitor Center Bookstore: 1-907-783-2326, x105)

Part 1: "The Mists of Time" on the legends, lives, and cultures of Alaska's earliest native peoples. Part 2: "Age of Discovery" on Russian America and the sea otter, known as "soft gold." Part 3: "Folly or Fortune" on "Seward's Folly." Part 4: "Adventures of a Pioneer" on a sourdough's recollections of the gold rush. Part 5: "The Silver Years" on the discovery of "black gold." (1985)

Battle for the Great Plains, video, 60 min. $59.95 purchase.

Narrated by Jane Fonda. A penetrating look at the many sides of the fight over the future of the Great Plains. A National Audubon Society Special.

The Donner Party, video, 90 min. $19.98 purchase. (From *The American Experience* series. PBS videos, 1-800-344-3332)

http://www.shop.pbs.org

The story of the men and women trapped for the winter of 1846 in the Sierra Nevadas of California after trying to take a shortcut to the valley below.

"Dakota Encounters," video, 13 min. $99 purchase, $45 rental. (New Day Films, 1-888-367-9154)

http://www.newday.com

Portrays the Dakota culture in the last days before European settlement. (n.d.)

"Ghost Dance," video, 9 min. $99 purchase, $45 rental. (New Day Films, 1-888-367-9154)

http://www.newday.com

The 1890 massacre at Wounded Knee is remembered through poetry, art, and the haunting beauty of the Dakota landscape. (n.d.)

Gold Fever, video, 60 min. $59.95 purchase. (From *The American Experience.* PBS series, 1-800-344-3332)

http://www.shop.pbs.org

What drove men and women from around the world to California? What did they encounter along the way? When they arrived? (1997)

Gold Rush, video, 60 min. $19.95 purchase. (Boettcher/Trinklein Television, Inc., 635 S. 8th St. Pocatello, Idaho 83201)

http://www.isu.edu/~trinmich/grvideo.html

http://www.shop.pbs.org

Story of the nineteenth century quest for gold in frontier California. Historical documentary of the discovery of gold, westward journey, fortunes made and lost, frontier women, conflict of cultures, environmental destruction, birth of San Francisco, and the long-term impact. Narrated by John Lithgow. (1998)

Last Stand at Little Big Horn, video, 60 min. $24.98 purchase. Total Marketing Services

http://www.totalmarketing.com/an/dbpage.pl/p/7653/jfv6214.91783

Looks at the conflict between the Sioux and the American settlers over western lands.

"Letters From America: The Life and Times of O.E. Rolvaag," video, 30 min., $99 purchase, $55 rental. (New Day Films, 1-888-367-9154)

http://www.newday.com/films/Letters_from_America.html

Immigrant author Ole Rolvaag captures the triumph and tragedy of American pioneering. (n.d.)

Lewis and Clark: The Journey of the Corps of Discovery, video, 240 min., $29.95 purchase. (PBS videos, Alexandria, Va., 1-800-344-3337)

http://www.shop.pbs.org

Drawing from the magnificent journals kept by the party and stories from Indian oral tradition, the film recreates the epic exploration of the Northwest by Meriweather Lewis and William Clark. (1997)

Lost in the Grand Canyon, video, 60 min. $59.95 purchase. (*The American Experience* series, 1-800-344-3332)

The story of John Wesley Powell's epic exploration of the Colorado River and the Grand Canyon. (1999)

"Plow That Broke the Plains," 16mm film, 25 min. $24 rental. (Center for Media and Independent Learning, University of California, Berkeley, #9204, 1-510-642-0460)

Documentary of the Great Plains from the defeat of the Indians to the Dust Bowl and the Depression of the 1930s. (1936)

"Paniolo O Hawai'i: Cowboys of the Far West," video, $21 purchase. (Film-Works Ltd., 1-808-585-9005)

http://filmworkspacific.com/order.html

Shows the history of Hawaii's *paniolos* (cowboys) and their culture in the cattle country of Hawaii. Narrated by Willie Nelson and others. (1997)

Romance and Reality: Yesterday's World of Work. Set of four videos. $177 purchase for set of videos and guides. (Hawkhill Associates, 1-800-422-4295)

http://www.hawkhill.com

Includes "Romance of the Lumberjack" (17 min.), "Voices of the Great Lakes" (18 min.),"Iron Mines and Men" (18 min.), "Thar She Blows" (22 min.) (1995-96)

"Searching for Paradise," video, 58 min. (From *The Land of the Eagle* series, Time-Life, No. 8 in series of 8 videos. Available on interlibrary loan from The Valley Library, Oregon State University, Corvallis, Ore.)

http://osulibrary.orst.edu/video/bio68.htm (E-mail: valley.circ@orst.edu)

The Chumash, Miwok, and Yokut tribes of California and the Spanish impact. (1991)

Surviving the Dust Bowl, video, 60 min. $19.98. (*The American Experience* series, PBS videos, Alexandria, Va., 1-800-344-3337)

http://www.shop.pbs.org

A new look at the ecological disaster which coincided with and contributed to the Great Depression of the 1930s. (1998)

"T.R. Country," video or film, 15 min. (National Park Service, Harper's Ferry Historical Association, 304-535-6881)

Chronicles the development of Teddy Roosevelt's environmental ethic as it developed during his visits to the Badlands of North Dakota. (n.d)

The West (directed by Ken Burns), set of 9 videos, 84 min. each, $149.98 purchase series. (PBS videos, Alexandria, Va., 1-800-344-3337)

http://www.shop.pbs.org

Each episode explores a different aspect of the West's history: Native American heritage, the Gold Rush, slavery, and immigration. Also looks at current resource use issues. (1996)

Westward the Course of Empire Takes Its Way, video, 90 min., $49.95 purchase. (*The Way West* series, PBS videos, Alexandria, Va., 1-800-344-3337)

http://www.shop.pbs.org

The westward expansion of American culture. (1994)

6. Urban Environments

"A Big Stink: City Sewer Systems," video, 29 min. $10.25 rental, $99 purchase. (Films for the Humanities and Sciences, 1-800-257-5126)

http://www.films.com

Focusing on London, this film traces the history of urban efforts to manage sewage.

"The City and the Environment," video, 23 min. $89.95 purchase. (Films for the Humanities and Sciences, 1-800-257-5126)

http://www.films.com

Focuses on three facets of the urban ecosystem: the underground infrastructure that enables a city to function; traffic and its management, and efforts to protect city trees from urban pollution. (n.d.)

"Frederick Law Olmsted and the Public Park in America," video, 58 min. $39.95 purchase. (BuyIndies.com)

http://www.buyindies.com/listings/1/3/FCTS-13248.html

Explores the creation of New York City's Central Park designed by Frederick Law Olmsted and the emergence of the profession of landscape architecture through reenactment of Olmsted's life. (1990)

Goin' to Chicago, video, 71 min. $195 purchase, 5 or more videos from California Newsreel are $99 each. (California Newsreel, 1-415-621-6196)

The great migration of African Americans from the rural south to the cities of the North and West, from the poverty of sharecropping to better paying jobs in Chicago. (1994)

"The Growth of Towns and Cities," video, 19 min. $75 rental, $129 purchase. (Films for the Humanities and Sciences, 1-800-257-5126)

http://www.films.com

This film looks at the problems caused by urbanization and industrialization. It also uses architectural evidence to trace local histories.

"Laid to Waste," video, 52 min., $75 rental, $225 purchase. (Center for Media and Independent Learning, University of California, Berkeley, 1-510-642-0460)

The West End of Chester, Pennsylvania, is bordered by a trash-to-steam incinerator, a sewage treatment plant, and a processing facility for hazardous medical waste. Its residents are largely African-American. An exploration of the environmental justice movement at the grassroots level. (1997)

"The Old Quabbin Valley: Politics and Conflict in Water Distribution," video, 28 min. $24.95 purchase (Lawrence Hott, Florentine Films, Distributed by Direct Cinema Limited, 1-800-345-6748)

Traces the history of Boston's water supply, the construction of the Quabbin Reservoir, the plan to divert the Connecticut River, and the east-west conflict in Massachusetts water policy.

"Taken for a Ride," video, 52 min. $55 rental. (New Day Films, 1-201-652-6590)

http://www.newday.com

This video argues that public transit declined in American cities because of the lobbying efforts of General Motors to dismantle streetcar lines. This was followed by a highway lobby in Washington that pushed through funding for construction of the nation's interstates. (n.d.)

Water and the Dream of the Engineers, video (or film), 80 min. $125 video rental. (Cine Research Associates, 170 Garden St., Cambridge, Mass. 02138; 1-617-442-9756)

http://www.cineresearch.org/films.htm

Four-part video (2 sections available separately). Part 1a, "Water for All," documents the history of modern water and sewer systems; Part 1b, "Water Wars: California," explores the social conflicts of water systems in California (available separately, 25 min. video rental $45); Part

2a, "Upstream-Downstream: New Orleans," places one system's struggle with chemical pollution in the context of history (available separately, 18 min. video rental $40); Part 2b, "The Sludge Dumpers," shows how the modern use of old sewer systems poses a threat to the water environment."

7. Conservation and Preservation

America's National Parks, video, $34.95 purchase. 2 cassettes, 85 min. ea. (Encounter Video, Inc. Portland, Ore. 1-800-677-7607)

http://www.encountervideo.com

Vol. 1: America's Adventurous and Majestic Parks; Vol. 2: America's Historic and Scenic Parks. Includes legends of "the ancient ones," the discoveries of traders and trappers, the trials of homesteaders and explorers, and the landscapes of the country's 53 national parks. (1995)

Cadillac Desert, set of four videos, 60–90 min. $99.95. (PMI/Home Vision Select, 1-800-343-4727)

http://www.shop.pbs.org

http://www.pbs.org/kteh/cadillacdesert/home.html

Based on Marc Reisner's best-selling account of the use and misuse of water in the American West. Told in four segments: "Mullholland's Dream," "An American Nile," "The Mercy of Nature," and "Last Oasis." (1997)

"The Cradle of Forestry in America," video, 18 min. $11.95. (Pisgah Forest, North Carolina, 1-877-457-4023)

http://www.cradleofforestry.com/store//storefront.html

The story of the efforts to make the Biltmore Estate and the Pisgah Forest into a national showpiece of managed forests. Features the work of George W. Vanderbilt, Frederick Law Olmsted, Gifford Pinchot, and Carl Schenck. (n.d.)

The Forest Through the Trees: Battle for the Redwoods, video, 60 min., $45 purchase. (Green TV, 1125 Hayes St. San Francisco, Calif. 94147; 1-415-255-4797) http://www.greentv.org/FTTT.htm (E-mail: fgreen@greentv.org)

The struggle to preserve the ancient redwood forests of the West Coast. (1990)

Hoover Dam, video, 60 min. $19.98 purchase. (*The American Experience* series, PBS videos, Alexandria, Va., 1-800-344-3337)

During the worst depression in American history, a group of engineers constructed one of the most spectacular and controversial dams in the world. (1997)

"John Muir: The Man, the Poet, the Legacy," video, 50 min. $24.95 purchase (Panorama International Productions, P.O. Box 1255, Beverly Hills, Ca. 90213)

Follows the life of John Muir from his birth in Scotland to his campaign to save Hetch Hetchy Valley through images of the places in which he lived and worked, accompanied by narration. (n.d.)

"On the Edge: Nature's Last Stand for Coast Redwoods," video, 33 min. rental $50, Purchase $150. (Produced by James Daniels for the California Department of Parks and Recreation and the Sempervirens Fund. Center for Media and Independent Learning, University of California, Berkeley, 1-510-642-0460).

Explores the attitudes and relationship of the California Indians to the redwoods, and shows how the Gold Rush and the ensuing growth of California resulted in the cutting of 95 percent of the redwoods by 1989. Traces the history of conservation efforts over the last century to protect the redwoods forests.

"Water Wars: The Battle of Mono Lake," film, 39 min. $35 rental. (Center for Media and Independent Learning, University of California, Berkeley, 1-510-642-0460)

Controversy over diversion of fresh water from California's Mono Lake to Los Angeles and its impact on nesting birds. (1984)

Wild by Law: Bob Marshall, Aldo Leopold, and Howard Zahniser, video, 60 min. $34.95 purchase. (Produced by Lawrence Hott and Diane Garry of Florentine Films, Distributed by Direct Cinema Limited, 1-800-345-6748)

http://www.directcinema.com

Dust Bowl, New Deal conservation, tourism, and wilderness issues from 1930–1964. Features Leopold and Zahniser's sons and daughters as well as historians and activists such as William Cronon, Roderick Nash, Wallace Stegner, Max Oelschlaeger, Baird Callicott, Floyd Dominy, and David Brower.

The Wilderness Idea, video, 60 min. $34.95 purchase. (Florentine Films, Distributed by Direct Cinema Limited, 1-800-345-6748)

http://www.directcinema.com

Gifford Pinchot, John Muir, and the battle over the Hetch Hetchy Valley of California (with Roderick Nash, William Cronon, Annette Kolodny, Michael Cohen, Stephen Fox, and others). (1989)

Yosemite: The Fate of Heaven, video, 60 min. $39.95 purchase. (Direct Cinema Limited, 1-800-345-6748)

http://www.directcinema.com

Narrated by Robert Redford. Shows the history and conquest of Yosemite by the militia and the problems Yosemite faces today. (1989)

8. Indian Land Policy

"The First and Last Frontier," video, 58 min. (From *The Land of the Eagle* series, Time-Life, No. 7 in series of 8 videos. Available on interlibrary loan from The Valley Library, Oregon State University, Corvallis, Ore.)

http://osulibrary.orst.edu/video/bio67.htm (E-mail: valley.circ@orst.edu)

The Tlingit, Aleut, and Eskimo encounter Russian fur traders and Yankee whalers in Alaska. (1991)

Last Stand at Little Big Horn, video, 60 min. $29.95 (Teacher's Video Co., 1-800-262-8837)

http://www.teachersvideo.com.

Views the conflict between Custer and Sitting Bull and Crazy Horse from both the settlers' and the Sioux's perspectives. (n.d.)

"Into the Shining Mountains," video, 58 min. (From *The Land of the Eagle* series, Time-Life, No. 5 in series of 8 videos. Available on interlibrary loan from The Valley Library, Oregon State University, Corvallis, Ore.)

http://osulibrary.orst.edu/video/bio65.htm (E-mail: valley.circ@orst.edu)

From the days of the Ute, Shoshone, and Blackfeet to the creation of Yellowstone National Park. (1991)

Koyaanisqatsi ("Life Out of Balance") Available from most video rental stores.

http://koyaanisqatsi.org/films/koyaanisqatsi.htm

Images of pristine nature and the developed industrialized world against the continual refrain of "Koyaanisqatsi." No additional dialog. (1975–82)

Make Prayers to the Raven, five 30-minute videos, $29.95 each, purchase; set $125 purchase. (University of Alaska KUAC TV, 1-907-474-7491)

Based on Richard Nelson's book *Make Prayers to the Raven*, on the Koyukon of Interior Alaska, their subsistence way of life, and their environmental ethics. Narrated by Barry Lopez.

9. The Rise of Ecology

"Ecological Realities: Natural Laws at Work," 14 min. video #37390, $40 rental; $150 purchase, or 16 mm film, #8174, $30 rental. (Center for Media and Independent Learning, University of California, Berkeley, 1-510-642-0460)

Flow of energy from the sun to plants and animals, life cycle of a marsh food chain, regulation of prairie population species. Concept of the food pyramid and recycling of mineral nutrients. (1976)

"Ecosystems," video, 34 min. $126 purchase. (Hawkhill Video, 1-800-422-4295)

http://www.hawkhill.com

Part 1, "From Nature Study to Ecology," traces the naturalist tradition from Henry David Thoreau to the quantitative tradition of Eugene and Howard Odum. Part 2, "The Web of Life," outlines concepts used by ecologists to study ecosystems today. Includes contributions of Darwin, Rachel Carson, and Aldo Leopold. (1999)

"The Gaia Hypothesis," video, 25 min., $129 purchase, $75 rental. (Films for the Humanities and Sciences, 1-800-257-5126) http://www.films.com

James Lovelock explains the development and evolution of his hypothesis, which considers the planet a self-regulating system of physical, chemical, and biological processes. (n.d.)

"Natural Resources," video, 38 min. $126 purchase. (Hawkhill Video, 1-800-422-4295)

http://www.hawkhill.com

Part 1, "The Record of the Past," shows how the concept of natural resources has changed over time, exemplified by whale oil and gold. Part 2,"The Promise of the Future," looks at current limitations on the world's resources. Includes the agricultural and industrial revolutions, acid rain, greenhouse effect, population, species extinction, fossil fuels and wind farms. (1999)

"A New England Pond," video, 27 min. $86 purchase. (Hawkhill Video, 1-800-422-4295)

http://www.hawkhill.com

Uses a small New England pond to teach basic concepts in ecosystem science, such as food webs, abiotic-producer-consumer-decomposer cycles, pond succession, and acid rain. (1998)

"Principles of Ecology," video, 23 min., $89.95 purchase. (Films for the Humanities and the Sciences, 1-800-257-5126)

http://www.films.com

Explains the underlying themes of the science, as well as exploring the Gaia hypothesis. (n.d.)

10. The Era of Environmentalism

Alaska: The Last Frontier? video. (Corporation for Public Broadcasting, 1-800-LEARNER)

While Alaska is perceived as this nation's last great frontier, the region, in fact, faces continuing struggles between its indigenous people, the lovers of its wilderness, and those who wish to develop its natural resources.

"Ancient Forests," video, 21 min. $50 rental; $150 purchase. (Center for Media and Independent Learning, University of California, Berkeley, 1-510-642-0460)

The story of the Pacific Northwest's threatened ecosystem. What do we stand to lose if we allow the logging of our last ancient forests? (1997)

"Do You Really Want to Live Like This?" video, $29.95 purchase. (From *The Race to Save the Planet* series, 1-800-LEARNER, The Annenberg/CPB Project, c/o Intellimation, P.O. Box 1922, Santa Barbara, Calif. 93116-1922)

The price of progress — smog, toxic wastes, and water pollution — as a product of Western industrialism. (1990)

Edward Abbey, A Voice in the Wilderness, video, 60 min. $18.96 purchase (Eric Temple Productions, P.O. Box 2284, South Burlington, Vt., 05407-2284)

http://shopping.yahoo.com/shop?d=v&id=1800205071

A biographical account of Edward Abby, his philosophy, and his work on behalf of national parks in Utah and the canyon lands of the southwestern United States. (1993)

"The Ever-Normal Granary," video, 24 min. $90 purchase (Part 5 of *America the Bountiful* Collection, set of 6, $395) (Visual Education Productions, 1-800-235-4146)

American agriculture in the 1930s, the Great Depression, and the Dust Bowl. Agricultural plans of Henry A. Wallace and FDR's New Deal. (n.d.)

The Fight in the Fields, video and book. (Paradigm Productions, 1-800-903-7804)

http://www.paradigmproductions.org/paradigm.html

Story of Cesar Chavez and the farmworkers of California in the 1960s through the 1980s.

"The Four Corners: A National Sacrifice Area?" video, 58 min. $25 rental; $49 purchase. (Bullfrog Films, 1-800-543-3764)

Producer Christopher McLeod documents the cultural and ecological impacts of coal stripmining, uranium mining, and oil shale development in Utah, Colorado, New Mexico, and Arizona — homeland of the Hopi and Navajo. (1983)

"Global Warming: Turning Up the Heat," video, 46 min. $75 rental. (Bull-frog Films, 1-800-543-3764)

Examines how human activity has contributed to the trend in global warming. Guests include noted scientists and government officials.

In the Light of Reverence. Video, 73 min. $95 rental; $275 purchase. (Bull-dog Films, P.O Box 149, Oley, Pa. 19547, 1-800-543-3764)

http://www.bullfrogfilms.com/catalog/ilr.html

Produced by Christopher McLeod. A portrait of land-use conflicts over Native American sacred sites on public and private land around the West. (2001)

Now or Never, video, 60 min. $39.95 purchase; $169 series. (Corporation for Public Broadcasting, 1-800-LEARNER)

http://www.learner.org/catalog/ordering/series66.html

http://www.pbs.org/wgbh/shop/ipsaveplanet.html

Hosted by Meryl Streep and narrated by Roy Scheider, this video is the last in a series entitled *Race to Save the Planet.* Looks at the efforts of individuals who today are trying to make a difference in the fight to protect the environment. (1990)

Paul Ehrlich and The Population Bomb, video, 60 min. $149 purchase, $75 rental. (Films for the Humanities and Sciences, 1-800-257-5126)

http://www.films.com

Ehrlich set out thirty years ago to convince the world that the planet's ecosystems would not be able to support unlimited population growth indefinitely. This film makes the same argument, using interviews with Ehrlich, and brings the viewer to the present.

"The Poisoned Dream: The Love Canal Nightmare," video, 48 min., $89.95 purchase. (Films for the Humanities and Sciences, 1-800-257-5126)

http://www.films.com

A "riveting exposé" of the efforts of three mothers to force the New York and federal governments to take action against the dangerously polluted community of Love Canal.

Rachel Carson's Silent Spring, video, 60 min., $59.95 purchase. (*The American Experience* series, PBS videos, Alexandria, Va., 1-800-344-3337. Also available on interlibrary loan from The Valley Library, Oregon State University, Corvallis, Ore.)

http://osulibrary.orst.edu/video/env9.htm (E-mail: valley.circ@orst.edu)

The story of Carson's enduring classic and the controversy surrounding its publication. (1993)

"River," video, 34 min., $28 rental. (#38084 Center for Media and Independent Learning, University of California, Berkeley, 1-510-642- 0460)

Classic documentary, produced in 1937 by the Farm Security Administration, about the exploitation and ruin of the Mississippi River basin through agricultural, industrial and urban expansion and subsequent efforts at reforestation and flood control via TVa. (1937)

Water Is for Fighting Over, video. (Corporation for Public Broadcasting, 1-800-LEARNER)

The Truckee River Basin, which bridges the border between Nevada and California just north of Lake Tahoe, is the site of perpetual battle between the need for water and the desire to protect our scenic rivers.

"Waste Not, Want Not," video, $29.95 purchase; $169 series. (From *The Race to Save the Planet* series, No. 8, 1-800-LEARNER, The Annenberg/CPB Project, c/o Intellimation, P.O. Box 1922, Santa Barbara, Calif., 93116-1922)

http://www.pbs.org/wgbh/shop/ipsaveplanet.html

Hosted by Meryl Streep and narrated by Roy Scheider. Garbage barges, toxic dumping, landfills, and sewage pollution of coastal waters and alternatives to disposal such as source reduction and recycling. (1990)

"When the Rivers Run Dry," video, 29 min. $150 purchase, $50 rental. (#37146, Center for Media and Independent Learning, University of California, Berkeley, 1-510-642-0460)

Explores the long-standing conflicts among Anglo, Hispanic, and Indian farmers for the limited water resources of the southwest. (1980)

Electronic Resources

1. General Environmental History Resources

American Environmental History Resources
    http://www2.h-net.msu.edu/~environ/resources.html

American Society for Environmental History
    http://www2.h-net.msu.edu/~environ/
    The Web site for the American Society for Environmental History, the national academic organization.

Directory of Environmental History Internet Sources
    http://academicinfo.net/ehist.html

Ecohistory
    http://www.ecohistory.org

Nature Transformed: The Environment in American History
    http://www.nhc.rtp.nc.us:8080/tserve/nattrans/nattrans.htm
    Teacher Serve Web site from the National Humanities Center.

The Environment and World History Web Sites
    http://www.dana.edu/~dwarman/envrhist.htm

Environmental History

http://www.lib.duke.edu/forest/ehmain.html

An on-line guide to the joint publication of the Forest History Society and the American Society of Environmental History.

Environmental History: An Introduction

http://h-net2.msu.edu/~aseh/syllabi/nore.htm

Environmental History Bibliography

http://www.cnr.berkeley.edu/departments/espm/env-hist

A Web version of the bibliography compiled by Jessica Teisch for the National Endowment for the Humanities.

Environmental History: Explore the Field

http://www.cnr.berkeley.edu/departments/espm/env-hist/

Environmental History Timeline

http://www.runet.edu/~wkovarik/hist/hist.html

A timeline of World Environmental History (copyright 1997 Bill Kovarik, Ph.D.) and a list of related Web links.

Environmental History for High School Students and Teachers

http://www.geocities.com/Athens/Delphi/5777/index.htm

The National Library for the Environment

http://www.cnie.org/nle

A project of the Committee for the National Institute for the Environment (CNIE), the library address the need for objective, scientifically sound information on environmental issues.

Resources in Environmental History

http://www.library.cmu.edu/bySubject/History/environment.html

A comprehensive site from Carnegie Mellon University with links to environmental history associations, organizations, institutes, biographical resources, etc.

What Is Environmental History?

http://www.wsu.edu:8080/~forrest/envhist.html

2. Environmental History Societies and Related Associations

American Society for Environmental History
  http://www2.h-net.msu.edu/~environ/

Environmental History Discussion Lists (H-ASEH, H-Sci-Med-Tech, etc.)
  http://www.asap.unimelb.edu.au/hstm/hstm_ema.htm

Environmental History Discussion Lists (various topics)
  http://www.h-net.msu.edu/~aseh/archives/threads/

Forest History Society
  http://www.lib.duke.edu/forest/

The Institute for Environmental History
  http://www.st-and.ac.uk/academic/history/envhist/

3. Archival Materials

American Memory: Digital Collections from the Library of Congress
  http://lcweb2.loc.gov/ammem/ammemhome.html

American Studies Web
  http://www.georgetown.edu/crossroads/asw/

Documenting the American South: North American Slave Narratives
  http://metalab.unc.edu/docsouth/neh/neh.html

Evolution of the Conservation Movement, 1850–1920
  http://lcweb2.loc.gov/ammem/amrvhtml/conshome.html

Exploring the West from Monticello
  http://www.lib.virginia.edu/exhibits/lewis_clark/
  Digital images of maps and navigational aids used in planning the Lewis and Clark expedition, as well as links to related resources.

Hetch Hetchy Controversy
  http://lcweb2.loc.gov/ammem/amrvhtml/consbib8.html

Nature and Wilderness as Recreational Resources
   http://lcweb2.loc.gov/ammem/amrvhtml/consbib7.html

The Plymouth Colony Archive Project
   http://www.people.Virginia.EDU/~jfd3a/

Smithsonian Institute
   http://www.si.edu/

U.S. Historical Documents
   http://www.law.ou.edu/hist/

Voices from the Dust Bowl
   http://lcweb2.loc.gov/ammem/afctshtml/tshome.html

4. Bibliographies

Directory to Bibliography in Environmental Ethics
   http://ecoethics.net/bib/

Environmental Publications (search)
   http://www2.links2go.com/topic/Environmental_Publications

Environmental History (journal of American Society for Environmental History)
   http://www.asap.unimelb.edu.au/hstm/data/529.htm

Environment and History (journal of American Society for Environmental History)
   http://www.erica.demon.co.uk/EH.html

Non Timber Forest Product Bibliography Online
   http://www.ifcae.org/cgibin/ntfp/db/dbsql/db.cgi?db=bib&uid=default

The United States History Index (bibliographies and searches)
   http://www.ukans.edu/history/VL/USA/index.html

5. Biographical Resources

Edward Abbey

> http://www.utsidan.se/abbey/

Ansel Adams

> http://www.zpub.com/sf/history/adams.html

Jane Addams

> http://www.nobel.se/peace/laureates/1931/addams-bio.html

Isabella Bird

> http://www.geocities.com/Wellesley/7385/

Rachel Carson

> http://onlineethics.org/moral/carson/main.html
>
> http://rachelcarson.org/

George Washington Carver

> http://www.princeton.edu/~mcbrown/display/carver.html

Barry Commoner

> http://www.sciam.com/interview/commoner/062397commoner.html

James Fenimore Cooper

> http://library.cmsu.edu/cooper/COOPER.HTM

Paul Ehrlich

> http://www.pbs.org/kqed/population_bomb/theshow/bio.html

Thomas Jefferson

> http://www.pbs.org/jefferson/frame5.htm
>
> http://etext.virginia.edu/jefferson/

Chief Joseph

> http://www.pbs.org/weta/thewest/people/a_c/chiefjoseph.htm

Aldo Leopold

http://www.wilderness.org/profiles/leopold.htm

George Perkins Marsh

http://www.virtualvermont.com/history/gmarsh.html

http://sun3.lib.uci.edu/~slca/microform/resources/f-g/g_023a.htm

John Muir

http://www.sierraclub.org/john_muir_exhibit/

Gifford Pinchot

http://www.pinchot.org/gt/gp.html

John Wesley Powell

http://www.powellmuseum.org/MajorPowell.html

Theodore Roosevelt

http://tlc.ai.org/tr.htm

Wallace Stegner

http://206.14.7.53/gic/Stegner/wallace.html

Henry David Thoreau

http://www.geocities.com/Athens/7687/1thorea.html

Frederick Jackson Turner

http://xroads.virginia.edu/~HYPER/TURNER/home.html

Stewart Udall

http://www.lbjlib.utexas.edu/johnson/archives.hom/oralhistory.hom
/udall/udall.asp

6. Environmental Organizations and Information Centers

American Nature Study Society

http://hometown.aol.com/anssonline/index.htm

Earth Island Institute

http://www.earthisland.org

The EcoJustice Network

    http://www.igc.apc.org/envjustice/pocdir/

The Environmental Information Center:

    http://www.tei.or.th/eic/

    The Center is a resource for public education campaigns, focusing on grass-roots organizing.

Environmental Organization Web Directory

    http://www.webdirectory.com/

History Television

    http://www.historytelevision.ca/

National Audubon Society

    http://www.audubon.org/

Nature Conservancy

    http://www.tnc.org

Natural History Resources (organizations, government agencies)

    http://www.ucmp.berkeley.edu/subway/subway.html

Rocky Mountain Institute

    http://www.rmi.org

The Sierra Club

    http://www.sierraclub.org/history/

    The club's history and successful battles to protect the nation's resources.

7. Environmental Philosophy and Ethics

Environmental Ethics

    http://www.wsu.edu:8080/~forrest/ethics.html

Directory to Bibliography in Environmental Ethics

    http://ecoethics.net/bib/

Figures in Environmental History and Ethics
   http://www.ac.wwu.edu/~gmyers/eheassign1.html

The International Association for Environmental Philosophy
   http://www.environmentalphilosophy.org/

Web Sites Devoted to Environmental Philosophy
   http://h-net2.msu.edu/~aseh/syllabi/isenberg.html

8. Government Agencies

Army Corps of Engineers
   http://www.usace.army.mil/

Bureau of Land Management
   http://www.blm.gov

Bureau of Reclamation
   http://www.usbr.gov

Environmental Protection Agency
   http://www.epa.gov

The Federal Web Locator
   http://www.infoctr.edu/fwl/

Geographic Information Systems
   http://info.er.usgs.gov/research/gis/title.html

History in the National Park Service: Ask a Historian
   http://www.cr.nps.gov/history/askhist.htm

The National Environmental Data Index
   http://www.nedi.gov/

National Park Service
   http://www.nps.gov

U.S. Fish and Wildlife Service
   http://www.fws.gov/index.html

U.S. Forest Service
   http://www.fs.fed.us/

U.S. Government Environmental Web Sites
   http://members.aol.com/fitzenviro/sites.html

9. Natural History

Alaskan Animals
   http://www.teelfamily.com/links/animals/

American Nature Study Society
   http://hometown.aol.com/anssonline/index.htm

Desert Fishes
   http://www.utexas.edu/ftp/depts/tnhc/.www/fish/dfc/dfc_topc.html

Endangered Species of Hawaii
   http://hisurf.com/~enchanted/

Salmon
   http://www.riverdale.k12.or.us/salmon.htm

U.S. Fish and Wildlife Service
   http://www.fws.gov/

10. Natural Resources

Agricultural History and Rural Studies
   http://www.iastate.edu/~history_info/ahahs/links.htm

   A page of links to archival resources, publications, and museums related to agricultural history in the United States.

International River Networks: River Basics

http://www.irn.org/basics/basic.shtml

Information on the history of dams from the perspective of an international environmental organization committed to protecting wild rivers.
Those interested in specific dams should search the Web using the
name of that dam, as there are innumerable sites devoted to individual
dams such as the Grand Coulee, the Hoover, and the Glen Canyon.

Mining History Association

http://www.lib.mtu.edu/mha/mha.htm

Association for those interested in the history of mining and metallurgy. Site contains links to archival resources and to other related organizations.

Sites of Ecological Interest

http://eagle.bio.unipr.it/EcoWWW.html

11. Regional Resources

Alaska Volcanos

http://www.avo.alaska.edu/html/resource.html

Albuquerque's Environmental Story

http://www.cabq.gov/aes/s1geol.html

American Environmental Photographs, 1891–1936: Images from the University of Chicago Library

http://memory.loc.gov/ammem/award97/icuhtml/aephome.html

The American Experience: Surviving the Dust Bowl

http://www.pbs.org/wgbh/amex/dustbowl/

Kansas Historical Geography

http://www.kgs.ukans.edu/Publications/primer/primer01.html

Los Angeles River Tour

http://www.lalc.k12.ca.us/target/units/river/riverweb.html

Maps of the Southwest

>   http://www.calacademy.org/research/library/elkus/stories/maps.htm

The Midwestern U.S. 16,000 Years Ago

>   http://www.museum.state.il.us/exhibits/larson/

Native American Environmental Curriculum Project

>   http://mesa.colorado.edu/~topper/nsfhome.htm

Northwest Hatcheries' History

>   http://jcomm.uoregon.edu/~josh/salmon/chrono.html

Pacific Northwest Logging History

>   http://www.aone.com/~robert/histlog.html

A Trip Through the Grand Canyon

>   http://www.azstarnet.com/grandcanyonriver/GCrt.html

United States Research Report Links

>   http://swasco.net/New%20Web%20Pages/saxton/usregions/intro.html
>
>   A site with links to various Web resources for historical research. Divided by geographical region.

WestWeb: Western Environmental History

>   http://scholar.library.csi.cuny.edu/westweb/pages/enviro.html
>
>   A collection of Web resources for those interested specifically in the history of the American West.

12. Environmental Justice Resources

Cesar Chavez

>   http://latino.sscnet.ucla.edu/research/chavez/
>
>   A Web site commemorating his life and achievements.

The Ecojustice Network

>   http://www.igc.apc.org/envjustice/

The Network addresses environmental issues facing communities of color in the United States and provides on-line services, informational resources, and training for activists and organizations involved in the environmental justice movement.

History of Environmental Justice

http://www-personal.umich.edu/~jrajzer/nre/history.html

13. Teaching Resources

EELink-Environmental Education on the Internet

http://www.eelink@eelink.umich.edu/

A site dedicated to assisting educators in integrating environmental issues and learning into their curricula. Intended for students, teachers and education professionals.

Environmental History for High School Students and Teachers

http://www.geocities.com/Athens/Delphi/5777/index.htm

Environmental Science Education Resources

http://www.educationindex.com/environ/

Green Teacher

http://www.greenteacher.com/

A Web version of the quarterly magazine by and for K–12 educators, includes articles, ready-to-use activities, and resource listings and reviews.

Figures in Environmental History and Ethics

http://www.ac.wwu.edu/~gmyers/eheassign1.html

Guide to Environmental Education

http://princetonol.com/groups/stonybrook/envedu.html

Resources for Teaching and Researching Northwest Environmental History

http://www.wsu.edu:8080/~forrest/

Science and Technology Curriculum Guide (Grades K–12)
   http://princetonol.com/groups/stonybrook/links.htm#teach

WestWeb: On Teaching Western History (Resources)
   http://scholar.library.csi.cuny.edu/westweb/pages/teach.html

14. Course Syllabi in American Environmental History

American Environmental and Cultural History (U.C. Berkeley)
   http://cnr.berkeley.edu/departments/espm/env-hist/espm160

American Environmental History (Brown University)
   http://h-net2.msu.edu/~aseh/syllabi/isenberg.html

American Environmental History (Western Washington University)
   http://h-net2.msu.edu/~aseh/syllabi/stewart.htm

American Environmental History (University of Arizona)
   http://h-net2.msu.edu/~aseh/syllabi/weiner2.htm

American Environmental History (University of Wisconsin, Madison)
   http://h-net2.msu.edu/~aseh/syllabi/crononl.htm

Environmental History of the Americas (University of Dayton)
   http://h-net2.msu.edu/~aseh/syllabi/bednarek.htm

Environmental History Syllabi
   http://www.h-net.msu.edu/~aseh/syllabi/

Global Issues in Environmental History (College of New Rochelle)
   http://www-rci.rutgers.edu/~collie/global.html

Natural Resources/Environmental Issues (Willamette University)
   http://www.willamette.edu/dept/newspub/media_resource
   /nat_resources.html

Nature and Culture in the American Landscape (Bowdoin College)
    http://library.bowdoin.edu/classes/hist/226.2/

N.E.H. Environmental History Instate at Kansas State University
    http://www-personal.ksu.edu/~jsherow/envirohp.htm

North American Environmental History (University of Kansas)
    http://h-net2.msu.edu/~aseh/syllabi/worster2.htm

Social and Environmental History (Washington State University)
    http://h-net2.msu.edu/~aseh/syllabi/sonnen.htm

Surveys of Environmental History Programs
    http://www.h-net.msu.edu/~aseh/surveys.html

15. General Environmental Education

EE-Link: Environmental Education on the Internet
    http://nceet.snre.umich.edu/

Environmental Education
    http://www.wsu.edu:8080/~forrest/education.html

Environmental History for High School Teachers
    http://www.geocities.com/Athens/Delphi/5777/index.htm

Interactive Mapping Ideas and Resources
    http://www.mapcruzin.com/mapideas.htm

Links for Students of History and the Social Sciences
    http://www.vwc.edu/wwwpages/dgraf/graf32a.htm

Resources for Teaching and Researching Northwest Environmental History
    http://www.wsu.edu:8080/~forrest/

The World-Wide Web Virtual Library: History
    http://history.cc.ukans.edu/history/WWW_history_main.html

16. Historical Overview Web Sites

The American Environment and Native-European Encounters,
1000–1875

Environmental History: Explore the Field

   http://www.cnr.berkeley.edu/departments/espm/env-hist

American Society for Environmental History resources

   http://www2.h-net.msu.edu/~environ/resources.html

Buffalo and American Indians

   http://www.nhc.rtp.nc.us:8080/tserve/nattrans/ntecoindian/essays
   /buffalo.htm

The Columbian Exchange

   http://www.nhc.rtp.nc.us:8080/tserve/nattrans/ntecoindian/essays
   /columbian.htm

Native American Religion

   http://www.nhc.rtp.nc.us:8080/tserve/eighteen/ekeyinfo/natrel.htm

Pueblo Indians of Mesa Verde

   http://www.swcolo.org/Tourism/Archaeology/MesaVerde.html

Religious traditions of the Micmac

   http://www.mun.ca/rels/hrollmann/native/micmac/micmac1.html

Squanto's history

   http://members.aol.com/calebj/squanto.html

Syllabus for a course on world environmental history

   http://www.mansfield.ohio-state.edu/faculty/dominick/history765.htm

University of Colorado environmental history resource links

   http://web.uccs.edu/~history/index/environhist.html

Wampanoag Horticulture

   http://www.plimoth.org/Library/Wampanoag/wamphort.htm

The New England Wilderness Transformed, 1600–1850

Cattle in Plymouth Colony

http://members.aol.com/calebj/cattle.html

The *Mayflower's* Voyage

http://members.aol.com/calebj/voyage.html

Who are the Puritans?

http://www.nhc.rtp.nc.us:8080/tserve/eighteen/ekeyinfo/puritan.htm

Puritan Origins of the American Wilderness Movement

http://www.nhc.rtp.nc.us:8080/tserve/nattrans/ntwilderness/essays/puritan.htm

William Bradford, History of Plymouth Plantation (1620–1647)

http://www.swarthmore.edu/SocSci/bdorsey1/41docs/14-bra.html

The Tobacco and Cotton South, 1600–1900

Antebellum White South Economy

http://cghs.dade.k12.fl.us/slavery/white_south/market.htm

Robert Beverley: *The History and Present State of Virginia* (1705–6)

http://www.stockton.edu/~gilmorew/0colhis/vlls-89.htm

The Civil War: An Environmental View

http://www.nhc.rtp.nc.us:8080/tserve/nattrans/ntuseland/essays/amcwar.htm

The Enterprise Boll Weevil Monument today

http://www.al.com/wacky/vandals.html

John White and Roanoke Island

http://www.nps.gov/fora/jwhite.htm

Nature and the Market Economy, 1750–1850

Crevecoeur and American Identity and the Letters

http://www.vcu.edu/engweb/crev.htm

http://xroads.virginia.edu/~HYPER/CREV/home.html

Harvard Forest Models

http://www.news.harvard.edu/gazette/1999/04.15/harvard.forest.html

Jefferson as reflection of America's racial ambiguity

http://www.theatlantic.com/issues/97mar/jeffer/jeffer.htm

Paintings of the Hudson River School

http://dfl.highlands.com/DFL_Painters/Index.html

Poems of Phillis Wheatley

http://www.geocities.com/Athens/Parthenon/5471/president
/pw_intro.htm

*Western Frontiers: The Settlement of California
and the Great Plains, 1820–1930*

African Americans in the Fur Trade West

http://www.coax.net/people/lwf/furtrade.htm

Cultural History of the South Yuba River Canyon

http://www.websterweb.com/Museums_Parks/syrp/syrc/index.html

The Gold Rush

http://www.sacbee.com/goldrush/

Pioneer and Emigrant Women

http://www.rootsweb.com/~nwa/pioneer.html

Urban Environments, 1850–1960

Jane Addams's Hull-House Museum and Biographies

http://www.uic.edu/jaddams/hull/

Environmental Justice and Chester, Pennsylvania

http://www.penweb.org/chester/

Garbage Timeline

http://www.astc.org/exhibitions/rotten/timeline.htm

Roads, Highways, and Ecosystems

> http://www.nhc.rtp.nc.us:8080/tserve/nattrans/ntuseland/essays/roads.htm

Conservation and Preservation, 1785–1950

The Challenge of the Arid West

> http://www.nhc.rtp.nc.us:8080/tserve/nattrans/ntwilderness/essays/aridwest.htm

Conservation chronology :1847–1920

> http://memory.loc.gov/ammem/amrvhtml/cnchron3.html

Northern Pacific Railroad establishes Yellowstone as a profit center

> http://xroads.virginia.edu/~MA96/RAILROAD/ystone.html

John Muir's 1909 pamphlet advocating preservation of Hetch Hetchy in Yosemite

> http://www.sfmuseum.org/john/muir.html

John Wesley Powell - brief biography

> http://www.desertusa.com/magnov97/nov_pap/du_jwpowell.html

The Puritan Origins of the American Wilderness Movement

> http://www.nhc.rtp.nc.us:8080/tserve/nattrans/ntwilderness/essays/puritan.htm

The regulation of grazing on public rangelands, past and present

> http://pantheon.yale.edu/~jalbert/cowsch4.htm

The History of Yosemite National Park

> http://www.terraquest.com/highsights/valley/yoshist.html

Indian Land Policy, 1800–1990

The Cherokee Trail of Tears

> http://rosecity.net/tears/
>
> http://ngeorgia.com/history/cherokeeindex.html

Current tensions between the Miccosukee Tribe and the National Park Service in Everglades National Park

http://commdocs.house.gov/committees/resources/hii46735.000
/hii46735_0.htm

The Flight of the Nez Perce

http://www.bitterroot.net/usdafs/NezPerceflight.html

History of the Southern Utes

http://www.southern-ute.nsn.us/

Klamath Indians and U.S. Forest Service: struggles over water and land

http://www.newswest.com/naissues/klamath/klamath.html

Photographs of Chief Plenty Coups and Crow villagers

http://www.uwyo.edu/ahc/digital/throssel/photographs.htm

The Sheepeaters

http://www.hanksville.org/daniel/misc/sheepeaters.html

The Rise of Ecology, 1890–1990

American environmental photographs 1891–1936

http://lcweb2.loc.gov/ammem/award97/icuhtml/aephome.html

Environmental Conflict in History — Middle Ages to Rachel Carson

http://geocities.com/RainForest/3621/HIST.HTM

Frederic Clements and forest succession

http://www.anu.edu.au/Forestry/silvinative/daniel/chapter2/2.2.html

Aldo Leopold: a profile

http://tws.org/profiles/leopold.htm

The Era of Environmentalism, 1940–2000

Rachel Carson and the Awakening of Environmental Consciousness

http://www.nhc.rtp.nc.us:8080/tserve/nattrans/ntwilderness/essays
/carson.htm

http://www.rachelcarson.org/

EcoJustice Network

http://www.igc.org/envjustice/

Environmental justice programs at the EPA

http://es.epa.gov/oeca/main/ej/ejustepa.html

Environmental philosophers

http://baobabcomputing.com/naturalphilosophy/

Hoover Dam: perspectives and images

http://xroads.virginia.edu/~MA98/haven/hoover/front2.html

Pesticide Information Center On-line

http://picol.cahe.wsu.edu/

Martha Ellen Stortz essay on environmental ethics

http://www.webofcreation.org/education/articles/stortz.htm

Bibliographical Essay

1. The American Environment and Native-European Encounters

Environmental history incorporates into history the characteristics of particular environments and the methods people have used to exploit or conserve forests, waters, soils, and animals. In *Ecological Imperialism: The Biological Expansion of Europe* (New York: Cambridge University Press, 1986), Alfred W. Crosby spells out his theory of the "portmanteau biota": the cooperative, if unconscious assistance that livestock, crops, and microbes provided Europeans as they explored and conquered native populations around the world. Following Crosby's foundational work, Jared Diamond published *Guns, Germs, and Steel: The Fates of Human Societies* (New York: W.W. Norton, 1998) to expound on the idea that the advance of civilizations was shaped primarily by the plenitude or paucity of domesticable animals and plants that happened to evolve on particular continents. The east-west orientation of Eurasia was superior to the north-south orientation of the Americas in the transmission of domesticates, technologies, and political systems.

Another important work that sets the stage for the transformation of the physical environment is Stephen Pyne's *Fire in America: A Cultural History of Wildland and Rural Fire* (Princeton: Princeton University Press, 1982). In this work, Pyne treats fire as a cultural phenomenon. The impact and consequences of fires depend not only upon environmental factors but also upon the cultural system in which the fire takes place. How people judge a fire, their understanding of its dynamics, and their ability to control it all depend on the values and knowledge of the community. Yi-Fu Tuan is a geographer whose work also makes an important contribution to the study of how humans perceive and understand their environments. His *Topophilia: A Study of Environmental Perception, Attitudes, and Values* (New York: Columbia University Press, 1974) and other related studies look at cultural and regional comparisons, as well as reflections of human environmental values in art, philosophy, and religion.

The ways in which American Indians have related to the land ecologically has been the subject of several books. Calvin Martin in *Keepers of the Game: Indian-Animal Relationships and the Fur Trade* (Berkeley: University of California Press, 1978) argues that Indians believed in animal spirits, taboos, and rituals that allowed for ordained killing of animals, who gave their permission to be killed in return for respectful treatment of their remains. Introduced diseases, Martin argues, undermined the power of the tribal shaman, leading to the spiritual breakdown of the community. The Micmac, Ojibwa, and other northeastern hunting cultures apostatized, betraying the animal spirits and breaking traditional taboos, thereby allowing the animals to be harvested for the fur trade. *Clothed-in-Fur and Other Tales: An Introduction to an Ojibwa World View*, edited by J. Baird Callicott and Thomas W. Overholt, (Washington, D.C.: University Press of America, 1982), reveals relationships between Indians and the land that preserved nature and its bounty. J. Donald Hughes in *American Indian Ecology* (El Paso: Texas Western University Press, 1983) likewise presents a highly positive view of American Indians as practicing ecologically viable methods of hunting, fishing, gathering, and horticulture within spiritual patterns that allowed for respect for the land and its nonhuman inhabitants.

The size of the aboriginal population has been extensively treated and disputed. Rejecting earlier estimates of a million people at the dawn of European settlement of North America, Henry F. Dobyns in 1966 used epidemic data to recalculate the aboriginal population at 10 to 12 million, and then revised it upward to 18 million in his 1983 book *Their Number Become Thinned: Native American Population Dynamics in Eastern North America* (Knoxville: University of Tennessee Press, in cooperation with the Newberry Library Center for the History of the American Indian, 1983). In *The Ecological Indian: Myth and History* (New York: Norton, 1999), Shepard Krech III reassesses the demographic evidence and methods used by Dobyns and others, settling on a pre-contact population of between 4 and 7 million. He also reevaluates the stereotype that all Indians lived in ecological harmony and balance with the land, arguing instead that a diversity of cultural adaptations and practices prevailed, which varied in the extent to which Indians left the environment in a usable state for future generations.

2. The New England Wilderness Transformed

The important and highly influential concept of wilderness was introduced to environmental history by Roderick Nash in 1967 and elaborated in subsequent editions, in 1973 and 1982. His classic book *Wilderness and the American Mind* (New Haven: Yale University Press, 1967) examines the United States' changing intellectual sentiment toward nature and wilderness from the colonial period to the 1950s, detailing the nation's evolving interest in wilderness experiences and outdoor recreation. One of the first places in which wilderness was encountered and transformed was New England. In his influential book *Changes in the Land: Indians, Colonists, and the Ecology of New England* (New York: Hill and Wang, 1983), William Cronon argues that the arrival of Europeans in colonial New England heralded a dramatic shift in the use of forests and other natural resources. At the heart of the conflict between Indians and Europeans lay their different attitudes toward land use, and the success of capitalism is inseparable from the elimination of the New England Indians. Cronon's work was followed in 1989 by Carolyn Merchant's *Ecological Revolutions: Nature, Gender, and Science in New England* (Chapel Hill: University of North Carolina Press, 1989), which explicates a theory of revolutions in land use, arguing that essential shifts twice transformed gender relations, economic production, and modes of consciousness in New England history — once with the arrival of the colonists in the seventeenth century, and once with the Market Revolution of the early nineteenth century.

3. The Tobacco and Cotton South

By comparison with forest history and wilderness preservation, the environmental history of the South has received little attention from environmental historians, but some excellent histories have appeared in the field. Originally published in 1926, the classic work on the Tobacco South is Avery O. Craven's *Soil Exhaustion as a Factor in the Agricultural History of Virginia and Maryland, 1606–1860* (reprint, Gloucester, Mass.: Peter Smith Publishers), which examines soil depletion as a product of frontier conditions. Timothy Silver sets out an exemplary multicultural environmental history of the early South that includes the ways that Indians, blacks, and whites all used the land and forests in A New Face on the Countryside: Indians, Colonists, and Slaves in the South Atlantic Forests, 1500–1800 (Cambridge: Cambridge University Press, 1990). Mart A. Stewart's environmental history of the deep South focuses on rice cultivation and labor processes in "What Nature Suffers to Groe": Life, Labor, and Landscape on the Georgia Coast, 1680–1920 (Athens: University of Georgia Press, 1996).

In This Land, This South: An Environmental History (Lexington: University of Kentucky Press, 1983), Albert E. Cowdrey argues that row crop monocultures of cotton and corn set up the conditions for soil erosion, pest outbreaks, and parasites. Carville Earle disputes the role of ordinary, small-time farmers in creating soil erosion in "The Myth of the Southern Soil Miner: Macrohistory, Agricultural Innovation, and Environmental Change," an essay that appeared in The Ends of the Earth, edited by Donald Worster (New York: Cambridge University Press, 1988). Pete Daniel, in Breaking the Land: The Transformation of Cotton, Tobacco, and Rice Cultures Since 1880 (Urbana: University of Illinois Press, 1985), looks at the sharecropping system that became widespread after the Civil War, consolidating power in the hands of bankers, merchants, and loan companies and perpetuating soil degradation. Cooperative extension agents aided farmers in dealing with problems of soil erosion and pest outbreaks.

4. Nature and the Market Economy

Another foundational concept for the formation of the American environment is that of the pastoral landscape and the idea of the land as a garden. Leo Marx's The Machine in the Garden: Technology and the Pastoral Ideal in America (New York: Oxford University Press, 1964) shows how Americans have historically imagined their nation in two very different ways: as a rural paradise and as an industrialized, mechanized nation. He examines the historical roots of these images and traces their impact on the course of the nation's history. Equally important in the area of artistic responses to the environment is Barbara Novak's Nature and Culture: American Landscape Painting, 1825–1875 (New York: Oxford University Press, 1980). Novak focuses on the governing idea of nature in American painting, while tracing its roots to European schools of art and the American artist's vision of the country as a movement between ideas such as nature and civilization, dark and light, rural and cultivated. William Cronon's innovative essay "Telling Tales on Canvas," in *Discovered Lands, Invented Pasts* (New Haven: Yale University Press, 1992), interprets American paintings as narrative moments that reveal the past, present, and future directions of the environment as influenced by human action.

John Stilgoe's *Common Landscape of America, 1580–1845* (New Haven: Yale University Press, 1982) likewise deals with landscape formation. He examines the European roots of the early settlements in America and then moves on to discuss the layering of regional and national icons and the imposition of turnpikes, fences, and the surveyor's grid on the American rural landscape in the eighteenth and nineteenth centuries.

The market economy that transformed nature into economic resources is explicated by Charles Sellers in his influential book *The Market Revolution: Jacksonian America, 1815–1846* (New York: Oxford University Press, 1991). Sellers looks at the impact of the concurrent transportation and market revolutions in reinforcing a competitive, get-ahead spirit consistent with the capitalist transformation of land and life that resulted in intensive economic development.

5. Western Frontiers: The Settlement of California and the Great Plains

Setting the stage for the interpretation of the westward movement is the foundational essay by Frederick Jackson Turner, "The Significance of the Frontier in American History," written in 1893 and published in the *Annual Report of the American Historical Association* (Washington, D.C., 1894). Turner argues that the environment stripped the settler of his European cultural garb and reformed his character as an American, born of free land and the democratic ideal. After a century, Turner's argument is still significant for the emphasis it places on the western environment in shaping American history and for the influence it has had on generations of scholars of the American West.

One of the persistent accounts of Great Plains environmental history has been the demise of the bison at the hands of the U.S. military and buffalo hunters, who came close to exterminating the millions of buffalo that once roamed the plains and on which Indian livelihood depended. But ecological factors set the stage for that rapid decline, according to Dan Flores in "Bison Ecology and Bison Diplomacy: The Southern Plains from 1800 to

1850" (*Journal of American History* 78 [September 1991]: 465–485). Horses introduced in the eighteenth and nineteenth centuries competed with bison, and the climate changed to arid, warmer conditions unfavorable to the grasses that supported bison reproduction. Andrew Isenberg's *Destruction of the Bison: An Environmental History, 1750–1920* (New York: Cambridge University Press, 2000) likewise uses ecological history to recast the bison narrative, arguing that Indians participated in the bison trade in exchange for guns, blankets, kettles, and liquor, and that there was no conspiracy between hide hunters and the military to decimate the bison.

Walter Prescott Webb's foundational work *The Great Plains* (Boston: Ginn, 1931) portrays the transformation of the arid, treeless, level Great Plains through technological innovations, including barbed wire, the six-shooter, the windmill, and the John Deere plow, showing how they allowed American settlers to dominate the Great Plains environment in an unprecedented fashion. Several works by Donald Worster, on the other hand, analyze changes in the Western environment as negatively affected by technology and the capitalist transformation of nature. In *Dust Bowl: The Southern Plains in the 1930s* (New York: Oxford University Press, 1979), Worster overturns the myth of the pioneer conquering the Great Plains by demonstrating how the homesteader confidence in providence, hard work, and capitalism ultimately led to environmental catastrophe, in the form of the Dust Bowl. His book *Rivers of Empire: Water, Aridity, and the Growth of the American West* (New York: Oxford University Press, 1985) looks at the efforts of the federal government, engineers, and capitalists to harness the arid West's rivers in order to convert the barren landscape into a farmer's paradise. Drawing on the writings of Karl Wittfogel concerning past hydraulic societies, Worster demonstrates the increased ecological and social costs of efforts to exploit human labor and the natural environment under the capitalist mode of production.

Richard White in *The Organic Machine: The Remaking of the Columbia River* (New York: Hill and Wang, 1995) seeks to bridge the gap between nature and culture through a history of human labor and the energy provided to humans by the environment. Rather than emphasizing recreational or philosophical relationships between man and nature, White uses the dams and salmon fisheries of the Columbia River to refocus environmental history on the role of labor and the working person's relationship with the natural world. Like White, Arthur McEvoy looks at fisheries on the West Coast, but instead emphasizes the role of law, the social structures of various groups of indigenous and immigrant peoples, and differing forms of environmental cognition. His influential book *The Fisherman's Problem: Ecol-*

ogy and Law in the California Fisheries, 1850–1980 (New York: Cambridge University Press, 1986) covers the evolution of fishing in California from the period of Native American settlement to the post–World War II period, showing how the environment and society have evolved dynamically, and that the modern environmental threat to the health of the fisheries lies in the social conditions of fishing and problems of regulating the industry.

6. Urban Environments

In Nature's Metropolis: Chicago and the Great West (New York: Norton, 1991), William Cronon tells the history of Chicago's dramatic growth in the late nineteenth century by examining the development of the markets for meat, lumber, and grain. Theodore Steinberg's Nature Incorporated: Industrialization and the Waters of New England (New York: Cambridge University Press, 1991) focuses on the problem of river development in the increasingly industrialized countryside of northern New England. Issues of urban environmental and workplace health are explicated by Christopher Sellers in Hazards of the Job: From Industrial Disease to Environmental Health (Chapel Hill: University of North Carolina Press, 1997).

Martin V. Melosi has written extensively on the urban environment. His Garbage in the Cities: Refuse, Reform, and the Environment, 1880–1980 (Austin: University of Texas Press, 1980) details the country's growing need to dispose of waste and the technological problems faced and overcome. His edited volume Pollution and Reform in American Cities, 1870–1930 (Austin: University of Texas Press, 1980) gathers together foundational articles on air, water, noise, and garbage pollution and attempts to initiate reform. Melosi's Coping with Abundance: Energy and Environment in Industrial America (New York: Knopf, 1985) looks at issues of energy in an industrializing nation, while The Sanitary City: Urban Infrastructure in America from Colonial Times to the Present (Baltimore: Johns Hopkins University Press, 2000) links problems of maintaining clean water supplies with discharging domestic wastes and industrial effluents. Another major contribution to understanding urban environmental problems has been made by Joel A. Tarr in The Search for the Ultimate Sink: Urban Pollution in Historical Perspective (Akron, Ohio: University of Akron Press, 1996), a collection of his essays on city wastes and their environmental consequences.

Kenneth Jackson's Crabgrass Frontier: the Suburbanization of the United States (New York: Oxford University Press, 1985) is an important work on the suburban environment. Jackson charts the evolution of transportation systems in relation to the socioeconomic composition of communities. Adam Rome's *The Bulldozer in the Countryside* (New York: Cambridge University Press, 2001) looks at suburban sprawl and the rise of environmentalism.

7. Conservation and Preservation

In his classic book on the roots of the conservation movement, *Conservation and the Gospel of Efficiency* (Cambridge: Harvard University Press, 1959), Samuel P. Hays argues that the roots of natural resource management during the Progressive Era lay in the period's devotion to efficiency, and in the role played by a rising cadre of experts who researched and promoted scientific methods of managing the country's forests, water, and rangelands. An overview of the recent environmental movement's development in the United States is offered by Donald Fleming in "Roots of the New Conservation Movement," which appeared in *Perspectives in American History* 6 (1972). Fleming focuses on major writers of the recent movement, including Rachel Carson and Aldo Leopold, as well as René Dubos, Barry Commoner, and Paul Ehrlich. Samuel Trask Dana and Sally K. Fairfax's *Forest and Range Policy, Its Development in the United States* (New York: McGraw-Hill, 1980) is a comprehensive treatment of the changing policies and laws pertaining to forest and range conservation, and includes a highly useful time line of significant dates in forest and environmental history.

The history of the concurrent preservation movement is elaborated by Stephen Fox in *The American Conservation Movement: John Muir and his Legacy* (Boston: Little, Brown, 1981). Fox traces the history of environmental activism in the United States with the underlying assumption that today's movement draws its inspiration from John Muir's philosophy of preservation. Paul Brooks's *Speaking for Nature: How Literary Naturalists from Henry Thoreau to Rachel Carson Have Shaped America* (Boston: Houghton, Mifflin, 1980) traces the history of a tradition of American writers whose descriptions of nature have helped fellow citizens to appreciate their homeland. In *Made from This Earth: American Women and Nature* (Chapel Hill: University of North Carolina Press, 1993), Vera Norwood examines American women's contributions to the study and preservation of nature from the early 1800s to the present. She argues that women have sought to validate their role in society and home through their work in nature.

The history of the creation of America's national parks, many out of what were considered lands too worthless for resource development, is recounted in Alfred Runte's *National Parks: The American Experience*, 3d ed. (Lincoln: University of Nebraska Press, 1997). In *How the Canyon Became Grand: A Short History* (New York: Viking Press, 1998), Steven Pyne provides an arresting analysis of the social construction of nature in the Grand Canyon, from worthless impediment to Spanish explorations, to the discovery of the canyon as a place of geological grandeur, to the iconographic indemnification of the canyon as wilderness. The preservation movement also included the saving of the giant charismatic redwood forests. Susan Schrepfer's *The Fight to Save the Redwoods: A History of Environmental Reform, 1917–1978* (Madison: University of Wisconsin Press, 1983) focuses on California's coastal redwoods and the struggle to save them from the ax and sawmill.

8. Indian Land Policy

Francis Paul Prucha's *The Indians in American Society: From the Revolutionary War to the Present* (Berkeley: University of California Press, 1985) is a valuable overview of two centuries of U.S. policy toward Indians, making the case for tribal sovereignty. John Ehle's *Trail of Tears: The Rise and Fall of the Cherokee Nation* (New York: Doubleday, 1988) details the heartbreaking removal of the Cherokees from their ancestral lands and their journey to reservations in Oklahoma. In *American Indian Water Rights and the Limits of Law* (Lawrence: University Press of Kansas, 1991), Lloyd Burton argues that, from the 1908 Winters decision that gave Indians rights of first appropriation to water on their reservations up through the 1960s, the courts have upheld Indian water rights, while Congress and the executive have ignored or subverted them. Mark David Spence in *Dispossessing the Wilderness: Indian Removal and the Making of the National Parks* (New York: Oxford University Press, 1999) looks at the ways Indians were removed from their tribal territories that became Yellowstone, Glacier, and Yosemite National Parks, as those lands came to symbolize sublime, pristine wilderness in which humans are visitors who do not remain. In *American Indians and National Parks* (Tucson: University of Arizona Press, 1998), Robert H. Keller and Michael F. Turek look at conflicts between Indians and Park administrations over hunting and gathering rights and over the maintenance of cultural traditions, while in *Indian Country, God's Country: Native Americans and the National Parks* (Washington, D.C.: Island Press, 2000), Philip Burnham details the effects of tourism on Indians in and around the parks.

9. The Rise of Ecology

Aldo Leopold was one of the nation's foremost twentieth-century conservationists. A *Sand County Almanac and Sketches Here and There* (New York: Oxford University Press, 1949) combines reflections on his efforts to restore an abandoned, eroded farm in Wisconsin with the articulation of a land ethic. Having spent his life practicing and teaching wildlife management, Leopold gradually began to see the benefits of allowing nature to manage itself. Susan Flader's *Thinking Like a Mountain: Aldo Leopold and the Evolution of an Ecological Attitude Toward Deer, Wolves, and Forests,* (Columbia: University of Missouri Press, 1974) tells the story of Leopold's work in game management and his conversion to an ecological way of thinking.

An ecological perspective developed also through plant and animal ecology. Ronald C. Tobey's *Saving the Prairies: The Life Cycle of the Founding School of American Plant Ecology, 1895–1955* (Berkeley: University of California Press, 1981) traces the evolution of institutionalized ecology in the midwest universities of the United States, focusing on the careers of C. E. Bessey, Frederic Clements, and, in Great Britain, A. G. Tansley. Donald Worster's influential *Nature's Economy: A History of Ecological Ideas* (Cambridge: Cambridge University Press), first released in 1977 with an updated edition in 1994, takes the perspective of individuals who studied nature between the eighteenth century and the present, examines the influence of social issues on their work, and traces the history of the development of ecology as a science. *The Background of Ecology: Concept and Theory* by ecologist Robert P. McIntosh (New York: Cambridge University Press, 1985) focuses on internal developments within the field of ecology itself.

10. The Era of Environmentalism

Rachel Carson's 1962 *Silent Spring* (Boston: Houghton, Mifflin) awakened a generation of environmentalists to the dangers of pesticide use and the fallacy of believing blindly in science and technology. Her book has remained a classic as much as for its inspired use of language as for its message. Linda J. Lear's prize-winning biography, *Rachel Carson: Witness for Nature* (New York: Holt, 1997), relates Carson's early life, her work in marine biology, and her struggles to alert the world to the problems of pesticide accumulation in the food chain.

Samuel P. Hays's sequel to his history of the early conservation movement deals with environmental values in the post–World War II period. In
*Beauty, Health, and Permanence: Environmental Politics in the United
States, 1955–1985* (New York: Cambridge University Press, 1987), Hays
speculates that changes in the values of the American public after World
War II account for the rise in environmentalism during the 1960s and
1970s. In A *History of Environmental Politics Since 1945* (Pittsburgh: University of Pittsburgh Press, 2000), he details the rise of environmental politics, economics, science, and technology. John Opie's *Nature's Nation: An
Environmental History of the United States* (Ft. Worth, Texas.: Harcourt
Brace, 1998), a comprehensive overview of the development of environmental history in the United States, is especially noteworthy for its up-to-
date history of the contemporary environmental movement. Robert Gottlieb's work defines new directions for environmental history. *Forcing the
Spring: The Transformation of the American Environmental Movement*
(Washington, D.C.: Island Press, 1993) combines the field's traditional focus on wilderness preservation with a new interest in the work of labor activists who have sought to improve the urban environment.

Bibliography

1. What Is Environmental History?

"Theories of Environmental History." *Environmental Review* 11 (special issue, winter 1987): 251–305.

Balee, William, ed. *Advances in Historical Ecology*. New York: Columbia University Press, 1998.

Cronon, William. "A Place for Stories: Nature, History, and Narrative." *Journal of American History* 78 (March 1992): 1347–76.

——. "The Uses of Environmental History." *Environmental History Review* 17 (fall 1993): 1–22.

Crosby, Alfred W. "The Past and Present of Environmental History." *American Historical Review* 100(4) (October 1995): 1177–90.

Diamond, Jared. *Guns, Germs, and Steel: The Fates of Human Societies*. New York: Norton, 1997.

Dilsaver, Lary and Craig Colton. *The American Environment: Interpretations of Past Geographies*. Lanham, Md.: Rowman and Littlefield, 1992.

Hays, Samuel P. *Explorations in Environmental History*. Pittsburgh: University of Pittsburgh Press, 1998.

Hughes, J. Donald. "Ecology and Development as Narrative Themes of World History." *Environmental History Review* 19 (spring 1995): 1–16.

——. *The Face of the Earth: Environment and World History*. Armonk, N.Y.: M. E. Sharpe, 2000.

Jamieson, Duncan R. "American Environmental History." *CHOICE* 32(1) (September 1994): 49–60.

Kline, Benjamin. *First Along the River: A Brief History of the U.S. Environmental Movement.* 2d ed. San Francisco: Acada Books, 2000.

Leach, Melissa, and Cathy Green. "Gender and Environmental History: From Representation of Women and Nature to Gender Analysis of Ecology and Politics. *Environment and History* 3 (October 1997): 343–70.

Leibhardt, Barbara. "Interpretation and Causal Analysis: Theories in Environmental History." *Environmental Review* 12(1) (1988): 23–36.

Lewis, Chris H. "Telling Stories About the Future: Environmental History and Apocalyptic Science." *Environmental History Review* 17 (fall 1993): 43–60.

McNeill, John. *Something New Under the Sun: An Environmental History of the Twentieth Century World.* New York: Norton, 2000.

Miller, James J. *An Environmental History of Northeast Florida.* Gainesville: University Press of Florida, 1998.

O'Connor, James. "What is Environmental History? Why Environmental History?" *Capitalism, Nature, Socialism* 8(2) (June 1997): 1–27.

Opie, John. *Nature's Nation: An Environmental History of the United States.* Fort Worth, Tex.: Harcourt Brace, 1998.

Rothman, Hal K. *The Greening of a Nation? Environmentalism in the United States Since 1945.* Fort Worth, Tex.: Harcourt Brace, 1998.

Russell, Emily Wyndham Barnett. *People and the Land Through Time: Linking Ecology and History.* New Haven: Yale University Press, 1997.

Stewart, Mart A. "Environmental History: Profile of a Developing Field." *History Teacher* 31 (May 1998): 351–68.

Whitney, Gordon G. *From Coastal Wilderness to Fruited Plain: A History of Environmental Change in Temperate North America from 1500 to the Present.* New York: Cambridge University Press, 1996.

Worster, Donald, ed. *The Ends of the Earth: Perspectives on Modern Environmental History.* Cambridge and New York: Cambridge University Press, 1988.

Worster, Donald. "Nature and the Disorder of History." *Environmental History Review* 18 (summer 1994): 1–15.

Worster, Donald, et al. "A Roundtable: Environmental History." *Journal of American History* 74(4) (March 1990): 1087–1147.

2. Anthologies and Bibliographies

Alglemeyer, Mary, and Eleanor R. Seagraves. *The Natural Environment: An Annotated Bibliography on Attitudes and Values.* Washington, D.C.: Smithsonian Institution Press, 1984.

Beegel, Susan F., Susan Shillinglaw, and Wesley N. Tiffney Jr., eds. *Steinbeck and the Environment: Interdisciplinary Approaches.* Tuscaloosa: University of Alabama Press, 1997.

Callicott, J. Baird, and Michael P. Nelson, eds. *The Great New Wilderness Debate: An Expansive Collection of Writings Defining Wilderness from John Muir to Gary Snyder.* Athens: University of Georgia Press, 1998.

Cronon, William, ed. *Uncommon Ground: Toward Reinventing Nature.* New York: Norton, 1995.

Cronon, William, George Miles, and Jay Gitlin, eds. *Under an Open Sky: Rethinking America's Western Past.* New York: Norton, 1992.

Davis, Richard C. *North American Forest History: A Guide to Archives and Manuscripts in the United States and Canada.* Santa Barbara, Calif.: Forest History Society, 1977.

———, ed. *Encyclopedia of American Forest and Conservation History.* 2 vols. New York: Macmillan, 1983.

Dilsaver, Lary M., and Craig E. Colten, eds. *The American Environment: Interpretations of Past Geographies.* Lanham, Md.: Rowman and Littlefield, 1992.

Etulain, Richard W., ed. *The American West in the Twentieth Century: A Bibliography.* Norman: University of Oklahoma Press, 1994.

Haycox, Steven W., and Mary Childers Mangusso, eds. *An Alaska Anthology: Interpreting the Past.* Seattle: University of Washington Press, 1996.

Hirt, Paul, and Dale Goble, eds. *Northwest Lands, Northwest Peoples.* Seattle: University of Washington Press, 1999.

Jackson, Wes, ed. *Meeting the Expectations of the Land: Essays in Sustainable Agriculture and Stewardship.* San Francisco: North Point Press, 1984.

Jaehn, Thomas. *The Environment in the Twentieth-Century American West: A Bibliography.* Albuquerque: Center for the American West, Department of History, University of New Mexico, 1990.

Kiy, Richard, and John D. Wirth. *Environmental Management on North America's Borders.* College Station: Texas A&M University Press, 1998.

Melosi, Martin V. *Bibliography on Urban Pollution Problems in American Cities from the Mid-nineteenth through the Mid-twentieth Centuries.* Monticello, Ill.: Vance Bibliographies, 1981.

Merchant, Carolyn. *Green versus Gold. Sources in California's Environmental History.* Washington, D.C.: Island Press, 1998.

Merchant, Carolyn, ed. *Major Problems in American Environmental History.* Lexington, Mass.: D. C. Heath, 1993.

——, ed. *Key Concepts in Critical Theory: Ecology.* Atlantic Highlands, N.J.: Humanities Press International, 1994.

Miller, Char and Hal Rothman, eds. *Out of the Woods: Essays in Environmental History.* Pittsburgh : University of Pittsburgh Press, 1997.

Meyerson, Joel, ed. *Henry David Thoreau: The Cambridge Companion.* New York: Cambridge University Press, 1995.

Nash, Roderick, ed. *American Environmentalism: Readings in Conservation History.* New York: McGraw-Hill, 1990.

Neiderheiser, Clodaugh M. *Forest History Sources of the United States and Canada; A Compilation of the Manuscript Sources of Forestry, Forest Industry, and Conservation History, Forest History Foundation.* St. Paul, Minn.: St. Paul Publications, 1956.

Papadakis, Elim. *Historical Dictionary of the Green Movement.* Lanham, Md.: Scarecrow Press, 1998.

Penna, Anthony, ed. *Nature's Bounty: Historical and Modern Environmental Perspectives.* Armonk, N.Y.: M. E. Sharpe, 1999.

Pinkett, Harold T. "Sources of American Forest and Conservation History." *Journal of Forest History* 25 (October 1981): 210–12.

Petulla, Joseph M. *American Environmental History.* 2nd ed. Columbus: Merrill, 1988.

Riebsame, William, ed. *Atlas of the New West: Portrait of a Changing Region.* New York: Norton, 1997.

Rothenberg, Marc. *The History of Science and Technology in the United States: a Critical and Selective Bibliography.* New York: Garland, 1982–1993.

Rothman, Hal K., ed. *Reopening the American West.* Tucson: University of Arizona Press, 1998.

Sherow, James E., ed. *A Sense of the American West: An Anthology of Environmental History.* Albuquerque: University of New Mexico Press: Published in cooperation with the University of New Mexico Center for the American West, 1998.

Soos, Frank, and Kesler Woodward, eds. *Under Northern Lights: Writers and Artists View the Alaskan Landscape.* Seattle: University of Washington Press, 2000.

Wadell, Craig, ed. *And No Birds Sing: Rhetorical Analyses of Rachel Carson's Silent Spring.* Carbondale, Ill.: Southern Illinois University Press, 2000.

Wall, Derek, ed. *Green History: A Reader in Environmental Literature, Philosophy and Politics.* New York: Routledge, 1994.

Wells, Edward R. *Historical Dictionary of North American Environmentalism.* Lanham, Md.: Scarecrow Press, 1997.

White, Richard and John M. Findlay. *Power and Place in the North American West.* Seattle: University of Washington Press, 1999.

Worster, Donald, ed. *American Environmentalism: The Formative Period, 1860–1915.* New York: Wiley, 1973.

3. Biographies and Autobiographical Writings

Abbey, Edward. *Desert Solitaire.* Salt Lake City: Peregrine Smith, Inc., 1981.

Albright, Horace M., and Frank A. Taylor. "How We Saved the Big Trees." *Saturday Evening Post* 225 (February 1953): 31–32, 107–108.

Audubon, John James. *Delineations of American Scenery and Character.* New York: G. A. Baker, 1926.

Austin, Mary, and John Muir. *Writing the Western Landscape.* Edited by Ann H. Zwinger. Boston: Beacon Press, 1994.

Backes, David. *A Wilderness Within: The Life of Sigurd F. Olson.* Minneapolis: University of Minnesota Press, 1997.

Black Elk (Oglala Lakota). *Black Elk Speaks.* Edited by John Neihardt. New York: Morrow, 1932.

Bonta, Marcia Myers, ed. *American Women Afield: Writings by Pioneering Women Naturalists.* College Station: Texas A&M University Press, 1995.

Bradford, William. *Of Plimoth Plantation.* Boston: Wright and Potter, 1901.

Breton, Mary Joy. *Women Pioneers for the Environment.* Boston: Northeastern University Press, 1998.

Burroughs, John, and John Muir, et al. *Alaska, the Harriman Expedition, 1899.* New York: Dover Publications, 1986.

Carson, Rachel. *Lost Woods: The Discovered Writing of Rachel Carson.* Edited and with an introduction by Linda Lear. Boston: Beacon Press, 1998.

Cather, Willa. *My Antonia,* 1918.

Cather, Willa. *O Pioneers,* 1913.

Chief Standing Bear. *Land of the Spotted Eagle.* Boston: Houghton Mifflin, 1933.

Clappe, Louise Amelia. *The Shirley Letters, Being Letters Written in 1851–1852 from the California Mines.* Salt Lake City: Peregrine Smith Books, 1985.

Crèvecoeur, J. Hector St. John de. *Letters from an American Farmer* [1782]. New York: E.P. Dutton, 1957.

Davies, Gilbert W., and Florice M. Frank, eds. *Forest Service Memories: Past Lives and Times in the United States Forest Service.* Hat Creek, Calif.: History Ink Books, 1997.

DeSanto, Jerry. "Foundation for a Park: Explorer and Geologist Bailey Willis in the Area of Glacier National Park." *Forest and Conservation History* 39(3) (July 1995).

Dorman, Robert L. *A Word for Nature: Four Pioneering Environmental Advocates, 1845–1913.* Chapel Hill: University of North Carolina Press, 1998.

Douglas, David. *The Oregon Journals of David Douglas: Of His Travels and Adventures Among the Traders and Indians in the Columbia, Willamette and Snake River Regions During the Years 1825, 1826, and 1827.* Ashland: Oregon Book Society, 1972.

Eliot, Charles William. *Charles Eliot, Landscape Architect* [1902]. With a new introduction by Keith N. Morgan. University of Massachusetts Press in association with the Library of American Landscape History, 1999.

Farquhar, Francis, ed. *Up and Down California in 1860–1864: The Journal of William H. Brewer.* Berkeley: University of California Press, 1974.

Fields, Leslie Leyland. *The Entangling Net: Alaska's Commercial Fishing Women Tell Their Lives.* Urbana: University of Illinois Press, 1997.

Flader, Susan. *Thinking like a Mountain: Aldo Leopold and the Evolution of an Ecological Attitude Toward Deer, Wolves, and Forests,* 1974.

Fortuine, Robert, ed. *The Alaska Diary of Adelbert von Chamisso, Naturalist on the Kotzebue Voyage, 1815–1818.* Anchorage, Alaska: Cook Inlet Historical Society, 1986.

Fox, Steven. *John Muir and His Legacy: The American Conservation Movement*. Boston: Little, Brown, 1981.

Frazier, Ian. *Great Plains*. New York: Penguin Books, 1989.

Frost, O. W., ed. *Bering and Chirikov: The American Voyages and Their Impact*. Anchorage: Alaska Historical Society, 1992.

Grinnell, George Bird. *Alaska 1899: Essays from the Harriman Expedition*. Seattle: University of Washington Press, 1995.

Guthrie, A. B. *Big Sky, Fair Land: The Environmental Essays of A. B. Guthrie*. Flagstaff, Ariz.: Northland Press, 1988.

Hamerstrom, Frances. *My Double Life: Memoirs of a Naturalist*. Madison: University of Wisconsin Press, 1994.

Holmes, Steven J. *The Young John Muir: An Environmental Biography*. Madison: University of Wisconsin Press, 1999.

Hutchings, James Mason. *Scenes of Wonder and Curiousity in California. Illustrated with Over One Hundred Engravings. A Tourist's Guide to the Yo-Semite Valley*. New York: A. Roman, 1872.

James, Edwin. *Account of an Expedition from Pittsburgh to the Rocky Mountains, Under the Command of Major Stephen H. Long, from the Notes of Major Long* [1822–23]. Barre, Mass.: Imprint Society, 1972.

John, Betty, ed. *Libby: The Alaskan Diaries and Letters of Libby Beaman, 1879–1880*. Boston: Houghton Mifflin, 1989.

Jones, Holway R. *John Muir and the Sierra Club: The Battle for Yosemite*. San Francisco: Sierra Club, 1965.

Kemble, Fanny. *Journal of a Residence on a Georgian Plantation in 1838–1839* [1863]. Edited by John A. Scott. New York: Knopf, 1961.

King, Clarence. *Mountaineering in the Sierra Nevada* [1872]. Edited by Francis P. Farquhar. New York: Norton, 1935.

Kowsky, Francis R. *Country, Park and City: The Architecture and Life of Calvert Vaux*. New York: Oxford University Press, 1998.

Langford, Nathaniel Pitt. *The Discovery of Yellowstone Park; Journal of the Washburn Expedition to the Yellowstone and Firehole Rivers in the Year 1870*. Foreword by Aubrey L. Haines. Lincoln: University of Nebraska Press, 1972 [1905].

Lear, Linda J. *Harold L. Ickes: The Aggressive Progressive, 1874–1933*. New York: Garland, 1981.

Lear, Linda J. *Rachel Carson: Witness for Nature*. New York: Holt, 1997.

Lewis, Meriwether. *The Journals of Lewis and Clark*. Edited by Frank Bergon. New York: Viking, 1989.

Lind, Anna M. "Women in Early Logging Camps: A Personal Reminiscence." *Journal of Forest History* 19 (July 1975): 128–35.

Lowenthal, David. *George Perkins Marsh: Prophet of Conservation.* Seattle: University of Washington Press, 2000.

Meine, Curt. *Aldo Leopold: His Life and Work.* Madison: University of Wisconsin Press, 1987.

Mercier, Francois Xavier. *Recollections of the Yukon: Memoirs from the Years 1868–1885.* Translated by Linda Finn Yarborough. Anchorage: Alaska Historical Society, 1986.

Miers, Earl Schenck. *Vitus Bering and James Cook Discover Alaska and Hawaii.* Newark, Del.: Published for the Friends of the Curtis Paper Company, 1960.

Miller, Peter. "John Wesley Powell: Vision for the West." *National Geographic* 185(4) (April 1994): 86–114.

Momaday, N. Scott. *The Way to Rainy Mountain.* Albuquerque: University of New Mexico Press, 1969.

Muir, John. *Steep Trails: California, Utah, Nevada, Washington, Oregon, the Grand Canyon.* Boston: Houghton Mifflin, 1918.

Muir, John, ed. *Picturesque California and the Region West of the Rocky Mountains, from Alaska to Mexico.* 2 vols. New York: J. Dewing, 1888.

Olmsted, Frederick Law. *The Slave States, Before the Civil War* [1861]. Edited by Henry Wish. New York: Capricorn, 1959.

——. *Landscape Into Cityscape: Frederick Law Olmsted's Plans for a Greater New York City.* Edited by Albert Fein. Ithaca, N.Y.: Cornell University Press, 1968.

——. *Creating Central Park, 1857–1861.* Edited by Charles E. Beveridge and David Schuyler. Vol. 3 of *The Papers of Frederick Law Olmsted* (1977, 1992). Baltimore, Md.: Johns Hopkins University Press, 1983.

——. *The California Frontier, 1863–1865.* Edited by Victoria Post Ranney et al. Vol. 5 of *The Papers of Frederick Law Olmsted* (1977, 1992). Baltimore: Johns Hopkins University Press, 1990.

Parsons, James J. "A Geographer Looks at the San Joaquin Valley." *Geographical Review* 76(4) (1986): 371–89.

Peyser, Joseph L., ed. *Letters from New France: The Upper Country, 1686–1783.* Urbana: University of Illinois Press, 1992.

Pinchot, Gifford. *Breaking New Ground.* New York: Harcourt Brace, 1947.

Powell, John Wesley. *The Exploration of the Colorado River and Its Canyons.* Edited with an introduction by Wallace Stegner, 1987.

Redondo, Margaret Proctor. "Valley of Iron: One Family's History of Madera Canyon." *Journal of Arizona History* 34 (autumn 1993): 233–74.

Roosevelt, Theodore. *Theodore Roosevelt's America: Selections from the Writings of the Oyster Bay Naturalist.* Edited by Farida A. Wiley. New York: Devin-Adair, 1955.

Rotella, Carlo. "Travels in a Subjective West: The Letters of Edwin James and Major Stephen Long's Scientific Expedition, 1819–1820." *Montana the Magazine of Western History* 41 (autumn 1991): 20–34.

Royce, Sarah. *A Frontier Lady: Recollections of the Gold Rush and Early California* [1932]. Lincoln: University of Nebraska Press, 1977.

Sauer, Carl O. *Land and Life: A Selection from the Writings of Carl Ortwin Sauer.* Edited by John Leighly. Berkeley: University of California Press, 1963.

Seton, Ernest Thompson. *Ernest Thompson Seton's America: Selections from the Writings of the Artist-Naturalist.* Edited by Farida A. Wiley. New York: Devin-Adair, 1954.

Silko, Leslie Marmon. *Almanac of the Dead: A Novel.* New York: Simon and Schuster, 1991.

Sobrero, Maria Carolina Isabella Luigia. *An Italian Baroness in Hawai'i: The Travel Diary of Gina Sobrero, Bride of Robert Wilcox, 1887.* Translated by Edgar C. Knowlton, 1887.

Spaulding, Jonathan. *Ansel Adams and the American Landscape: A Biography.* Berkeley: University of California Press, 1992.

Slate, Frederick. *Biographical Memoir of Eugene Woldemar Hilgard, 1833–1916.* Washington, D.C.: National Academy of Sciences, 1919.

Stegner, Wallace. *The American West as Living Space.* Ann Arbor: University of Michigan Press, 1987.

Stratton, David H. *Tempest Over Teapot Dome: The Story of Albert B. Fall.* Norman: University of Oklahoma Press, 1998.

Strong, Douglas H. *Dreamers and Defenders: American Conservationists.* Lincoln: University of Nebraska Press, 1984.

Taylor, David, ed. *South Carolina Naturalists: An Anthology, 1700–1860.* Columbia: University of South Carolina Press, 1998.

Thoreau, Henry David. *Walden.* Boston: Ticknor and Fields, 1854.

Todd, John. *The Sunset Land; or the Great Pacific Slope.* Boston: Lee and Shepard, 1870.

Twain, Mark. *Roughing It.* New York: Literary Classics of the United States, 1984.

Twining, Charles E. *George S. Long: Timber Statesman*. Seattle: University of Washington Press, 1994.

———. *F. K. Weyerhaeuser: A Biography*. St. Paul: Minnesota Historical Society Press, 1997.

Unger, Douglas. *Leaving the Land*. New York: Ballantine, 1984.

Van Stone, James W., ed. *Russian Exploration in Southwest Alaska: The Travel Journals of Peter Korsakovskiy (1818) and Ivan Ya. Vasilev (1829)*. Translated by David H. Kraus. Fairbanks: University of Alaska Press, 1988.

Winks, Robin W. *Laurance S. Rockefeller: Catalyst for Conservation*. Washington, D.C.: Island Press, 1997.

Welsh, Stanley L. *John Charles Frémont: Botanical Explorer*. St. Louis: Missouri Botanical Garden Press, 1998.

Worster, Donald. *A River Running West: The Life of John Wesley Powell*. New York: Oxford, 2001.

Wyss, Max, Robert Y. Koyanagi, and Doak C. Cox, eds. *The Lyman Hawaiian Earthquake Diary, 1833–1917*. Washington, D.C.: Government Printing Office, 1992.

Yochelson, Ellis L. *Charles Doolittle Walcott, Paleontologist*. Kent, Ohio: Kent State University Press, 1998.

4. African Americans and the Environment

General

Bullard, Robert D. *Dumping in Dixie: Race, Class, and Environmental Quality*, 1990.

Bryant, Bunyon, ed. *Environmental Justice: Issues, Policies, and Solutions*. Washington, D.C.: Island Press, 1995.

Hurley, Andrew. "The Social Bases of Environmental Change in Gary, Indiana, 1945–1980." *Environmental Review* 12 (1988): 1–19.

———. *Environmental Inequalities: Class, Race, and Industrial Pollution in Gary, Indiana, 1945–1980*. Chapel Hill: University of North Carolina Press, 1995.

Mowrey, Marc, and Tim Redmond. *Not in Our Backyard: The People and Events That Shaped America's Modern Environmental Movement*. New York: Morrow, 1993.

Smith, E. Valerie. "The Black Corps of Engineers and the Construction of the Alaska (ALCAN) Highway." *Negro History Bulletin* 51(1) (December 1993): 22–38.

Szasz, Andrew. *EcoPopulism: Toxic Waste and the Movement for Environmental Justice, Social Movements, Protest, and Contention.* Vol. 1. Minneapolis: University of Minnesota Press, 1994.

Takaki, Ronald T. *Iron Cages: Race and Culture in Nineteenth-Century America.* New York: Knopf, 1979.

Weber, Devra. *Dark Sweat, White Gold: California Farm Workers, Cotton, and the New Deal.* Berkeley: University of California Press, 1994.

Slavery and Agriculture

Aiken, Charles S. *The Cotton Plantation South Since the Civil War.* Baltimore: Johns Hopkins University Press, 1998.

Bennett, Hugh H. *The Soils and Agriculture of the Southern States.* New York: Macmillan, 1921.

Brandfon, Robert L. *Cotton Kingdom of the New South: A History of the Yazoo Mississippi Delta from Reconstruction to the Twentieth Century.* Cambridge: Harvard University Press, 1967.

Breen, T. H. *Tobacco Culture: The Mentality of the Great Tidewater Planters on the Eve of Revolution.* Princeton: Princeton University Press, 1985.

Carr, Lois Green, Russell R. Menard, and Lorena S. Walsh. *Robert Cole's World: Agriculture and Society in Early Maryland.* Chapel Hill: Published by the University of North Carolina Press for the Institute of Early American History and Culture (Williamsburg, Va.), 1991.

Clemens, Paul G. E. *The Atlantic Economy and Colonial Maryland's Eastern Shore: From Tobacco to Grain.* Ithaca: Cornell University Press, 1980.

Cowdrey, Albert E. *This Land, This South: An Environmental History.* Lexington: University of Kentucky Press, 1983.

Craven, Alvery O. *Soil Exhaustion as a Factor in the Agricultural History of Virginia and Maryland, 1606–1860.* Urbana: University of Illinois, 1926.

Daniel, Pete. *Breaking the Land: The Transformation of Cotton, Tobacco, and Rice Cultures Since 1880.* Urbana: University of Illinois Press, 1985.

Genovese, Eugene D. *The Political Economy of Slavery: Studies in the Economy and Society of the Old South.* New York: Pantheon Books, 1967.

Gray, Lewis C. *History of Agriculture in the Southern United States to 1860.* Gloucester, Mass.: Peter Smith, 1958.

Gregg, William. *Essays on Domestic Industry: or, An Inquiry into the Expediency of Establishing Cotton Manufactures in South-Carolina.* Charleston, S.C.: Burges and James, 1845.

Hilgard, E. W. *General Discussion of the Cotton Production of the United States.* Washington, D.C.: U.S. Census Office, 1884.

Hunter, W. D. "The Boll-weevil Problem." *Farmers' Bulletin,* no. 1329. Washington, D.C.: U.S. Department of Agriculture, 1923.

Hurt, R. Douglas. *Agriculture and Slavery in Missouri's Little Dixie.* Columbia: University of Missouri Press, 1992.

Jefferson, Thomas. *Notes on the State of Virgina.* London: J. Stockdale, 1787.

Kulikoff, Allan. *Tobacco and Slaves: The Development of Southern Cultures in the Chesapeake, 1680–1800.* Chapel Hill: Published by the University of North Carolina Press for the Institute of Early American History and Culture (Williamsburg, Va.), 1986.

Littlefield, Daniel. *Rice and Slaves: Ethnicity and the Slave Trade in Colonial South Carolina.* Baton Rouge: Louisiana State University Press, 1981.

Menard, Russell R. "The Tobacco Industry in the Chesapeake Colonies, 1617–1730: An Interpretation." *Research in Economic History* 5 (1980): 109–77.

Moore, John Hebron. *The Emergence of the Cotton Kingdom in the Old Southwest: Mississippi, 1770–1860.* Baton Rouge: Louisiana State University Press, 1988.

Orton, Thomas H. "The New Water Management Era and the Return of Southwest Cotton to the Old South." *Agricultural History* 66(2) (spring 1992): 307–27.

Otto, John Solomon. *The Southern Frontiers, 1607–1860: The Agricultural Evolution of the Colonial and Antebellum South.* New York: Greenwood Press, 1989.

Reidy, Joseph P. *From Slavery to Agrarian Capitalism in the Cotton Plantation South: Central Georgia, 1800–1880.* Chapel Hill: University of North Carolina Press, 1992.

Russell, Dick. "Environmental Racism: Minority Communities and Their Battle Against Toxics." *The Amicus Journal* 11 (spring 1989): 22–32.

Seabrook, Whitemarsh Benjamin. *A Memoir on the Origin, Cultivation and Uses of Cotton, from the Earliest Ages to the Present Time, with Especial Reference to the Sea-Island Cotton Plant, Including the Improvements in Its Cultivation.* Charleston, N.C.: Printed by Miller and Browne, 1844.

Silver, Timothy. *A New Face on the Countryside: Indians, Colonists, and Slaves in the South Atlantic Forests, 1500–1800*. New York: Cambridge University Press, 1990.

Stewart, Mart A. *"What Nature Suffers to Groe": Life, Labor, and Landscape on the Georgia Coast, 1680–1920*. Athens: University of Georgia Press, 1996.

Wolfenbarger, D. A., L. D. Hatfield, and E. V. Gage, eds. *The Tobacco Budworm and Bollworm in Cotton in the Mid-South, Southwestern United States and Mexico*. Dallas: Southwestern Entomological Society, 1991.

5. American Indian Land Use

Source Materials

Berkhofer, Robert F. *The American Indian: Essays from the Pacific Historical Review*. Santa Barbara, Calif.: Clio Books, 1974.

Dobyns, Henry F. *Their Number Become Thinned: Native American Population Dynamics in Eastern North America*. Knoxville: University of Tennessee Press in cooperation with the Newberry Library Center for the History of the American Indian, 1983.

Dobyns, Henry F., et al. *Indians of the Southwest: A Critical Bibliography*. Bloomington, Ill.: Published by Indiana University Press for the Newberry Library.

Edwards, Everett E. *A Bibliography on the Agriculture of the American Indians*. Washington, D.C.: U.S. Department of Agriculture, 1942.

Heizer, Robert F., Karen Nissen, and Edward Castillo. *California Indian History, A Classified and Annotated Guide to Source Materials*. Ramona, Calif.: Ballena Press, 1975.

Heizer, Robert F., and T. K. Whipple, eds. *The California Indians: A Source Book*. Berkeley: University of California Press, 1971.

Kroeber, A. L. *Handbook of the Indians of California*. Washington, D.C.: Government Printing Office; Smithsonian Institute Bureau of American Ethnology, 1925.

Salisbury, Neal. *The Indians of New England: A Critical Bibliography*. Bloomington: Published by Indiana University Press for the Newberry Library, 1982.

White, Richard. *"It's Your Misfortune and None of My Own": A New History of the American West*. Norman: University of Oklahoma Press, 1991.

Land Use

Anderson, M. Kat, Michael G. Barbour, and Valerie Whitworth. "A World of Balance and Plenty: Land, Plants, Animals, and Humans in a Pre-European California." *California History* 76 (summer and fall 1997): 12–47.

Baker, Emerson W. "'A Scratch with a Bear's Paw': Anglo-Indian Land Deeds in Early Maine." *Ethnohistory* 36 (summer 1989): 235–56.

Beck, David R.M. "The Importance of Sturgeon in Menominee Indian History." *Wisconsin Magazine of History* 79 (autumn 1995): 32–48.

Beckham, Stephen Dow. *The Indians of Western Oregon: This Land Was Theirs.* Coos Bay, Ore.: Arago Books, 1977.

Berry, Kate A. "Race for Water? Native Americans, Eurocentrism, and Western Water Policy." In *Environmental Injustices, Political Struggles: Race, Class, and the Environment,* edited by David E. Camacho. Durham, N.C.: Duke University Press, 1998.

Bilharz, Joy Ann. *The Allegany Senecas and Kinzua Dam: Forced Relocation Through Two Generations.* Lincoln: University of Nebraska Press, 1998.

Blackburn, T., and K. Anderson, eds. *Before the Wilderness: Environmental Management by Native Californians.* Menlo Park, Calif.: Ballene Press, 1993.

Boxberger, Daniel L. *To Fish in Common: The Ethnohistory of Lummi Indian Salmon Fishing.* Seattle: University of Washington Press, 1999.

Boyd, Robert. *The Coming of the Spirit of Pestilence: Introduced Infectious Diseases and Population Decline Among Northwest Indians, 1774–1874.* Seattle: University of Washington Press, 1999.

———, ed. *Indians, Fire, and the Land in the Pacific Northwest.* Corvallis: Oregon State University Press, 1999.

Brandenburg, Jim, et al. "The Land They Knew: A Portfolio" (1491: America Before Columbus). *National Geographic* 180(4) (October 1991): 14–100.

Brodeur, Paul. *Restitution, the Land Claims of the Mashpee, Passamaquoddy, and Penobscot Indians of New England.* Boston: Northeastern University Press, 1985.

Brown, Jennifer S.H., W.J. Eccles, and Donald P. Heldman, eds. *The Fur Trade Revisited: Selected Papers of the Sixth North American Fur Trade Conference, Mackinac Island, Michigan, 1991.* East Lansing: Michigan State University Press, 1994.

Brown, Kenneth. *Four Corners: History, Land, and People of the Desert Southwest*. New York: Harper Collins Publishers, 1995.

Brugge, David M. *The Navajo-Hopi Land Dispute: An American Tragedy*. Albuquerque: University of New Mexico Press, 1994.

Burnham, Philip. *Indian Country, God's Country: Native Americans and the National Parks*. Washington, D.C. Island Press, 2000.

Burton, Lloyd. *American Indian Water Rights and the Limits of Law*. Lawrence: University Press of Kansas, 1991.

Carlos, Ann M., and Frank D. Lewis. "Indians, the Beaver, and the Bay: The Economics of Depletion in the Lands of the Hudson's Bay Company, 1700–1763." *Journal of Economic History* 53(3) (September 1993): 465–95.

Carlson, Paul H. "Indian Agriculture, Changing Subsistence Patterns, and the Environment on the Southern Great Plains." *Agricultural History* 66(2) (spring 1992): 52–60.

Carlson, Richard G., ed. *Rooted Like the Ash Trees: New England Indians and the Land*. Naugatuck, Conn.: Eagle Wing Press, 1987.

Catton, Theodore. *Inhabited Wilderness: Indians, Eskimos, and National Parks in Alaska*. Albuquerque: University of New Mexico Press, 1997.

Connor, Sheila. *New England Natives: A Celebration of People and Trees*. Cambridge, Massachusetts: Harvard University Press, 1994.

Coombs, G., and F. Plog. "The Conversion of the Chumash Indians: An Ecological Interpretation." *Human Ecology* 5 (1977): 309–28.

Couro, Ted. *San Diego County Indians as Farmers and Wage Earners*. Pomona, Calif.: Ballena Press, 1975.

Cronon, William. *Changes in the Land: Indians, Colonists, and the Ecology of New England*. New York: Hill and Wang, 1983.

Day, Gordon M. "The Indian as an Ecological Factor in the Northeastern Forests." *Ecology* 34 (April 1953): 329–46.

Ehle, John. *Trail of Tears: The Rise and Fall of the Cherokee Nation*. New York: Doubleday, 1988.

Fausz, Frederick. "Profits, Pelts, and Power: English Culture in the Early Chesapeake, 1620–1652." *Maryland Historian* 14 (fall/winter 1983): 14–30.

Fish, Suzanne K., and Paul R. Fish. "Prehistoric Desert Farmers of the Southwest." *Annual Review of Anthropology* 23 (annual 1994): 83–109.

Fletcher, Thomas C. *Paiute, Prospector, Pioneer: The Bodie-Mono Lake Area in the Nineteenth Century*. Lee Vining, Calif.: Artemisia Press, 1987.

Gates, Paul Wallace. *The Rape of Indian Lands*. New York: Arno Press, 1979.

Gilbert, Joan. *The Trail of Tears Across Missouri*. Columbia: University of Missouri Press, 1996.

Gobalet, K, W., and T. L. Jones. "Prehistoric Native American Fisheries of the Central California Coast." *Transactions of the American Fisheries Society* 124(6) (November 1995): 813–23.

Gould, R.A. "Ecology and Adaptive Response Among the Tolowa Indians of Northwestern California." *Journal of California Anthropology* 2 (1975): 148–70.

Grant, Frank R., et al. *The Forests of Anishinabe: A History of Minnesota Chippewa Tribal Forestry, 1854–1991*. Missoula, Mont.: U.S. Department of the Interior, Bureau of Indian Affairs, Branch of Forestry, 1992.

Hack, John Tilton. *The Changing Physical Environment of the Hopi Indians of Arizona*. Millwood, N.Y.: Kraus Reprint, 1974.

Hauptman, Laurence M. "The Iroquois Indians and the Rise of the Empire State: Ditches, Defense, and Dispossession." *New York History* 79 (October 1998): 325–58.

Haycox, Stephen W. "Economic Development and Indian Land Rights in Modern Alaska: the 1947 Tongass Timber Act." *Western Historical Quarterly* 21(1) (February 1990): 21–47.

Heizer, Robert F., and Albert B. Elsasser. *The Natural World of the California Indians*. Berkeley: University of California Press, 1980.

Holland Braund, Kathryn E. *Deerskins and Duffels: Creek Indian Trade with Anglo-America, 1685–1815*. Lincoln: University of Nebraska Press, 1993.

Hughes, J. Donald. *American Indian Ecology*. El Paso: Texas Western Press, 1983.

Hurt, R. Douglas. *Indian Agriculture in America: Prehistory to the Present*. Lawrence: University Press of Kansas, 1987.

Isenberg, Andrew Christian. "Indians, Whites, and the Buffalo: An Ecological History of the Great Plains, 1750–1900." Northwestern University, 1993.

Jennings, Francis. *The Invasion of America: Indians, Colonialism, and the Cant of Conquest*. Chapel Hill: Published by the University of North Carolina Press for the Institute of Early American History and Culture (Williamsburg, Va.), 1975.

Keller, Robert H. and Michael F. Turek. *American Indians and National Parks*. Tucson: University of Arizona Press, 1998.

Kame'eleihiwa, Lilikala. *Native Land and Foreign Desires: A History of Land Tenure Change in Hawai'i from Traditional Times Until the 1848 Mahele*. Honolulu: Bishop Museum Press, 1992.

Krech, Shepard, III. *The Ecological Indian: Myth and History.* New York: Norton, 1999.

——, ed. *Indians, Animals, and the Fur Trade: A Critique of Keepers of the Game.* Athens: University of Georgia Press, 1981.

——, ed. *The Subarctic Fur Trade: Native Social and Economic Adaptations.* Vancouver: University of British Columbia Press, 1984.

Lewis, David Rich. "Changing Subsistence, Changing Reservation Environments: The Hupa, 1850–1980s." *Agricultural History* 66 (spring 1992): 34–51.

——. "Still Native: The Significance of Native Americans in the History of the Twentieth-Century American West." *Western Historical Quarterly* 24(2) (May 1993): 203–228.

——. *Neither Wolf Nor Dog: American Indians, Environment, and Agrarian Change.* New York: Oxford University Press, 1994.

Lewis, G. Malcolm, ed. *Cartographic Encounters: Perspectives on Native American Mapmaking and Map Use.* Chicago: University of Chicago Press, 1998.

Lewis, Henry T. *Patterns of Indian Burning in California: Ecology and Ethnohistory,* 1973.

Luther Standing Bear (Sioux). *Land of the Spotted Eagle.* Edited by E.A. Brininstool. Boston: Houghton Mifflin, 1933.

MacLeish, William H. *The Day Before America: Changing the Nature of a Continent.* Boston: Houghton Mifflin, 1994.

Margolin, Malcolm. *The Ohlone Way.* Berkeley: Heydey, 1978.

Marsden, Susan, and Robert Galois. "The Tsimshian, the Hudson's Bay Company, and the Geopolitics of the Northwest Coast Fur Trade, 1787–1840." *Canadian Geographer* 39(2) (summer 1995): 169–84.

Martin, Calvin. *Keepers of the Game: Indian-Animal Relationships and the Fur Trade.* Berkeley: University of California Press, 1978.

Mathien, Frances Joan, ed. *Environment and Subsistence of Chaco Canyon, New Mexico,* Publications in Archaeology. Albuquerque, N.M.: National Park Service, U.S. Department of the Interior, 1985.

McEvoy, Arthur. *The Fisherman's Problem: Ecology and Law in the California Fisheries.* New York: Cambridge University Press, 1986.

McGuire, Thomas R., William B. Lord, and Mary G. Wallace, eds. *Indian Water in the New West.* Tucson: University of Arizona Press, 1993.

Merchant, Carolyn. *Ecological Revolutions: Nature, Gender, and Science in New England.* Chapel Hill: University of North Carolina Press, 1989.

Merrell, James. *The Indians' New World: Catawbas and Their Neighbors from European Contact Through the Era of Removal*. New York: Norton, 1989.

Meyer, Melissa L. "'We Can Not Get a Living as We Used To': Dispossession and the White Earth Anishinaabeg, 1889–1920." *American Historical Review* 96 (April 1991): 368–94.

———. *The White Earth Tragedy: Ethnicity and Dispossession at a Minnesota Anishinaabe Reservation, 1889–1920*. Lincoln: University of Nebraska Press, 1994.

Nabhan, Gary Paul. *Enduring Seeds: Native American Agriculture and Wild Plant Conservation*. San Francisco: North Point Press, 1989.

Nelson, Richard K. *Make Prayers to the Raven: A Koyukon View of the Northern Forest*. Chicago: University of Chicago Press, 1983.

Ray, Arthur J. *Indians in the Fur Trade: Their Role as Trappers, Hunters, and Middlemen in the Lands Southwest of Hudson Bay, 1660–1870*. Toronto: University of Toronto Press, 1974.

Reno, Philip. *Mother Earth, Father Sky, and Economic Development: Navajo Resources and Their Use*. Albuquerque: University of New Mexico Press, 1981.

Rothman, Hal. *Navajo National Monument: A Place and its People: An Administrative history*. Santa Fe, N.M. : Southwest Regional Office, Division of History, 1991.

Russell, Emily W.B. "Indian-Set Fires in the Forests of the Northeastern United States." *Ecology* 64 (February 1983): 77–88.

Satz, Ronald N. "Chippewa Treaty Rights: The Reserved Rights of Wisconsin's Chippewa Indians in Historical Perspective." *Transactions* 79(1) (1991): xix–25.

Sawa, Martin. *Land Use Planning on California Indian Reservations*. Monticello, Ill.: Vance Bibliographies, 1978.

Silver, Timothy. *A New Face on the Countryside: Indians, Colonists, and Slaves in the South Atlantic Forests, 1500–1800*. Cambridge: Cambridge University Press, 1990.

Smiley, H. D. "The Fur Trade in Retrospect." *Journal of the West* 30 (October 1991): 52–63.

Solnit, Rebecca. *Savage Dreams: A Journey into the Hidden Wars of the American West*. San Francisco: Sierra Club Books, 1994.

Spence, Mark. "Dispossessing the Wilderness: Yosemite Indians and the National Park Ideal, 1864–1930." *Pacific Historical Review* 65 (February 1996): 27–59.

Spence, Mark David. *Dispossessing the Wilderness: Indian Removal and the Making of the National Parks*. New York: Oxford University Press, 1999.

Spicer, Edward Holland. *Cycles of Conquest: The Impact of Spain, Mexico, and the United States on the Indians of the Southwest, 1533–1960.* Tucson: University of Arizona Press, 1962.

Stanford, Dennis J., and Jane S. Day, eds. *Ice Age Hunters of the Rockies.* Denver: Denver Museum of Natural History, 1992.

Sunder, John E. *The Fur Trade on the Upper Missouri, 1840–1865.* Norman: University of Oklahoma Press, 1993.

Vecsey, Christopher, and William A. Starna, eds. *Iroquois Land Claims.* Syracuse, N.Y.: Syracuse University Press, 1988.

Ward, R. C. "The Spirits Will Leave: Preventing the Desecration and Destruction of Native American Sacred Sites on Federal Land." *Ecology Law Quarterly* 19(4) (1992): 795–846.

West, Elliott. *The Contested Plains: Indians, Goldseekers, and the Rush to Colorado.* Lawrence: University of Kansas Press, 1998.

White, Richard. *The Roots of Dependency: Subsistence, Environment, and Social Change Among the Choctaws, Pawnees, and Navajos.* Lincoln: University of Nebraska Press, 1983.

———. "Native Americans and the Environment." In *Scholars and the Indian Experience,* edited by W. R. Swagerty. Bloomington: Indiana University Press, 1984, pp. 179–204.

———. *The Middle Ground: Indians, Empires, and Republics in the Great Lakes Region, 1650–1815.* Cambridge: Cambridge University Press, 1991.

Wills, Wirt Henry. *Early Prehistoric Agriculture in the American Southwest.* Santa Fe, N.M.: School of American Research Press, 1988.

Wood, W. Raymond. "An Introduction to the History of the Fur Trade on the Northern Plains." *North Dakota History* 61 (summer 1994): 2–6.

6. American Indian Religion

Adamson, Joni. *American Indian Literature, Environmental Justice, and Ecocriticism: The Middle Place.* Tucson: University of Arizona Press, 2001.

Albanese, Catherine. *Nature Religion in America: From the Algonkian Indians to the New Age.* Chicago: University of Chicago Press, 1990.

Anderson, Douglas D. "Sailing an Ancient Landbridge: A Journey into Eskimo Prehistory, from Alaska to the Russian Far East." *Archaeology* 47(4) (July/August 1994): 50–54.

Arden, Harvey. *Travels in a Stone Canoe: The Return to the Wisdomkeepers.* New York: Simon and Schuster, 1998.

Atkinson, Mary Jourdan. *Indians of the Southwest*. 4th ed. San Antonio, Tex.: Naylor, 1963.

Baldwin, Gordon Curtis. *The Apache Indians: Raiders of the Southwest*. New York: Four Winds Press, 1978.

Bandelier, Adolph Francis Alphonse. *A History of the Southwest: A Study of the Civilization and Conversion of the Indians in Southwestern United States and Northwestern Mexico from the Earliest Times to 17—*. Citta del Vaticano: Biblioteca Apostolica Vaticana, 1969.

Benson, Henry C. *Life Among the Choctow Indians, and Sketches of the South-west*. Cincinnati: L. Swormstedt and A. Poe, 1860.

Berkhofer, Robert F. *The White Man's Indian: Images of the American Indian from Columbus to the Present*. New York: Knopf: Random House, 1978.

Black Elk (Oglala Lakota). *Black Elk Speaks*. Edited by John Neihardt. New York: Morrow, 1932.

Bodfish, Waldo. *Kusiq: An Eskimo Life History from the Arctic Coast of Alaska*. Fairbanks: University of Alaska Press, 1991.

Booth, A., and H. Jacobs. "Ties That Bind: Native American Beliefs as a Foundation for Environmental Consciousness." *Environmental Ethics* 12 (1990): 27–43.

Brice, Wallace A. *History of Fort Wayne, from the Earliest Known Accounts of this Point, to the Present Period. Embracing an Extended View of the Aboriginal Tribes of the Northwest, Including, More Especially, the Miamies*. Fort Wayne, Ind.: D. W. Jones and Son, printers, 1868.

Brightman, Robert Alain. *Grateful Prey: Rock Cree Human-Animal Relationships*. Berkeley: University of California Press, 1993.

Brown, Joseph Epes. *Animals of the Soul: Sacred Animals of the Oglala Sioux*. Rockport, Mass.: Element, 1992.

Brown, Robin C. *Florida's First People: 12,000 Years of Human History*: Pineapple Press, 1994.

Bushnell, O. A. *The Gifts of Civilization: Germs and Genocide in Hawai'i*. Honolulu: University of Hawaii Press, 1993.

Callicott, J. Baird, and Thomas W. Overholt, eds. *Clothed-in-Fur and Other Tales: An Introduction to an Ojibwa World View*. Washington, D.C.: University Press of America, 1982.

Calloway, Colin G. *The Western Abenakis of Vermont, 1600–1800: War, Migration, and the Survival of an Indian People*. Norman: University of Oklahoma Press, 1994.

Catlin, George. *North American Indians* [1844]. Edited by Peter Matthiessen. New York: Viking, 1989.

Cheek, Lawrence E. A.D. 1250: Ancient Peoples of the Southwest. Phoenix: Arizona Highways, 1994.

Chief Standing Bear. Land of the Spotted Eagle. Boston: Houghton Mifflin, 1933.

Clark, Galen. Indians of the Yosemite Valley and Vicinity, their [sic] History, Customs and Traditions. 4th edition. Yosemite Valley, Calif.: G. Clark, 1910.

Clifton, James A., ed. The Invented Indian: Cultural Fictions and Government Policies: Transaction Publishers, 1994.

Collier, John. On the Gleaming Way: Navajos, Eastern Pueblos, Zunis, Hopis, Apaches, and Their Land. Denver, Colo.: Sage Books, 1949.

Cook, Sherburne F. The Conflict Between the California Indian and White Civilization. Berkeley and Los Angeles: University of California Press, 1943.

——. The Population of the California Indians, 1769–1970. Berkeley and Los Angeles: University of California Press, 1976.

Dale, Edward E. The Indians of the Southwest: A Century of Development Under the United States. Norman: University of Oklahoma Press, 1976.

Douglas, David. The Oregon Journals of David Douglas: Of His Travels and Adventures Among the Traders and Indians in the Columbia, Willamette and Snake River Regions During the Years 1825, 1826, and 1827. Ashland: Oregon Book Society, 1972.

Douthit, Nathan. A Guide to Oregon South Coast History: Including an Account of the Jedediah Smith Exploring Expedition of 1828 and Its Relations with the Indians. Coos Bay, Ore.: River West Books, 1986.

DuBois, Constance Goddard. The Condition of the Mission Indians of Southern California. Philadelphia: Office of the Indian Rights Association, 1901.

Dutton, Bertha P. The Pueblos. Englewood Cliffs, N.J.: Prentice-Hall, 1976.

——. The Rancheria, Ute, and Southern Paiute Peoples. Englewood Cliffs, N.J.: Prentice-Hall, 1976.

Eastman, Charles Alexander. The Soul of the Indian; An Interpretation by Charles Alexander Eastman (Ohiyesa). Boston: Houghton Mifflin, 1911.

Foreman, Grant. Indians and Pioneers: The Story of the American Southwest Before 1830. Norman: University of Oklahoma Press, 1936.

Fortuine, Robert. Chills and Fever: Health and Disease in the Early History of Alaska. Fairbanks: University of Alaska Press, 1989.

Graves, Michael W., and David J. Addison. "The Polynesian Settlement of the Hawaiian Archipelago: Integrating Models and Methods in Archaeological Interpretation." *World Archaeology* 26(3) (February 1995): 380–400.

Gutierrez, Ramon. *When Jesus Came the Corn Mothers Went Away.* Stanford: Stanford University Press, 1991.

Hall, Robert L. *An Archaeology of the Soul: North American Indian Belief and Ritual.* Urbana: University of Illinois Press, 1997.

Harmon, Alexandra. "Lines in Sand: Shifting Boundaries between Indians and Non-Indians in the Puget Sound Region." *Western Historical Quarterly* 26(4) (winter 1995): 429–54.

Hauptman, Laurence M., and James D. Wherry, eds. *The Pequots in Southern New England.* Norman: University of Oklahoma Press, 1990.

Heizer, Robert F., and Gordon W. Hewes. *Animal Ceremonialism in Central California in the Light of Archaeology.* Lancaster, Penn.: n.p., 1940.

Hines, Gustavus. *Life on the Plains of the Pacific: Oregon, Its History, Condition and Prospects, Containing a Description of the Geography, Climate and Productions, with Personal Adventures Among the Indians.* New York: Miller, Orton and Mulligan, 1857.

Jennings, Francis. *The Founders of America: How Indians Discovered the Land, Pioneered in It, and Created Great Classical Civilizations; How They Were Plunged into a Dark Age by Invasion and Conquest, and How They Are Reviving.* New York: Norton, 1993.

John, Elizabeth Ann Harper. *Storms Brewed in Other Men's Worlds: The Confrontation of Indians, Spanish, and French in the Southwest, 1540–1795.* College Station: Texas A&M University Press, 1975.

Josephy, Alvin M., Jr., ed. *America in 1492: The World of the Indian Peoples Before the Arrival of Columbus.* New York: Knopf (distributed by Random House), 1992.

Keegan, Marcia. *Mother Earth, Father Sky: Navajo and Pueblo Indians of the Southwest.* New York: Grossman, 1974.

Kirch, Patrick Vinton, and Marshall Sahlins. *Anahulu: The Anthropology of History in the Kingdom of Hawaii.* Chicago: University of Chicago Press, 1992.

Knoop, Anna Marie. *The Federal Indian Policy in the Sacramento Valley, 1846–1860.* Berkeley, Calif.: n.p., 1941.

Kroeber, Theodora. *Ishi in Two Worlds.* Berkeley: University of California Press, 1961.

Leupp, Francis Ellington. *Notes of a Summer Tour Among the Indians of the Southwest.* Philadelphia: Indian Rights Association, 1897.

Lopez, Barry. *The Rediscovery of North America.* Lexington: University Press of Kentucky, 1990.

Merchant, Carolyn. *Ecological Revolutions: Nature, Gender, and Science in New England*. Chapel Hill: University of North Carolina Press, 1989.

Momaday, N. Scott. *The Way to Rainy Mountain*. Albuquerque: University of New Mexico Press, 1969.

Moodie, D. Wayne, A. J. W. Catchpole, and Kerry Abel. "Northern Athapaskan Oral Traditions and the White River Volcano." *Ethnohistory* 39(2) (spring 1992): 148–72.

Nairne, Thomas. *Nairne's Muskhogean Journals: The 1708 Expedition to the Mississippi River*. Edited, with an introduction, by Alexander Moore. Jackson: University Press of Mississippi, 1988.

Nelson, Richard. "Understanding Eskimo Science: Traditional Hunters' Insights into the Natural World Are Worth Rediscovering." *Audubon* 95(5) (September/October 1993): 102–108.

Nobbe, George. "Native Culture: A New Discovery Rewrites the History of Alaska's Alutiq Eskimos." *Omni* 17(6) (March 1995): 28.

Nordyke, Eleanor C. *The Peopling of Hawaii*. 2d ed. Honolulu: University of Hawaii Press, 1989.

Pearce, Roy Harvey. *Savagism and Civilization: A Study of the Indian and American Mind*. Baltimore: Johns Hopkins Press, 1965.

Plenty-coups (Crow). *Plenty-coups: Chief of the Crows*. Edited by Frank B. Linderman. New York: John Day, 1930.

Powers, William R., and John F. Hoffecker. "Late Pleistocene Settlement in the Nenana Valley, Central Alaska." *American Antiquity* 54(2) (April 1989): 263–88.

Ray, Dorothy Jean. *The Eskimos of Bering Strait, 1650–1898*. Seattle: University of Washington Press, 1992.

Remy, Jules. *Contributions of a Venerable Savage to the Ancient History of the Hawaiian Islands*. Translated by William T. Brigham. Boston: A. A. Kingman, 1868.

Reynolds, Brad, and Don Doll. "Athapaskans Along the Yukon." *National Geographic* 177(2) (February 1990): 44–70.

Riccio, Thomas. "A Message from the Eagle Mother: The Messenger's Feast of the Inupiat Eskimo." *TDR* 37(1) (spring 1993): 115–47.

Salisbury, Neil. *Manitou and Providence: Indians, Europeans, and the Making of New England, 1500–1643*. New York: Oxford University Press, 1982.

Schlesier, Karl H., ed. *Plains Indians, A.D. 500–1500: The Archaeological Past of Historic Groups*. Norman: University of Oklahoma Press, 1994.

Silko, Leslie Marmon. "Landscape, History, and the Pueblo Imagination." *Antaeus* 57 (autumn 1986): 83–94.

Silko, Leslie Marmon. *Ceremony.* New York: Penguin, 1977.

Speck, Frank G. *Penobscot Man.* Philadelphia: University of Pennsylvania Press, 1940.

Stannard, David E. *Before the Horror: The Population of Hawai'i on the Eve of Western Contact.* Honolulu: Social Science Research Institute, University of Hawaii, 1989.

Tanner, Adrian. *Bringing Home Animals: Religious Ideology and Mode of Production of the Mistassini Cree Hunters.* New York: St. Martin's Press, 1979.

Trumbull, Henry. *History of the Discovery of America: Of the Landing of Our Forefathers at Plymouth, and of Their Most Remarkable Engagement with the Indians in New-England.* Norwich, Conn.: Printed by James Springer for the author, 1812.

Turner, Frederick Jackson. *The Character and Influence of the Indian Trade in Wisconsin: A Study of the Trading Post as an Institution,* Johns Hopkins University Studies in Historical and Political Science. Baltimore: Johns Hopkins Press, 1891.

Wright, Gary A. *People of the High Country: Jackson Hole Before the Settlers.* New York: P. Lang, 1984.

7. Asian Americans and the Environment

Bunje, Emil Theodore Hieronymus. *The Story of Japanese Farming in California.* Berkeley, Calif.: n.p., 1937.

California Farmers Cooperative Association. *Japanese Immigration and the Japanese in California.* San Francisco: California Farmers Co-operative Assocations, 1920.

Chan, Sucheng. *The Bittersweet Soil: The Chinese in California Agriculture, 1860–1910.* Berkeley: University of California Press, 1986.

Ch'iu, P'ing. *Chinese labor in California, 1850–1880, an Economic Study.* Madison: State Historical Society of Wisconsin for the Department of History, University of Wisconsin, 1963.

Daniels, Roger. *The Politics of Prejudice, the Anti-Japanese Movement in California, and the Struggle for Japanese Exclusion.* Berkeley: University of California Press, 1962.

Galarza, Ernesto. *Farm Workers and Agribusiness in California, 1947–1960.* Notre Dame, Ind.: University of Notre Dame Press, 1977.

Goldberg, George. *East meets West; the Story of the Chinese and Japanese in California.* New York: Harcourt Brace Jovanovich, 1970.

Hirata, Lucie Cheng. *Labor Immigration under Capitalism: Asian Workers in the United States before World War II.* Berkeley: University of California Press, 1984.

Holliday, J. S. *The World Rushed In: The California Gold Rush Experience.* New York: Simon and Schuster, 1981.

Hom, Gloria Sun, ed. *Chinese Argonauts: An Anthology of the Chinese Contributions to the Historical Development of Santa Clara County.* Los Altos Hills, Calif.: Foothill Community College, 1971.

Hundley, Norris, Jr., ed. *The Asian American: The Historical Experience.* Santa Barbara, Calif.: Clio Books, 1976.

Japanese Agricultural Association. *The Japanese Farmers in California.* San Francisco: The Japanese Agricultural Association, 1918.

Lim, Roger T. *The Chinese in San Francisco and the Mining Region of California, 1848–1858.* 1979.

Lukes, Timothy J., and Gary Y. Okihiro. *Japanese Legacy: Farming and Community Life in California's Santa Clara Valley.* Cupertino, Calif.: California History Center, De Anza College, 1985.

Lyndon, Sandy. *Chinese Gold: The Chinese in the Monterey Bay Region.* Capitola, Calif.: Capitola Book Company, 1985.

McWilliams, Carey. *Factories in the Field: The Story of Migratory Farm Labor in California.* Santa Barbara: Peregrine Publishers, 1971.

———. *Ill Fares the Land: Migrants and Migratory Labor in the United States.* Boston: Little, Brown, 1939.

———. *The California Revolution.* New York: Grossman Publishers, 1968.

Minke, Pauline. *Chinese in the Mother Lode, 1850–1870.* San Francisco: R and E Research Associates.

Minnick, Sylvia Sun. *Samfow: The San Joaquin Chinese Legacy.* Fresno, Ca.: Panorama West, 1988.

Ritter, Eric W. *The Historic Archaeology of a Chinese Mining Venture near Igo in Northern California.* Redding, Calif.: Bureau of Land Management, Ukiah District, Redding Resource Area, 1986.

Saloutos, T. "The Immigrant in Pacific Coast Agriculture, 1880–1950." *Agricultural History* 49 (1975): 182–201.

Saxton, Alexander. *The Indispensable Enemy: A Study of the Anti-Chinese Movement in California.* Berkeley: University of California Press, 1971.

Sheafer, Silvia Anne. *Chinese and the Gold Rush.* Whittier, Calif.: Journal Publications, 1979.

Sylva, Seville A. *Foreigners in the California Gold Rush*. San Francisco: R and E Research Associates, 1972.

Takaki, Ronald T. *A Different Mirror: A History of Multicultural America*. Boston: Little, Brown, 1993.

Tsurutani, Hisashi. *America-bound: The Japanese and the Opening of the American West*. Tokyo: The Japan Times, 1989.

Yen, Tzu-kuei. *Chinese Workers and the First Transcontinental Railroad of the United States of America*, 1977.

8. Environmental Philosophy and Landscape Perception

Albanese, Catherine. *Nature Religion in America: From the Algonkian Indians to the New Age*. Chicago: University of Chicago Press, 1990.

Ashworth, William. *Left Hand of Eden: Meditations on Nature and Human Nature*. Corvallis: Oregon State University Press, 1999.

Bookchin, Murray. *The Ecology of Freedom*. Palo Alto, Calif.: Cheshire Books, 1982.

Botkin, Daniel B. *Discordant Harmonies: A New Ecology for the Twenty-first Century*. New York: Oxford University Press, 1990.

——. *No Man's Garden: Thoreau and a New Vision for Civilization and Nature*. Washington, D.C.: Island Press, 2001.

Buell, Lawrence. *The Environmental Imagination: Thoreau, Nature Writing, and the Formation of American Culture*. Cambridge, Massachusetts: The Belknap Press of Harvard University Press, 1995.

Callicott, J. Baird. *Companion to a Sand County Almanac*. Madison: University of Wisconsin Press, 1987.

——. *In Defense of the Land Ethic: Essays in Environmental Philosophy*. Albany: State University of New York Press, 1989.

——. "Sustainability in Historical-Philosophical Context." *George Wright Forum* 10(4) (1993): 26–33.

——. "A Brief History of the American Land Ethic Since 1492." *Inner Voice* 6 (January/February 1994): 5–7.

——. *Beyond the Land Ethic: More Essays in Environmental Philosophy*. Albany: State University of New York Press, 1999.

Callicott, J. Baird, and Michael P. Nelson, eds., *The Great New Wilderness Debate*. Athens: University of Georgia Press, 1998.

Cohen, Michael. *The Pathless Way: John Muir and American Wilderness*. Madison: University of Wisconsin Press, 1984.

Commoner, Barry. *The Closing Circle*. New York: Knopf, 1971.

Cramer, Phillip F. *Deep Environmental Politics: The Role of Radical Environmentalism in Crafting American Environmental Policy*. Westport, Conn.: Praeger, 1998.

Cuomo, Chris J. *Feminism and Ecological Communities: An Ethic of Flourishing*. New York: Routledge, 1998.

DesJardins, Joseph. *Environmental Ethics: Concepts, Policy, and Theory*. Mountain View, Calif.: Mayfield, 1999.

Ehrlich, Paul. *Human Natures: Genes, Cultures, and the Human Prospect*. Washington, D.C. Island Press, 2000.

Engel, Leonard, ed. *The Big Empty: Essays on Western Landscapes as Narrative*. Albuquerque: University of New Mexico Press, 1994.

Fowler, Robert Booth. *The Greening of Protestant Thought*. Chapel Hill: University of North Carolina Press, 1995.

Freyfogle, Eric T. *Bounded People, Boundless Lands: Envisioning a New Land Ethic*. Washington, D.C.: Island Press, 1998.

Glacken, Clarence J. *Traces on the Rhodian Shore*. Berkeley: University of California Press, 1967.

Gobster, Paul H. "Aldo Leopold's 'Ecological Esthetic': Integrating Esthetic and Biodiversity Values." *Journal of Forestry* 93 (February 1995): 6–10.

Goin, Peter. *Humanature*. Austin: University of Texas Press, 1996.

Grusin, Richard. "Thoreau, Extravagance, and the Economy of Nature." *American Literary History* 5 (spring 1993): 30–50.

Gunter, Pete A.Y., and Max Oelschlaeger. *Texas Land Ethics*. Austin: University of Texas Press, 1997.

Harrison, Peter. *The Bible, Protestantism, and the Rise of Natural Science*. New York: Cambridge University Press, 1998.

Harrison, Robert Pogue. *Forests: The Shadow of Civilization*. Chicago: University of Chicago Press, 1992.

Hyde, Anne F. "Cultural Filters: The Significance of Perception in the History of the American West." *Western Historical Quarterly* 24 (August 1993): 351–74.

Kirby, Jack Temple. *Poquosin: A Study of Rural Landscape and Society*. Chapel Hill: University of North Carolina Press, 1995.

Kolodny, Annette. *The Lay of the Land: Metaphor and Experience in American Life and Letters*. Chapel Hill: University of North Carolina Press, 1975.

Kula, Erhun. *History of Environmental Economic Thought*. London: New York : Routledge, 1998.

Leopold, Aldo. *Game Management*. New York: Scribner's, 1933.

———. *A Sand County Almanac*. New York: Oxford University Press, 1949.

———. *Round River: From the Journal of Aldo Leopold*. Edited by Luna B. Leopold. New York: Oxford University Press, 1953.

———. *The River of the Mother of God and Other Essays*. Edited by J. Baird Callicott and Susan L. Flader. Madison: University of Wisconsin Press, 1991.

———. *For the Health of the Land: Previously Unpublished Essays and Other Writings*. Edited by J. Baird Callicott and Eric T. Freyfogle. Washington, D.C.: Island Press for Shearwater Books, 1999.

Leopold, Luna B. *A View of the River*. Cambridge: Harvard University, 1994.

Machor, James. *Pastoral Cities: Urban Ideals and the Symbolic Landscape of America*. Madison: University of Wisconsin Press, 1987.

Marx, Leo. *The Machine in the Garden: Technology and the Pastoral Ideal in America*. New York: Oxford University Press, 1964.

McEvoy, Arthur F. "Toward an Interactive Theory of Nature and Culture: Ecology, Production, and Cognition in the California Fishing Industry." *Environmental Review* 11 (winter 1987): 289–305.

McGreevy, Patrick V. *Imagining Niagara: The Meaning and Making of Niagara Falls*. Amherst: University of Massachusetts Press, 1994.

McGregor, Robert Kuhn. "Deriving a Biocentric History: Evidence From the Journal of Henry David Thoreau." *Environmental Review* 12(2) (summer 1988): 117–24.

Meine, Curt. *The Essential Aldo Leopold: Quotations and Commentaries*. Madison: University of Wisconsin Press, 1999.

Merchant, Carolyn. *The Death of Nature: Women, Ecology, and the Scientific Revolution*. New York: Harper and Row, 1980.

———. "The Theoretical Structure of Ecological Revolutions." *Environmental Review* 11 (winter 1987): 265–74.

———. *Earthcare: Women and the Environment*. New York: Routledge, 1996.

Mighetto, Lisa. *Wild Animals and American Environmental Ethics*. Tucson: University of Arizona Press, 1991.

Miller, Alan. *Gaia Connections*. Lanham, Md.: Rowman and Littlefield, 1991.

Muir, John. *Letters from Alaska*. Edited by Robert Engberg and Bruce Merrell. Madison: University of Wisconsin, 1993.

Nash, Roderick. *Wilderness and the American Mind.* New Haven: Yale University Press, 1973 (rev. ed.), 1982 (3d ed.).

——. *The Rights of Nature: A History of Environmental Ethics.* Madison: University of Wisconsin Press, 1989.

Norwood, Vera. *Made from This Earth: American Women and Nature.* Chapel Hill: University of North Carolina Press, 1993.

Novak, Barbara. *Nature and Culture: American Landscape Painting, 1825–1875.* New York: Oxford University Press, 1980.

Oelschlaeger, Max. *The Idea of Wilderness from Prehistory to the Age of Ecology.* New Haven: Yale University Press, 1991.

——. *The Wilderness Condition: Essays on Environment and Civilization.* San Francisco : Sierra Club Books, 1992.

——. *Caring for Creation: An Ecumenical Approach to the Environmental Crisis.* New Haven: Yale University Press, 1994.

Olsen, Brett J. "Wallace Stegner and the Environmental Ethic: Environmentalism as a Rejection of Western Myth." *Western American Literature* 29 (summer 1994): 123–42.

Price, Jennifer. *Flight Maps: Adventures with Nature in Modern America.* New York: Basic Books, 1999.

Ross, Andrew. *The Chicago Gangster Theory of Life: Nature's Debt to Society.* New York: Verso, 1994.

Rothenberg, David, and Marta Ulvaeus, eds. *The World and the Wild.* Tucson: University of Arizona Press, 2001.

Russell, Edmund P., III "The Strange Career of DDT: Experts, Federal Capacity, and Environmentalism in World War II." *Technology and Culture* 40(4) (October 1999): 770–96.

Schama, Simon. *Landscape and Memory.* New York: Knopf, 1995.

Schmitt, Peter J. *Back to Nature: The Arcadian Myth in Urban America.* New York: Oxford University Press, 1969.

Sessions, George, ed. *Deep Ecology for the Twenty-first Century.* Boston: Shambala, 1995.

Slovic, Scott. *Getting Over the Color Green: Contemporary Environmental Literature of the Southwest.* Tucson: University of Arizona Press, 2001.

Smith, Henry Nash. *Virgin Land.* Cambridge: Harvard University Press, 1950.

Steinberg, Theodore. *The Unnatural History of Natural Disaster in America.* New York: Oxford University Press, 2000.

Stoll, Mark. *Protestantism, Capitalism, and Nature in America.* Albuquerque: University of New Mexico Press, 1997.

Thoreau, Henry David. *Walden*. Boston: Ticknor and Fields, 1854.

Tichi, Cecelia. *New World, New Earth: Environmental Reform in American Literature from the Puritans through Whitman*. New Haven: Yale University Press, 1979.

Vogel, Steven. *Against Nature: The Concept of Nature in Critical Theory*. Albany: State University of New York Press, 1996.

Wells, Robert N., Jr. ed. *Law, Values, and the Environment: A Reader and Selective Bibliography*. Lanham, Md.: Scarecrow Press, 1996.

Westra, Laura. *An Environmental Proposal for Ethics: The Principle of Integrity*. Lanham, Md.: Rowman and Littlefield, 1994.

Worster, Donald. *Nature's Economy: The Roots of Ecology*. San Francisco: Sierra Club Books, 1977.

——. *Under Western Skies: Nature and History in the American West*. New York: Oxford University Press, 1992.

——. *An Unsettled Country: Changing Landscapes of the American West*. Albuquerque: University of New Mexico Press, 1994.

9. The Environmental Movement

Bookchin, Murray. *The Ecology of Freedom*. Palo Alto, Calif.: Cheshire Books, 1982.

Bramwell, Anna. *The Fading of the Greens: The Decline of Environmental Politics in the West*. New Haven: Yale University Press, 1994.

Carson, Rachel. *Silent Spring*. Boston: Houghton Mifflin, 1962; Cambridge: Riverside Press, 1962.

Cawley, R. McGreggor. *Federal Land, Western Anger: The Sagebrush Rebellion and Environmental Politics*. Lawrence: University Press of Kansas, 1993.

Commoner, Barry. *The Closing Circle*. New York: Knopf, 1971.

Dowie, Mark. *Losing Ground: American Environmentalism at the Close of the Twentieth Century*. Cambridge: MIT Press, 1995.

Dunlap, Riley E., and Angela G. Mertig, eds. *American Environmentalism: The U.S. Environmental Movement, 1970–1990*. Philadelphia: Taylor and Francis, 1992.

Dunlap, Thomas R. *DDT: Scientists, Citizens, and Public Policy*. Princeton: Princeton University Press, 1981.

Ehrlich, Paul. *The Population Bomb*. New York: Ballantine, 1968.

Faber, Daniel, ed. *The Struggle for Ecological Democracy: Environmental Justice Movements in the United States.* New York: Guilford Press, 1998.

Fairfax, Sally K. and Darla Guenzler. *Conservation Trusts.* Lawrence: University Press of Kansas, 2001.

Gibbs, Lois M. *Love Canal: My Story.* Albany: State University of New York Press, 1982.

Gottlieb, Robert. *Forcing the Spring: The Transformation of the American Environmental Movement.* Washington, D.C.: Island Press, 1993.

——. "Reconstructing Environmentalism: Complex Movements, Diverse Roots." *Environmental History Review* 17 (winter 1993): 1–19.

Hays, Samuel P. *Conservation and the Gospel of Efficiency: The Progressive Conservation Movement, 1890–1920.* Cambridge: Harvard University Press, 1959.

——. *Beauty, Health, and Permanence: Environmental Politics in the United States, 1955–1985.* Cambridge, New York: Cambridge University Press, 1987.

——. *A History of Environmental Politics since 1945.* Pittsburgh: University of Pittsburgh Press, 2000.

Helvarg, David. *The War Against the Greens: The "Wise-Use" Movement, the New Right and Anti-Environmental Violence.* San Francisco: Sierra Club Books, 1994.

Hirt, Paul, and Dale Goble, eds. *Northwest Lands, Northwest Peoples.* Seattle: University of Washington Press, 1999.

Huffman, Thomas R. "Defining the Origins of Environmentalism in Wisconsin: A Study in Politics and Culture." *Environmental History Review* 16 (fall 1992): 47–69.

Kiy, Richard, and John D. Wirth, eds. *Environmental Management on North America's Borders.* College Station: Texas A&M University Press, 1998.

Little, Charles E. *Green Fields Forever: The Conservation Tillage Revolution in America.* Washington, D.C.: Island Press, 1987.

Manes, Christopher. *Green Rage.* Boston: Little, Brown, 1990.

Matthews, Anne. *Where the Buffalo Roam.* New York: Grove Weidenfeld, 1992.

McEvoy, Arthur. *The Fisherman's Problem: Ecology and Law in the California Fisheries.* New York: Cambridge University Press, 1986.

McKibben, Bill. *The End of Nature.* New York: Random House, 1989.

McNeill, John. *Something New Under the Sun: An Environmental History of the Twentieth Century World*. New York: Norton, 2000.

McPhee, John. *The Control of Nature*. New York: Farrar, Straus, Giroux, 1989.

Melosi, Martin V., ed. *Pollution and Reform in American Cities, 1870–1930*. Austin: University of Texas Press, 1980.

Merchant, Carolyn. *Earthcare: Women and the Environment*. New York: Routledge, 1996.

Mighetto, Lisa. *Wild Animals and American Environmental Ethics*. Tucson: University of Arizona Press, 1991.

Mowrey, Marc, and Tim Redmond. *Not in Our Backyard: The People and Events That Shaped America's Modern Environmental Movement*. New York: Morrow, 1993.

Nash, Roderick. *Wilderness and the American Mind*. 3rd ed. New Haven: Yale University Press, 1982.

O'Brien, Bob. *Our National Parks and the Search for Sustainability*. Austin.: University of Texas Press, 1999.

Rome, Adam. *The Bulldozer in the Countryside: Suburban Sprawl and the Rise of American Environmentalism*. New York: Cambridge University Press, 2001.

Rothman, Hal. *Devil's Bargains: Tourism in the Twentieth-century American West*. Lawrence: University Press of Kansas, 1998.

———. *The Greening of a Nation?: Environmentalism in the United States Since 1945*. Fort Worth: Harcourt Brace College Publishers, 1998.

Sale, Kirkpatrick. *The Green Revolution: The American Environmental Movement, 1962–1992*. Edited by Eric Foner. New York: Hill and Wang, 1993.

Scarce, Rik. *Eco-Warriors: Understanding the Radical Environmental Movement*. Chicago: Noble Press, 1990.

Schaffer, Daniel. *Environment and TVA: Toward a Regional Plan for the Tennessee Valley, 1930s*. Norris, Tenn.: Tennessee Valley Authority, Cultural Resources Program, 1984.

Shabecoff, Philip. *A Fierce Green Fire: The American Environmental Movement*. New York: Hill and Wang, 1993.

Short, C. Brandt. *Ronald Reagan and the Public Lands: America's Conservation Debate, 1979–1984*. College Station: Texas A&M University Press, 1989.

Switzer, Jacqueline Vaughn. *Green Backlash: The History and Politics of Environmental Opposition in the United States*. Boulder, Colo.: Lynne Rienner Publishers, Inc., 1997.

Udall, Stewart. *The Quiet Crisis*. New York: Holt, Rinehart and Winston, 1963.

Wellock, Thomas Raymond. *Critical Masses: Opposition to Nuclear Power in California, 1958–1978*. Madison: University of Wisconsin Press, 1998.

Worster, Donald, ed. *American Environmentalism: The Formative Period, 1860–1915*. New York: Wiley, 1973.

10. History of Ecology

Botkin, Daniel B. *Discordant Harmonies: A New Ecology for the Twenty-first Century*. New York: Oxford University Press, 1990.

Bramwell, Anna. *Ecology in the Twentieth Century: A History*. New Haven: Yale University Press, 1989.

Crosby, Alfred W. *The Columbian Exchange: Biological and Cultural Consequences of 1492*. Westport, Conn.: Greenwood Press, 1972.

——. *Ecological Imperialism: The Biological Expansion of Europe*. New York: Cambridge University Press, 1986.

——. *Germs, Seeds, and Animals: Studies in Ecological History*. New York: M. E. Sharpe, 1994.

Diamond, Jared. *Germs, Guns, and Steel: The Fates of Human Societies*. New York: Norton, 1998.

Dobbs, David. *The Great Gulf: Fishermen, Scientists, and the Struggle to Revive the World's Greatest Fishery*. Washington, D.C.: Island Press, 2000.

Egan, Dave, and Evelyn A. Howell, eds. *The Historical Ecology Handbook: A Restorationist Guide to Reference Ecosystems*. Washington, D.C.: Island Press, 2001.

Egerton, Frank. "Ecological Studies and Observations Before 1900." In *Issues and Ideas in America*, edited by Benjamin Taylor and Thurman White. Norman: University of Oklahoma Press, 1976, pp. 311–51.

Evans, Howard Ensign. *Pioneer Naturalists: The Discovery and Naming of North American Plants and Animals*. New York: Holt, 1993.

Flader, Susan. *Thinking Like a Mountain: Aldo Leopold and the Evolution of an Ecological Attitude Toward Deer, Wolves, and Forests*. Columbia: University of Missouri Press, 1974.

Flores, Dan. "Place: An Argument for Bioregional History." *Environmental History Review* 18 (winter 1994): 1–18.

Goetzmann, William H. *Exploration and Empire: The Explorer and the Scientist in the Winning of the American West.* New York: Norton, 1966.

Hamerstrom, Frances. *My Double Life: Memoirs of a Naturalist.* Madison: University of Wisconsin Press, 1994.

Lowenthal, David. *George Perkins Marsh: Prophet of Conservation.* Seattle: University of Washington Press, 2000.

Malin, James. *History and Ecology: Studies of the Grassland.* Edited by Robert P. Swierenga. Lincoln: University of Nebraska Press, 1984.

McIntosh, Robert P. *The Background of Ecology: Concept and Theory.* New York : Cambridge University Press, 1985.

Merchant, Carolyn, ed. *Key Concepts in Critical Theory: Ecology.* Atlantic Highlands, N.J.: Humanities Press International, 1994.

Opie, John, ed. *Americans and Environment: The Controversy Over Ecology.* Lexington, Mass.: D.C. Heath, 1971.

Real, Leslie A., and James H. Brown. *Foundations of Ecology: Classic Papers with Commentaries.* Chicago: University of Chicago Press, 1991.

Russell, Edmund. *War and Nature: Fighting Humans and Insects with Chemicals from World War I to* Silent Spring. Tucson: University of Arizona Press, 2001.

Smith, Michael L. 1987. *Pacific Visions: California Scientists and the Environment, 1850–1915.* New Haven: Yale University Press, 1987.

Tobey, Ronald C. *Saving the Prairies: The Life Cycle of the Founding School of American Plant Ecology, 1895–1955.* Berkeley: University of California Press, 1981.

Worster, Donald. *Nature's Economy: The Roots of Ecology.* San Francisco: Sierra Club Books, 1977.

11. History of Environmental Science

Bartlett, Robert V. *The Reserve Mining Controversy: Science, Technology, and Environmental Quality.* Bloomington: Indiana University Press, 1980.

Botkin, Daniel B. *Environmental Science: Earth as a Living Planet.* New York: Wiley, 1995.

Bowler, Peter J., ed. *The Norton History of the Environmental Sciences.* New York: Norton, 1993.

Caldwell, Lynton K. *Between Two Worlds: Science, the Environmental Movement, and Policy Choice.* New York: Cambridge University Press, 1990.

Carson, Rachel. *Silent Spring*. Boston: Houghton Mifflin Company, 1962; Cambridge: Riverside Press, 1962.

Commoner, Barry. *The Closing Circle*. New York: Knopf, 1972.

Cowdrey, Albert E. "Pioneering Environmental Law: The Army Corps of Engineers and the Refuse Act." *Pacific Historical Review* 44 (1975): 331–49.

Doutt, Richard L. "The Historical Development of Biological Control." In *Biological Control of Insect Pests and Weeds*, edited by Paul DeBach. New York: Reinhold, 1964, pp. 3–20.

Ehrlich, Paul R. *Betrayal of Science and Reason*. Washington, D.C.: Island Press, 1996.

Elliott, Clark A. *History of Science in the United States: a Chronology and Research Guide*. New York: Garland, 1996.

Hilgard, Eugene W. *Alkali Lands, Irrigation and Drainage in their Mutual Relations*. Sacramento, Calif.: State Printer, 1892.

———. *Agriculture for Schools of the Pacific Slope*. New York: Macmillan, 1910.

Jenny, Hans. *E. W. Hilgard and the Birth of Modern Soil Science*. Pisa, Italy: Collanda della Revista "Agrochimica," 1961.

Kloppenburg Jr., Jack Ralph. *First the Seed: The Political Economy of Plant Biotechnology, 1492–2000*. Cambridge: Cambridge University Press, 1988.

McEvoy, Arthur F. "Science, Culture, and Politics in U.S. Natural Resources Management." *Journal of the History of Biology* 25 (fall 1992): 469–86.

McEvoy, Arthur. *The Fisherman's Problem: Ecology and Law in the California Fisheries*. New York: Cambridge University Press, 1986.

Mowrey, Marc, and Tim Redmond. *Not in Our Backyard: The People and Events that Shaped America's Modern Environmental Movement*. New York: Morrow, 1993.

Perkins, John H. *Insects, Experts, and the Insecticide Crisis: The Quest for New Pest Management Strategies*. New York: Plenum Press, 1982.

Perkins, Priscilla C. *Scientific Information in the Decision to Dam Glen Canyon*. Los Angeles: University of California, 1975.

Rabbitt, Mary C. *Minerals, Lands, and Geology for the Common Defence and General Welfare: A History of Public Lands, Federal Science and Mapping Policy*. Washington, D.C.: U.S. Geological Survey, 1986.

Shortland, Michael, ed. *Science and Nature: Essays in the History of the Environmental Sciences*. Stanford in the Vale: British Society for the History of Science, 1993.

Wells, George Stevens. *Garden in the West; A Dramatic Account of Science in Agriculture.* New York: Dodd, Mead, 1969.

Whorton, James. *Before Silent Spring: Pesticides and Public Health in Pre-DDT America.* Princeton: Princeton University Press, 1974.

12. Conservation History and Legislation

Abbott, Gordon, Jr. *Saving Special Places: A Centennial History of the Trustees of Reservations: Pioneer of the Land Trust Movement.* Ipswich, Mass.: Ipswich Press, 1993.

Baird, W. David. "The American West and the Nixon Presidency, 1969–1974." *Journal of the West* 34 (April 1995): 83–90.

Brooks, Paul. *Speaking for Nature: How Literary Naturalists from Henry Thoreau to Rachel Carson Have Shaped America.* Boston: Houghton Mifflin, 1980.

Carr, Ethan. *Wilderness by Design: Landscape Architecture and the National Park Service.* Lincoln: University of Nebraska Press, 1998.

Catton, Theodore. *Inhabited Wilderness: Indians, Eskimos, and National Parks in Alaska.* 1st ed. Albuquerque: University of New Mexico Press, 1997.

Clements, Kendrick A. "Politics and the Park: San Francisco's Fight for Hetch Hetchy, 1908–1913." *Pacific Historical Review* 48 (1979): 185–215.

———. *Hoover, Conservation, and Consumerism: Engineering the Good Life.* Lawrence: University Press of Kansas, 2000.

Coates, Peter A. *The Trans-Alaska Pipeline Controversy: Technology, Conservation, and the Frontier.* Bethlehem, Penn.: Lehigh University Press, 1991.

Conard, Rebecca. *Places of Quiet Beauty: Parks, Preserves, and Environmentalism.* Iowa City: University of Iowa Press, 1997.

Cone, Joseph, and Sandy Ridlington, eds. *The Northwest Salmon Crisis: A Documentary History.* Corvallis: Oregon State University Press, 1996.

Cortner, Hanna J., and Margaret A. Moote. *The Politics of Ecosystem Management.* Washington, D.C.: Island Press, 1999.

Dana, Samuel T. and Sally K. Fairfax. *Forest and Range Policy, Its Development in the United States.* 2d ed. New York: McGraw-Hill, 1980.

DeVoto, Bernard. "Conservation: Down and on the Way Out." *Harper's Monthly* 209 (August 1954): 66–74.

Dilsaver, Lary M., ed. *America's National Park System: The Critical Documents*. Lanham, Md.: Rowman and Littlefield, 1994.

Dilsaver, Lary M., and Craig E. Colten, eds. *The American Environment: Interpretations of Past Geographies*. Lanham, Md.: Rowman and Littlefield, 1992.

Dilsaver, Lary M., and Douglas H. Strong. "Sequoia and Kings Canyon National Parks: One Hundred Years of Preservation and Resource Management." *California History* 69 (summer 1990): 98–117.

Dilsaver, Lary M., and William C. Tweed. *Challenge of the Big Trees: A Resource History of Sequoia and Kings Canyon National Parks*. Three Rivers, Calif.: Sequoia Natural History Association, 1990.

Dorsey, Kurkpatrick. *The Dawn of Conservation Diplomacy: U.S.-Canadian Wildlife Protection Treaties in the Progressive Era*. Foreward by William Cronon. Seattle: University of Washington Press, 1998.

Dunlap, Thomas R. *Nature and the English Diaspora: Environment and History in the United States, Canada, Australia, and New Zealand*. Cambridge: Cambridge University Press, 1999.

Durbin, Kathie. *Tongass: Pulp Politics and the Fight for the Alaska Rain Forest*. Corvallis: Oregon State University Press, 1999.

Fairfax, Sally K., and Darla Guenzler. *Conservation Trusts*. Lawrence: University Press of Kansas, 2001.

Fairfax, Sally K., and Carolyn Yale. *Federal Lands: A Guide to Planning, Management, and State Revenues*. Washington, D.C.: Island Press, 1987.

Fox, Steven. *John Muir and His Legacy: The American Conservation Movement*. Boston: Little, Brown, 1981.

Gangewere, Robert, ed. *The Exploited Eden: Literature on the American Environment*. New York: Harper and Row, 1972.

Gates, Paul W. *Fifty Million Acres: Conflicts Over Kansas Land Policy, 1854–1890*. New York: Atherton Press, 1966.

———. *History of Public Land Law Development*. Washington, D.C.: Government Printing Office, 1968.

Gilliam, Harold. *Between the Devil and the Deep Blue Bay: The Struggle to Save San Francisco Bay*. San Francisco: Chronicle Books, 1969.

Grant, Madison. *Saving the Redwoods, An Account of the Movement during 1919 to Preserve the Redwoods of California*. New York: New York Zoological Society, 1919.

Grant, Richard A. "The Fight for the California Desert: Conserve or Destroy?" *Cry California* 8(1) (1972–73): 4–12.

Hamilton, David E. *From New Day to New Deal: American Farm Policy from Hoover to Roosevelt, 1928–1933.* Chapel Hill: University of North Carolina Press, 1991.

Harvey, Mark W. T. *Echo Park and the American Cconservation Movement.* Seattle: University of Washington Press, 2000.

Hays, Samuel P. *Conservation and the Gospel of Efficiency: The Progressive Conservation Movement, 1890–1920.* Cambridge: Harvard University Press, 1959.

———. *Beauty, Health, and Permanence: Environmental Politics in the United States, 1955–1985.* Cambridge: Cambridge University Press, 1987.

———. *A History of Environmental Politics since 1945.* Pittsburgh: University of Pittsburgh Press, 2000.

Heiman, Michael. "Production Confronts Consumption: Landscape Perception and Social Conflicts in the Hudson Valley." *Environment and Planning D: Society and Space* 7 (1989): 165–73.

Helms, Douglas. "Conserving the Plains: The Soil Conservation Service in the Great Plains." *Agricultural History* 64(2) (spring 1990): 58–73.

Helms, Douglas, and Susan L. Flader, eds. *The History of Soil and Water Conservation.* Berkeley: Published by the University of California Press for the Agricultural History (Washington, D.C.), 1985.

Hirt, Paul, and Dale Goble, eds. *Northwest Lands, Northwest Peoples.* Seattle: University of Washington Press, 1999.

Huffman, Thomas R. *Protectors of the Land and Water: Environmentalism in Wisconsin, 1961–1968.* Chapel Hill: University of North Carolina Press, 1994.

Humberger, Charles E. "The Civilian Conservation Corps in Nebraska: Memoirs of Company 762." *Nebraska History* 75 (winter 1994): 292–300.

Huth, Hans. "Yosemite: The Story of an Idea." *Sierra Club Bulletin* 33 (March 1948): 47–78.

———. *Nature and the American: Three Centuries of Changing Attitudes.* Berkeley: University of California Press, 1957.

Jacoby, Karl. *Crimes Against Nature: Squatters, Poachers, Thieves, and the Hidden History of American Conservation.* Berkeley: University of California Press, 2001.

Jameson, John. "From Dude Ranches to Haciendas: Master Planning at Big Bend National Park, Texas." *Forest and Conservation History* 38(3) (July 1994): 108–17.

Jones, Holway R. "Mysterious Origin of the Yosemite Park Bill." *Sierra Club Bulletin* 48 (December 1963): 69–79.

——. *John Muir and the Sierra Club: The Battle for Yosemite.* San Francisco: Sierra Club, 1965.

Judd, Richard W. "Grass-Roots Conservation in Eastern Coastal Maine: Monopoly and the Moral Economy of Weir Fishing, 1893–1911." *Environmental History Review* 12 (summer 1988): 81–103.

——. "Reshaping Maine's Landscape: Rural Culture, Tourism, and Conservation, 1890–1929." *Journal of Forest History* 32 (October 1988): 180–90.

Kaufman, Polly Welts. *National Parks and the Woman's Voice: A History.* 1st ed. Albuquerque: University of New Mexico Press, 1997.

Klyza, Christopher McGrory. *Who Controls Public Lands?: Mining, Forestry, and Grazing Policies, 1870–1990.* Chapel Hill: University of North Carolina Press, 1996.

Licht, Daniel S. "The Great Plains: America's Best Chance for Ecosystem Restoration." *Wild Earth* 4(2) (summer 1994): 47–53.

Lovett, Francis N. *National Parks : Rights and the Common Good.* Lanham, Md.: Rowman and Littlefield, 1998.

Manning, Samuel F. *New England Masts and the King's Broad Arrow.* London: National Maritime Museum, 1979.

Marsh, George Perkins. *Man and Nature.* Edited by David Lowenthal. Cambridge: The Belknap Press of Harvard University Press, 1965.

Marshall, Robert. *The People's Forests.* New York: H. Smith and R. Haas, 1933.

Matthews, Greg, and Neal Kephart. "The Park at the Crossroads of History." *Wisconsin Natural Resources* 16 (August 1992): 22–27.

Mayo, Lawrence S. "The Kings Woods. Proceedings of the Massachusetts Historical Society." *54* (October 1920–June 1921): 50–61.

McClelland, Linda Flint. *Building the National Parks: Historic Landscape Design and Construction.* Baltimore: Johns Hopkins University Press, 1998.

McCloskey, Michael, and Albert Hill. *Mineral King: Mass Recreation Versus Park Protection in the Sierra.* San Francisco: Sierra Club, 1971.

Meyer, Judith L. *The Spirit of Yellowstone: The Cultural Evolution of a National Park.* Lanham, Md.: Rowman and Littlefield, 1996.

Muir, John. "The Hetch-Hetchy Valley: A National Question." *Sierra Club Bulletin* 16(5) (1910): 263–69.

——. "The Creation of Yosemite National Park." *Sierra Club Bulletin* 29 (October 1944): 49–60.

Nash, Roderick. "John Muir, William Kent, and the Conservation Schism." *Pacific Historical Review* 36 (November 1967): 423–33.

——. "The American Invention of National Parks." *American Quarterly* 22 (fall 1970): 726–35.

——. *Wilderness and the American Mind.* 3d ed. New Haven: Yale University Press, 1982.

——, ed. *American Environmentalism: Readings in Conservation History.* New York: McGraw-Hill, 1990.

National Geographic Society. "Great Basin National Park." *National Geographic* 175(1) (January 1989): 72–76.

Oelschlaeger, Max. *The Idea of Wilderness from Prehistory to the Age of Ecology.* New Haven: Yale University Press, 1991.

——. *The Wilderness Condition: Essays on Environment and Civilization.* San Francisco : Sierra Club Books, 1992.

Ogden, Kate Nearpass. "Sublime Vistas and Scenic Backdrops: Nineteenth-Century Painters and Photographers at Yosemite." *California History* 69 (summer 1990): 134–53.

Opie, John. *The Law of the Land: Two Hundred Years of American Farm Land Policy.* Lincoln: University of Nebraska Press, 1987.

Patton, Thomas W. "Forestry and Politics: Franklin D. Roosevelt as Governor of New York." *New York History* 75 (October 1994): 397–418.

Pinchot, Gifford. *The Fight for Conservation.* New York: Doubleday, Page, 1910.

——. "How Conservation Began in the United States." *Agricultural History* 11 (October 1937): 255–65.

Pinkett, Harold T. "Sources of American Forest and Conservation History." *Journal of Forest History* 25 (October 1981): 210–12.

Pyne, Steven. *How the Canyon Became Grand: A Short History.* New York: Viking, 1998.

Reilly, P. T. "The Lost World of Glen Canyon." *Utah Historical Quarterly* 63 (spring 1995): 122–34.

Robbins, William G. and Foster, James C. *Land in the American West: Private Claims and the Common Good.* Seattle: University of Washington Press, 2000.

Ronda, James P., ed. *Thomas Jefferson and the Changing West: From Conquest to Conservation.* Albuquerque: University of New Mexico Press; St. Louis: Missouri Historical Society Press, 1997.

Rothman, Hal. *On Rims and Ridges: The Los Alamos Area Since 1880*. Lincoln: University of Nebraska Press, 1992.

Runte, Alfred. "Beyond the Spectacular: The Niagara Falls Preservation Campaign." *New York Historical Society Quarterly* 57(1) (1973): 30–50.

———. *Yosemite: The Embattled Wilderness*. Lincoln: University of Nebraska Press, 1990.

———. *National Parks: The American Experience*. 3d ed. Lincoln: University of Nebraska Press, 1997.

Russell, Franklin. "The Vermont Prophet: George Perkins Marsh." *Horizon* 10(31) (1968): 16–23.

Schrepfer, Susan. *The Fight to Save the Redwoods: A History of Environmental Reform, 1917–1978*. Madison: University of Wisconsin Press, 1983.

Sherman, Michael. "The CCC in Vermont." *Vermont History News* (November-December 1994): 74–77.

Smith, Thomas G. "John Kennedy, Steward Udall, and New Frontier Conservation." *Pacific Northwest Quarterly* 64 (August 1995): 329–52.

Spence, Mark David. *Dispossessing theWilderness: Indian Removal and the Making of the National Parks*. New York: Oxford University Press, 1999.

Stoll, Mark. *Protestantism, Capitalism, and Nature in America*. Albuquerque: University of New Mexico Press, 1997.

Strong, Douglas H. "The Sierra Forest Preserve: The Movement to Preserve the San Joaquin Valley Watershed." *California Historical Society Quarterly* 46(1) (1967): 3–17.

———. *Dreamers and Defenders: American Conservationists*. Lincoln: University of Nebraska Press, 1984.

Stroud, Patricia Tyson. "Forerunner of American Conservation: Naturalist Thomas Say." *Forest and Conservation History* 39 (October 1995): 184–90.

Thomas, Christine L. "One Hundred Twenty Years of Citizen Involvement with the Wisconsin Natural Resources Board." *Environmental History Review* 15 (spring 1991): 61–81.

Tyrrell, Ian. *True Gardens of the Gods: Californian-Australian Environmental Reform, 1860–1930*. Berkeley: University of California Press, 1999.

Wellock, Thomas. "The Battle for Bodega Bay: The Sierra Club and Nuclear Power, 1958–1964." *California History* 71 (summer 1992): 192–211, 289–91.

Young, James T. "The Origins of New Deal Agricultural Policy." *Policy Studies Journal* 21(2) (1993): 190–209.

13. Agricultural History

Atack, Jeremy. "Tenants and Yeomen in the Nineteenth Century." *Agricultural History* 62(3) (1988): 6–32.

Baltensperger, Bradley H. "Farm Consolidation in the Northern and Central States of the Great Plains." *Great Plains Quarterly* 7(4) (fall 1987): 256–65.

Barrio, Raymond. *The Plum Plum Pickers.* Tempe: Bilingual Press, 1969.

Beeman, Randal S. and James A. Pritchard. *A Green and Permanent Land: Ecology and Agriculture in the Twentieth Century.* Lawrence: University Press of Kansas, 2001.

Betts, Raymond F. "'Sweet Meditation Through This Pleasant Country': Foreign Appraisals of the Landscape of Kentucky in the Early Years of the Commonwealth." *Register of the Kentucky Historical Society* 90 (1992): 26–44.

Bidwell, Percy Wells. *Rural Economy in New England at the Beginning of the Nineteenth Century.* New Haven, Conn.: n.p., 1916. Reprint, Clifton, N.J.: Augustus M. Kelley, 1972.

Black, John D. *The Rural Economy of New England: A Regional Study.* Cambridge: Harvard University Press, 1950.

Bogue, Allan G. *From Prairie to Corn Belt: Farming on the Illinois and Iowa Prairies in the Nineteenth Century,* 1963.

Bonnifeld, Paul. *The Dust Bowl: Men, Dirt, and Depression.* Albuquerque: University of New Mexico Press, 1979.

Burchell, R. A. "Opportunity and the Frontier: Wealth Holding in Twenty-Six Northern California Counties, 1848–1880." *Western Historical Quarterly* 18 (April 1987): 177–96.

Campbell, Susan M. *Sugar in Hawaii: A Guide to Historical Resources.* Edited by Linda K. Menton. Honolulu: Humanities Program of the State Foundation on Culture and the Arts in Cooperation with the Hawaiian Historical Society, 1986.

Chernykh, E. L. "Agriculture of Upper California: A Long Lost Account of Farming in California as Recorded by a Russian Observer at Fort Ross in 1841." *Pacific Historian* 11(1) (1967): 10–28.

Coke, J. Earl, and Ann Foley Scheuring. *Reminiscences of People and Change in California Agriculture, 1900–1975.* Davis: University of California Press, 1976.

Cowdrey, Albert E. *This Land, This South: An Environmental History*. Lexington: University of Kentucky Press, 1983.

Cronise, Titus Fey. *The Agricultural and Other Resources of California*. San Francisco: A. Roman, 1870.

Cronon, William. *Nature's Metropolis: Chicago and the Great West*. New York: Norton, 1991.

———. "A Place for Stories: Nature, History, and Narrative." *Journal of American History* 78 (March 1992): 1347–76.

Cutler, Hugh Carson. *Corn, Cucurbits, and Cotton from Glen Canyon*. Salt Lake City: University of Utah Press, 1966.

Danbom, David B. "Romantic Agrarianism in Twentieth-Century America." *Agricultural History* 65(4) (fall 1991): 1–12.

———. *Born in the Country: A History of Rural America*. Baltimore: Johns Hopkins University Press, 1995.

Dasmann, Raymond F. *California's Changing Environment*. San Francisco: Boyd and Fraser, 1981.

Day, Clarence A. *A History of Maine Agriculture, 1604–1860*. University of Maine Studies, 2d ser., no. 68. Orono: University of Maine Press, 1954.

Dean, Samuel. *The New-England Farmer; or Georgical Dictionary: Containing a Compendious Account of the Ways and Methods in Which the Important Art of Husbandry, in All Its Various Branches, Is, or May Be, Practised to the Greatest Advantage in This Country* [1790]. Boston: Wells and Lilly, 1822.

Derr, Mark. *Some Kind of Paradise: A Chronicle of Man and Land in Florida*. Gainesville: University Press of Florida, 1998.

Duggar, John Frederick. *Southern Field Crops, Exclusive of Forage Plants*. New York: Macmillan, 1911.

Earle, Carville. "The Myth of the Southern Soil Miner: Macrohistory, Agricultural Innovation, and Environmental Change." In *The Ends of the Earth*, edited by Donald Worster. New York: Cambridge University Press, 1988, pp. 175–210.

Eliot, Jared. *Essays Upon Field Husbandry in New England, and Other Papers, 1748–1762*. Edited by Harry J. Carman and Rexford G. Tugwell. New York: Columbia University Press, 1934.

Faragher, John Mack. *Sugar Creek: Life on the Illinois Prairie*. New Haven: Yale University Press, 1986.

Gates, Paul W. *The Farmer's Age: Agriculture, 1815–1860*. New York: Holt, Rinehart and Winston, 1960.

——. *Agriculture and the Civil War.* New York: Knopf, 1965.

——. "Public Land Disposal in California." *Agricultural History* 49 (January 1975): 158–78.

Gates, Paul W., ed. *The Fruits of Land Speculation.* New York: Arno Press, 1979.

Gregory, James. *American Exodus: The Dust Bowl Migration and Okie Culture in California.* New York: Oxford University Press, 1989.

Hamilton, Patrick. *The Resources of Arizona: Its Mineral, Farming and Grazing Lands, Towns and Mining Camps, Its Rivers, Mountains, Plains, and Mesas* ... Prescott, Ariz: n.p., 1881.

Hargreaves, Mary Wilma M. *Dry Farming in the Northern Great Plains, 1900–1925,* 1957.

Harvey, Mark W. T. "North Dakota, the Northern Plains, and the Missouri Valley Authority." *North Dakota History* 59 (summer 1992): 28–39.

Hedrick, Ulysses P. *A History of Agriculture in the State of New York.* Albany: Printed by J. B. Lyon Company for the New York State Agricultural Society, 1933.

Hewes, Leslie. *The Suitcase-Farming Frontier: A Study in the Historical Geography of the Central Great Plains.* Lincoln: University of Nebraska Press, 1973.

Hilgard, Eugene W. *Report on the Physical and Agricultural Features of the State of California, with a Discussion of the Present and Future of Cotton Production in the State.* San Francisco: Pacific Rural Press Office, 1884.

Hornbeck, David. "Land Tenure and Rancho Expansion in Alta California, 1784–1846." *Journal of Historical Geography* 4(4) (1978): 371–90.

Hurt, R. Douglas. *The Dust Bowl: An Agricultural and Social History.* Chicago: Nelson-Hall, 1981.

——. *American Farm Tools: From Hand-Power to Steam-Power.* Manhattan, Kans.: Sunflower University Press, 1982.

——. "Federal Land Reclamation in the Dust Bowl." *Great Plains Quarterly* 6(2) (spring 1986): 94–106.

——. *Agricultural Technology in the Twentieth Century.* Manhattan, Kans.: Sunflower University Press, 1991.

——. *American Agriculture: A Brief History.* Ames: Iowa State University Press, 1994.

——. *American Farms: Exploring Their History.* Malabar, Fla.: Krieger, 1996.

Ingraham, Joseph H., ed. *The Sunny South: Embracing Five Years' Experience of a Northern Governess in the Land of Sugar and Cotton.* Philadelphia: G. G. Evans, 1860.

Kahane, Joyce D. *The Sugar Industry in Hawaii*. Honolulu: Legislative Reference Bureau, 1987.

Knobloch, Frieda. *The Culture of Wilderness: Agriculture as Colonization in the American West*. Chapel Hill: University of North Carolina Press, 1996.

Kolodny, Annette. *The Land Before Her: Fantasy and Experience of the American Frontiers, 1630–1869*. Chapel Hill: University of North Carolina Press, 1984.

Lange, Dorothea, and Paul S. Taylor. *An American Exodus: A Record of Human Erosion*. New York: Reynal and Hitchcock, 1930.

Lehman, Tim. *Public Values, Private Lands: Farmland Preservation Policy, 1933–1985*. Chapel Hill: University of North Carolina Press, 1995.

Lewis, Michael E. "National Grasslands in the Dust Bowl." *Geographical Review* 79(2) (April 1989): 161–71.

Little, Charles E. *Challenge of the Land*. New York: Open Space Action Institute, 1968.

———. "Dusty Old Dust." In *Hope for the Land*. New Brunswick, N.J.: Rutgers University Press, 1992.

Lokke, Janet. "'Like a Bright Tree of Life'; Farmland Settlement of the Sacramento River Delta." *California History* 59(3) (1980): 222–39.

Lookingbill, Brad D. *Dust Bowl, USA: Depression America and the Ecological Imagination, 1929–1941*. Seattle: University of Washington Press, 2001.

Lyons, Norbert. *The McCormick Reaper Legend*. New York: Exposition Press, 1955.

Malin, James. *Winter Wheat in the Golden Belt of Kansas*. Lawrence: University of Kansas Press, 1944.

———. *The Grassland of North America: Prolegomena to Its History*. Lawrence, Kans.: James C. Malin (Ann Arbor, Mich.: Edwards Brothers), 1947.

May, Dean L. *Three Frontiers: Family, Land, and Society in the American West, 1850–1900*. New York: Cambridge University Press, 1994.

McClelland, Peter D. *Sowing Modernity: America's First Agricultural Revolution*. Ithaca, N.Y.: Cornell University Press, 1997.

McCorvie, Mary R., and Christopher L. Lant. "Drainage District Formation and the Loss of Midwestern Wetlands, 1850–1930." *Agricultural History* 67 (fall 1993): 13–39.

McDean, Harry C. "Dust Bowl Historiography." *Great Plains Quarterly* 6(2) (spring 1986): 117–26.

McDermott, John D. "No Small Potatoes: Problems of Food and Health at Fort Laramie, 1849–1859." *Nebraska History* 79 (winter 1998): 162–70.

McGowan, William P. "Industrializing the Land of Lono: Sugar Plantation Managers and Workers in Hawaii, 1900–1920." *Agricultural History* 69(2) (spring 1995): 177–201.

McMath, Robert C., Jr. "Populism in Two Counties: Agrarian Protest in the Great Plains and Prairie Provinces." *Agricultural History* 69(4) (fall 1995): 517–47.

McNall, Neil A. "Lumbering and Agriculture in the Southern Tier." In McNall *An Agricultural History of the Genesee Valley, 1790–1860.* Philadelphia: University of Pennsylvania Press, 1952.

McWilliams, Carey. *Factories in the Field: The Story of Migratory Farm Labor in California.* Santa Barbara: Peregrine Publishers, 1971.

Meinig, Donald W. *The Great Columbia Plain: A Historical Geography, 1805–1910.* Seattle: University of Washington Press, 1968, 1995.

Merchant, Carolyn. *Ecological Revolutions: Nature, Gender, and Science in New England.* Chapel Hill: University of North Carolina Press, 1989.

Miner, Craig. *Harvesting the High Plains: John Kriss and the Business of Wheat Farming, 1920–1950.* Lawrence: University Press of Kansas, 1998.

Moles, Jerry A. "Who Tills the Soil? Mexican-American Workers Replace the Small Farmer in California: An Example from Colusa County." *Human Organization* 38(1) (1979): 20–27.

Mourt's Relation. *A Relation or Iournall of the Beginning and Proceedings of the English Plantation Setled at Plimoth in New England, by Certaine English Aduenturers Both Merchants and Others.* London: Printed for Iohn Bellamie, 1622.

Nakayama, Mona. *Maritime Industries of Hawaii: A Guide to Historical Resources.* Honolulu: Foundation on Culture and the Arts in Cooperation with the Hawaiian Historical Society, 1987.

Olmstead, A. L., and P. Rhode. "An Overview of California Agricultural Mechanization, 1870–1930." *Agricultural History* 62 (September 1988): 86–112.

Page, B., and R. Walker. "From Settlement to Fordism: The Agro-Industrial Revolution in the American Mid-West." *Economic Geography* 67 (1991): 281–315.

Parfit, Michael. "The Dust Bowl." *Smithsonian* 20(3) (June 1989): 46–54.

Phillips, Sarah T. "Lessons from the Dust Bowl: Dryland Agriculture and Soil Erosion in the United States and South Africa, 1900–1950." *Environmental History* 4(2) (April 1999): 245–66.

Pichaske, David R., ed. *Late Harvest: Rural American Writing.* New York: Paragon House, 1992.

Pisani, Donald J. *From the Family Farm to Agribusiness: The Irrigation Crusade in California and the West, 1850–1931.* Berkeley: University of California Press, 1984.

Popper, Deborah E., and Frank J. Popper. "The Great Plains: From Dust to Dust." *Planning* (December 1987): 12–18.

Rasmussen, Barbara. *Absentee Landowning and Exploitation in West Virginia, 1760–1920.* Lexington: University Press of Kentucky, 1994.

Raup, Hugh M. "The View From John Sanderson's Farm: A Perspective for the Use of the Land." *Forest History* 10 (April 1966): 2–11.

Richardson, Jean. *Partnerships in Communities: Reweaving the Fabric of Rural America.* Washington, D.C.: Island Press, 2000.

Riney-Kehrberg, Pamela. *Rooted in Dust: Surviving Drought and Depression in Southwestern Kansas.* Lawrence: University Press of Kansas, 1994.

Robbins, William G. *Landscapes of Promise: The Oregon Story, 1800–1940.* Seattle: University of Washington Press, 1997.

———. "Willamette Eden: The Ambiguous Legacy." *Oregon Historical Quarterly* 99 (summer 1998): 189–218.

Robinson, W. W. *Land in California: The Story of Mission Lands, Ranchos, Squatters, Mining Claims, Railroad Grants, Land Script, Homesteads.* 1948. Reprint, 1979. Berkeley: University of California Press.

Rosenberg, Norman J. "Adaptations to Adversity: Agriculture, Climate and the Great Plains of North America." *Great Plains Quarterly* 6(3) (summer 1986): 202–17.

Rothenberg, Winifred B. "The Market and Massachusetts Farmers, 1750–1855." *Journal of Economic History* 41 (1981): 283–314.

Rowntree, Lester B. "Drought During California's Mission Period, 1769–1834." *Journal of California and Great Basin Anthropology* 7(1) (1985): 7–20.

Ruiz, Vicki. *Cannery Women, Cannery Lives: Mexican Women, Unionization, and the California Food Processing Industry, 1930–1950.* Albuquerque: University of New Mexico Press, 1987.

Russell, Howard S. *A Long, Deep Furrow: Three Centuries of Farming in New England.* Hanover, N.H.: University Press of New England, 1976.

Sachs, Aaron. "Dust to Dust." *World Watch* 7 (January/February 1994): 32–35.

Saloutos, T. "The Immigrant in Pacific Coast Agriculture, 1880–1950." *Agricultural History* 49 (1975): 182–201.

Schumacher, Max George. *The Northern Farmer and His Markets During the Late Colonial Period* [1948]. New York: Arno Press, 1975.

Schuyler, Michael W. "New Deal Farm Policy in the Middle West: A Retrospective View." *Journal of the West* 33(4) (October 1994): 52–63.

Steinbeck, John. *The Grapes of Wrath*. New York: Penguin, 1992.

Stewart, Mart A. *"What Nature Suffers to Groe": Life, Labor, and Landscape on the Georgia Coast, 1680–1920*. Athens: University of Georgia Press, 1996.

Stoll, Steven. *The Fruits of Natural Advantage: Making the Industrial Countryside in California*. Berkeley: University of California Press, 1998.

Svobida, Lawrence. *Farming the Dust Bowl: A First-Hand Account from Kansas*. Lawrence: University Press of Kansas, 1968.

Thompson, Henry A. "The Life of a Vermont Farmer and Lumberman: The Diaries of Henry A. Thompson of Grafton and Saxton River." *Vermont History* 42 (spring 1974): 89–139.

Tocque, Philip. *A Peep at Uncle Sam's Farm, Workshop, Fisheries, and C.* Boston: C. H. Pierce, 1851.

Turner, Stan. *The Years of Harvest: A History of the Tule Lake Basin*. Eugene, Ore.: (95 E. 49th Ave., Eugene, Ore. 97465), 1987.

Ulrich, Hugh. *Losing Ground: Agricultural Policy and the Decline of the American Farm*. Chicago: Chicago Review Press, 1989.

Vaught, David. "An Orchardist's Point of View: Harvest Labor Relations on a California Almond Ranch, 1892–1921." *Agricultural History* 69(4) (fall 1995): 563–92.

Veregge, N. "Sense of Place in the Prairie Environment: Settlement and Ecology in Geary County, Kansas." *Great Plains Quarterly* 15(2) (spring 1995): 117–32.

Wacker, Peter O., and Paul G.E. Clemens. *Land Use in Early New Jersey: A Historical Geography*. Newark: New Jersey Historical Society, 1994.

Wagner-Wright, Sandra. *History of the Macadamia Nut Industry in Hawai'i, 1881–1981*. Lewiston, N.Y.: E. Mellen Press, 1995.

Walters, William D., Jr., and Jonathan Smith. "Woodland and Prairie Settlement in Illinois, 1830–70." *Forest and Conservation History* 36 (January 1992): 15–21.

Warren, John Quincy Adams. *California Ranchos and Farms, 1846–1862, Including the Letters of John Quincy Adams Warren of 1861, Being Largely Devoted to Livestock, Wheat Farming, Fruit Raising, and the Wine Industry*. Madison: State Historical Society of Wisconsin, 1967.

Weber, Devra. *Dark Sweat, White Gold: California Farm Workers, Cotton, and the New Deal*. Berkeley: University of California Press, 1994.

Wells, George Stevens. *Garden in the West: A Dramatic Account of Science in Agriculture.* New York: Dodd, Mead, 1969.

Wickson, Edward James. *One Thousand Questions in California Agriculture Answered.* San Francisco: Pacific Rural Press, 1914.

Worster, Donald. *Dust Bowl: The Southern Plains in the 1930s.* New York: Oxford University Press, 1979.

———. "The Dirty Thirties: A Study in Agricultural Capitalism." *Great Plains Quarterly* 6(2) (1986): 107–16.

14. Forest History

Agee, James K. *Fire Ecology of Pacific Northwest Forests.* Washington, D.C.: Island Press, 1993.

———. "Forest Fire History and Ecology of the Intermountain West." *Inner Voice* 7 (March/April 1995): 6–7.

Ayers, Harvard, Jenny Hager, and Charles E. Little, eds. *An Appalachian Tragedy: Air Pollution and Tree Death in the Eastern Forests of North America.* San Francisco: Sierra Club Books, 1998.

Backes, David. "Wilderness Visions: Arthur Carhart's 1922 Proposal for the Quetico-Superior Wilderness." *Forest and Conservation History* 35 (July 1991): 128–37.

Bajema, Carl. "Timber Express." *Michigan History* 77 (November/December 1993): 42–46.

Baldwin, Henry I. "The Trees of Nantucket." *American Forests and Forest Life* 34 (November 1928): 664–65, 684.

Barron, Hal S. *Mixed Harvest: The Second Great Transformation in the Rural North, 1870–1930.* Chapel Hill: University of North Carolina Press, 1997.

Belknap, Jeremy. *The History of New Hampshire.* 3 vols. Boston: Bradford and Read, 1784.

Bell, Mary T. *Cutting Across Time: Logging, Rafting, and Milling the Forests of Lake Superior.* Schroeder, Minn.: Schroeder Area Historical Society, 1999.

Berger, Jonathan, and John W. Sinton. *Water, Earth, and Fire: Land Use and Environmental Planning in the New Jersey Pine Barrens.* Baltimore: Johns Hopkins University Press, 1985.

Blair, James P., and Rowe Findley. "Will We Save Our Own?" (Pacific Northwest's Endangered Old-Growth Forests). *National Geographic* 178(3) (September 1990): 106–37.

Bliss, John C. "Evidence of Ethnicity: Management Styles of Forest Owners in Wisconsin." *Forest and Conservation History* 36 (April 1992): 63–72.

Bonyhady, Tim. "Artists with Axes." *Environment and History* 1 (June 1995): 221–39.

Booth, Douglas E. *Valuing Nature: The Decline and Preservation of Old-Growth Forests.* Lanham, Maryland: Rowman and Littlefield, 1994.

Boswell, Sharon A., Lorelea Hudson, and Nancy F. Renk. *Historic Overview of the Kootenai National Forest.* Vol. 1. Edited by Christian J. Miss. Seattle: Northwest Archaeological Associates, 1994.

Boylen, Bruce. *Old-Growth Forests and the Spotted Owl.* Sacramento, Calif.: California State Library, State Information and Reference Center, 1991.

Bridenbaugh, Carl. "Yankee Use and Abuse of the Forest in the Building of New England, 1620–1660." In Bridenbaugh *Early Americans.* New York: Oxford University Press, 1981, pp. 92–120.

Brooks, Charles E. "Overrun with Bushes: Frontier Land Development and the Forest History of the Holland Purchase, 1800–1850." *Forest and Conservation History* 39(1) (January 1995).

Brown, Beverly A. *In Timber Country: Working People's Stories of Environmental Conflict and Urban Flight.* Philadelphia: Temple University Press, 1995.

Brown, James H. "The Role of Fire in Altering the Species Composition of Forests in Rhode Island." *Ecology* 41 (April 1960): 310–16.

Browne, Charles C., and Howard B. Reed, eds. *Visions, Toil and Promises: Man in Vermont's Forests.* St. Johnsbury, Vt.: Fairbanks Museum and Planetarium, 1985.

Buckner, Edward. "Alabama's Forests — A Timeline." *Alabama's Treasured Forests* 13 (fall 1994): 14–15.

———. "Prehistory of the Southern Forest." *Forest Farmer* 54 (July/August 1995): 20–22.

Bunting, Robert. "Abundance and the Forests of the Douglas-Fir Bioregion, 1840–1920." *Environmental History Review* 18 (winter 1994): 41–62.

———. *The Pacific Raincoast: Environment and Culture in an American Eden, 1778–1900.* Lawrence: University Press of Kansas, 1997.

Burgess, Sherwood D. "The Forgotten Redwoods of the East Bay." *California Historical Society Quarterly* 30 (March 1951): 1–14.

———. "Lumbering in Hispanic California." *California Historical Society Quarterly* 41 (September 1962): 237–48.

Burns, Anna C. *A History of the Kisatchie National Forest*. Pineville, La.: U.S. Department of Agriculture, Forest Service, Southern Region, Kisatchie National Forest, 1994.

Campbell, Faith Thompson, and Scott E. Schlarbaum. *Fading Forests: North American Trees and the Threat of Exotic Pests*. New York: Natural Resources Defense Council, 1994.

Carlson, Douglas. "A Succession of Trees." *Georgia Review* 48 (summer 1994): 299–310.

Carlton, William R. "New England Masts and the King's Navy." *New England's Quarterly* 12 (March 1939): 4–18.

Carranco, Lynwood F., and John T. Labbe. *Logging the Redwoods*. Caldwell, Idaho: Caxton Printers, 1975.

Carroll, Charles Francis. *The Timber Economy of Puritan New England*. Providence, R.I.: Brown University Press, 1973.

Carroll, Matthew S. *Community and the Northwestern Logger: Continuities and Changes in the Era of the Spotted Owl*. Boulder, Colo.: Westview Press, 1995.

Carter, Jane Levis. *The Paper Makers: Early Pennsylvanians and their Water Mills*. Kennett Square, Penn.: KNA Press Inc., 1982.

Chaney, Ralph W. *Redwoods of the Past*. Berkeley: Save-the-Redwoods League, 1964.

Chase, Doris Harter. *They Pushed Back the Forest*. Colfax, Calif.: Chase, 1959.

Clark, Galen. *The Big Trees of California*. Yosemite Valley, Calif.: Press of Reflex Publishing Company, 1907.

Clow, Richmond L. "Timber Users, Timber Savers: Homestake Mining Company and the First Regulated Timber Harvest." *South Dakota History* 22 (fall 1992): 213–37.

Cohen, Michael P. *A Garden of Bristlecones: Tales of Change in the Great Basin*. Reno: University of Nevada Press, 1998.

Cohn, Lisa. "The Forests of the Lower Columbia River Region: A Case History of Use and Resilience." *Forest Perspectives* 4 (summer 1994): 5–8.

Collister, L. C. "Fiftieth Anniversary of the Santa Fe's Experiment with Eucalyptus." *Cross Tie Bulletin* 39 (February 1958): 9–10.

Cook, Annabel Kirschner. "Increasing Poverty in Timber-Dependent Areas in Western Washington." *Society and Natural Resources* 8 (March–April 1995): 97–109.

Cook, Don. "Oak Orchard: A Long Term Success Story." *Conservationist* 49 (October 1994): 2–7.

Cooper, Ellwood. *Forest Culture and Eucalyptus Trees*. San Francisco: Cubery, 1876.

Council, Inland Empire Public Lands. *Legacy of Congress's 1864 Northern Pacific Railroad Land Grant*. Spokane, Wash.: Inland Empire Public Lands Council, 1995.

Covington, W. Wallace, and Margaret M. Moore. "Southwestern Ponderosa Forest Structure: Changes Since Euro-American Settlement." *Journal of Forestry* 92 (January 1994): 39–47.

Cox, Thomas R. *Mills and Markets: A History of the Pacific Coast Lumber Industry to 1900*. Seattle: University of Washington Press, 1974.

Cox, Thomas R., Robert S. Maxwell, Philip Drennon Thomas, and Joseph J. Malone. *This Well-Wooded Land: Americans and Their Forests from Colonial Times to the Present*. Lincoln: University of Nebraska Press, 1985.

Cramer, Marianne. "Urban Renewal: Restoring the Vision of Olmsted and Vaux in Central Park's Woodlands." *Restoration and Management Notes* 11 (winter 1993): 106–16.

Critchfield, Richard. *Trees, Why Do You Wait? America's Changing Rural Culture*. Washington, D.C.: Island Press, 1991.

Cronon, William. *Nature's Metropolis: Chicago and the Great West*. New York: Norton, 1991.

Dana, Samuel T., and Sally K. Fairfax. *Forest and Range Policy, Its Development in the United States*. 2d ed. New York: McGraw-Hill, 1980.

Dasmann, Raymond F. *California's Changing Environment*. San Francisco: Boyd and Fraser, 1981.

Davies, Gilbert W., and Florice M. Frank, eds. *Forest Service Memories: Past Lives and Times in the United States Forest Service*. Hat Creek, Calif.: History Ink Books, 1997.

DeBell, Dean S., and Robert O. Curtis. "Silviculture and New Forestry in the Pacific Northwest." *Journal of Forestry* 91 (December 1993): 25–30.

DeBuys, William. *Enchantment and Exploitation: The Life and Hard Times of a New Mexico Mountain Range*. Albuquerque: University of New Mexico Press, 1985.

Devall, Bill, ed. *Clearcut: The Tragedy of Industrial Forestry*. San Francisco: Sierra Club Books/Earth Island Press, 1993.

Dickerson, Lynn. "Tree Farming in the Roanoke Valley." *Virginia Forests* 51 (spring 1995): 17–21.

Dilsaver, Lary M. "Resource Conflict in the High Sierra." In *The Mountainous West: Explorations in Historical Geography*, edited by Lary M. Dilsaver and William Wyckoff. Lincoln: University of Nebraska Press, 1995, pp. 281–302.

Donaldson, Alfred Lee. *A History of the Adirondacks*. 2 vols. New York: Century, 1921.

Draffan, George. *Annotated Bibliography on Railroad Land Grants, the Northern Pacific Railroad, and Its Corporate Descendants*. Spokane, Wash.: Inland Empire Public Lands Council, 1995.

———. *Annotated Bibliography on Timber Corporations Derived from the Northern Pacific Railroad Land Grant: Weyerhaeuser, Boise Cascade, Potlatch, and Plum Creek*. Spokane, Wash.: Inland Empire Public Lands Council, 1995.

Drobney, Jeffrey A. "Where Palm and Pine Are Blowing: Convict Labor in the North Florida Turpentine Industry, 1877–1923." *Florida Historical Quarterly* 72 (April 1994): 411–34.

———. "The Transformation of Work in the North Florida Timber Industry, 1890–1910." *Gulf Coast Historical Review* 10 (fall 1994): 93–110.

Dunwiddie, Peter W. "Forest and Health: The Shaping of the Vegetation on Nantucket Island." *Journal of Forest History* 33 (June 1989): 126–33.

Eames, Ninetta. "Staging in the Mendocino Redwoods." *Overland Monthly* 20 (August/September 1892): 113–131; 265–84.

Eddy, John Mathewson. *In the Redwood's Realm: By-ways of Wild Nature and Highways of Industry as Found in Humboldt Co., Cal*. San Francisco: D. S. Stanley, 1893.

Eichorn, Arthur Francis. *The Mt. Shasta Story; Being a Concise History of the Famous California Mountain*. Mount Shasta, Calif.: Mount Shasta Herald, 1957.

Ellsworth, Rodney Sydes. *The Giant Sequoia*. Oakland, Calif.: J.D. Berger, 1924.

Epstein, Mitch, and Bruce Stutz. "Stands of Time: The Old-Growth Forests of the Pacific Northwest..." *Audubon* 95(1) (January/February 1993): 62–78.

Everest, F. Alton. *Tales of High Clackamas Country: An Anecdotal History of Experiences on the Lakes Ranger District of the Mount Hood National Forest, 1930–1935*. Sandy, Ore.: St. Paul's Press, 1993.

Everett, Nigel. *The Tory View of Landscape*. New Haven: Yale University Press for the Paul Mellon Centre for Studies in British Art, 1994.

Ficken, Robert E. *The Forested Land: A History of Lumbering in Western Washington.* Durham, N.C.: Forest History Society, 1987.

Fiorelli, Edward A., and James M. Rossi. "The Place of Bad Woods: Staten Island's High Rock Park." *Conservationist* 38 (July/August 1983): 15–21.

Flader, Susan, ed. *The Great Lakes Forests: An Environmental and Social History.* Minneapolis: University of Minnesota Press, 1983.

Flippen, John Brooks. "The Nixon Administration, Timber, and the Call of the Wild." *Environmental History Review* 19 (summer 1995): 37–54.

Foster, Charles H.W., ed. *Stepping Back to Look Forward: A History of the Massachusetts Forest.* Petersham, Mass.: Distributed by Harvard University for Harvard Forest, 1998.

Foster, David R. "Land-Use History (1730–1990) and Vegetation Dynamics in Central New England." *Journal of Ecology* 80(4) (December 1992): 753–72.

Fredriksen, R.L. *Erosion and Sedimentation Following Road Construction and Timber Harvest on Unstable Soils in Three Small Western Oregon Watersheds.* Portland, Ore.: Pacific Northwest Forest and Range Experiment Station, U.S. Department of Agriculture, 1970.

Fritz, Emanuel. *The Story Told by a Fallen Redwood.* 3rd ed. Berkeley, Ca.: Save-the-Redwoods League, 1939.

——. *The Development of Industrial Forestry in California.* Seattle: University of Washington College of Forestry, 1960.

Fritzell, Peter A. "The Wilderness and the Garden: Metaphors for the American Landscape." *Forest History* 12 (April 1968): 16–23.

Geores, Martha. *Common Ground: The Struggle for Ownership of the Black Hills National Forest.* Lanham, Md. : Rowman and Littlefield, 1996.

Gerland, Jonathan. "Adjusting to Change: Railroads Brought a New Way of Life for East Texans." *Crosscut* 3–4 (first quarter 1995).

Goin, Peter. *Stopping Time: A Rephotographic Survey of Lake Tahoe.* Albuquerque: University of New Mexico Press, 1992.

Gordon, Burton L. *Monterey Bay Area: Natural History and Cultural Imprints.* Pacific Grove, Calif.: Boxwood Press, 1974.

Gough, Robert J. "Richard T. Ely and the Development of the Wisconsin Cutover." *Wisconsin Magazine of History* 75 (autumn 1991): 3–38.

Graham, Joseph N., et al. *Environment, Vegetation, and Regeneration After Timber Harvest in the Hungry-Pickett Area of Southwest Oregon.* Portland, Ore.: U.S. Department of Agriculture, Forest Service, Pacific Northwest Forest and Range Experiment Station, 1982.

Greber, Brian J. *Conservation Plans for the Northern Spotted Owl and Other Forest Management Proposals in Oregon: The Economics of Changing Timber Availability.* Corvallis, Ore.: Forest Research Laboratory, College of Forestry, Oregon State University, 1990.

Guthrie, Keith. "Blackwater Canyon: Showpiece of a World-Class Logging Operation." *Log Train* 10 (fall 1993): 6–12.

———. "By Rail to Stony River Dam." *Log Train* 11 (winter 1995): 8–14.

Hafen, LeRoy R., ed. *Fur Traders, Trappers, and Mountain Men of the Upper Missouri.* Lincoln: University of Nebraska Press, 1995.

Hansen, Kathy, William Wyckoff, and Jeff Banfield. "Shifting Forests: Historical Grazing and Forest Invasion in Southwestern Montana." *Forest and Conservation History* 39(2) (April 1995): 66–76.

Harper, Roland M. "Changes in the Forest Area of New England in Three Centuries." *Journal of Forestry* 16 (April 1918): 442–52.

Harrison, Robert Pogue. *Forests: The Shadow of Civilization.* Chicago: University of Chicago Press, 1992.

Harshberger, John W. "Nature and Man in the Pocono Mountain Region, Pennsylvania." *Bulletin of the Philadelphia Geographical Society* 13 (April 1915): 64–71.

Harvard Forest. *The Harvard Forest Models.* Petersham, Mass.: Harvard Forest, 1941.

Hawley, Ralph C., and Austin F. Hawes. *Forestry in New England: A Handbook of Eastern Forest Management.* New York: Wiley, 1912.

Hess, Terry L. "McKean County: Where the Gold is Green." *Pennsylvania Heritage* 9 (1983): 2–8.

Hicks, Ray R., Jr. *Ecology and Management of Central Hardwood Forests.* New York: Wiley, 1998.

Hines, Gustavus. *Wild Life in Oregon: Being a Stirring Recital of Actual Scenes of Daring and Peril Among the Gigantic Forests and Terrific Rapids of the Columbia River (the Mississippi of the Pacific Slope).* New York: R. Worthington, 1889.

Hirt, Paul W. *A Conspiracy of Optimism: Management of the National Forests since World War Two.* Lincoln: University of Nebraska Press, 1994.

Hirt, Paul, and Dale Goble, eds. *Northwest Lands, Northwest Peoples.* Seattle: University of Washington Press, 1999.

Hobbs, John E. "The Beginnings of Lumbering as an Industry in the New World, and First Efforts at Forest Protection: A Historical Study." *Forest Quarterly* 4 (March 1906): 14–23.

Horton, Tom. "Longleaf Pine: A Southern Revival." *Audubon* 97(2) (March/April 1995): 74–82.

Huggard, Christopher J. and Arthur R. Gómez. *Forests Under Fire: A Century of Ecosystem Mismanagement in the Southwest.* Tucson: University of Arizona Press, 2001.

Hurt, Bert. "Sawmill History of the Sierra National Forest, California." *Timberman* 44 (March 1943): 10–13, 30–32.

Hyde, John. *Wonderland, or, The Pacific Northwest and Alaska: With a Description of the Country Traversed by the Northern Pacific Railroad.* St. Paul, Minn.: Northern Pacific Railroad, 1888.

Hyman, Lou. "Biological Diversity in Alabama Forests: A Historical Perspective." *Alabama's Treasured Forests* 12 (fall 1993): 10–11.

Illick, Joseph S. *Pennsylvania Trees.* Harrisburg: Pennsylvania Department of Forestry, 1914.

Irland, Lloyd C. *Wildlands and Woodlots: The Story. of New England's Forests.* Hanover, N.H.: University Press of New England, 1982.

Jacobson, George L., Jr., and Ann Dieffenbacher-Krall. "White Pine and Climate Change: Insights from the Past." *Journal of Forestry* 93 (July 1995): 39–42.

James, David A. *Grisdale, Last of the Logging Camps: A Photo Story of Simpson Camps from 1890–1986.* Belfair, Wash.: Mason County Historical Society, 1986.

Jeffrey, David. "Yellowstone: The Great Fires of 1988." *National Geographic* 175(2) (February 1989): 252–77.

Jensen, Derrick, George Draffan, and John Osborn. *Railroads and Clearcuts: Legacy of Congress's 1864 Northern Pacific Railroad Land Grant.* Spokane, Wash.: Inland Empire Public Lands Council, 1995.

Johanneck, Donald P. *A History of Lumbering in the San Bernardino Mountains.* Redlands, Calif.: San Bernardino County Museum, 1975.

Johnson, Benjamin Heber. "Conservation, Subsistence, and Class at the Birth of Superior National Forest." *Environmental History* 4(1) (January 1999): 80–99.

Johnson, Melvin C. "The Road to Possum Walk: Immigrants and the East Texas Sawmill Culture." *Crosscut* (Fourth Quarter 1994): 3–4.

Johnston, Hank, and James Law. *Railroads of the Yosemite Valley.* Long Beach, Calif.: Johnston-Howe Publications, 1963.

——. *Thunder in the Mountains: The Life and Times of Madera Sugar Pine.* Los Angeles: Trans-Anglo Books, 1968.

Johnston, Jeremy. "Preserving the Beasts of Waste and Desolation: Theodore Roosevelt and Predator Control in Yellowstone National Park." *George Wright Forum* 15 (4 November 1998): 19–26.

Jonas, Gerald. *The Living Earth Book of North American Trees*. Pleasantville, N.Y.: Reader's Digest Association, 1993.

Judd, Richard W. "Reshaping Maine's Landscape: Rural Culture, Tourism, and Conservation, 1890–1929." *Journal of Forest History* 32 (October 1988): 180–90.

Judd, Richard W., Edwin A. Churchill, and Joel W. Eastman, eds. *Maine: The Pine Tree State from Prehistory to the Present*. Orono, Maine: University of Maine Press, 1995.

Kauffman, Erle. "The Trees of Potowomut: An Early Private Tree Planting Experiment Has Given America a Unique Man-Made Forest." *American Forests and Forest Life* 35 (June 1929): 329–32.

Kilpinen, Jon T. "Finnish Cultural Landscapes in the Pacific Northwest." *Pacific Northwest Quarterly* 86 (winter 1994/95): 25–34.

Kingsley, Ronald F. "Chestnut Grove: An Early 19th-Century Lime Burning Industry in the Connecticut Western Reserve." *North American Archaeologist* 14(1) (1993): 71–85.

Kirby, Jack Temple. *Poquosin: A Study of Rural Landscape and Society*. Chapel Hill: University of North Carolina Press, 1995.

Klein, Daniel B., and John Majewski. "Plank Road Fever in Antebellum America: New York State Origins." *New York History* 75 (January 1994): 39–65.

Klyza, Christopher McGrory. *Who Controls Public Lands?: Mining, Forestry, and Grazing Policies, 1870–1990*. Chapel Hill: University of North Carolina Press, 1996.

LaLande, Jeff. *An Environmental History of the Little Applegate River Watershed, Jackson County, Oregon*. Medford, Oregon: U.S. Department of Agriculture, Forest Service, Rogue River National Forest, 1995.

Langston, Nancy. *Forest Dreams, Forest Nightmares: The Paradox of Old Growth in the Inland West*. Seattle: University of Washington Press, 1995.

Lee, Francis B. "Forests of Colonial Jersey as the Settlers Found Them." *Forester* 1 (1895): 30.

Lehmann, Richard A. "The Laws of the Land." *Wisconsin Natural Resources* 17 (June 1993): 4–7.

Lehmkuhl, John F., et al. *Historical and Current Forest Landscapes of Eastern Oregon and Washington. Part I: Vegetation Pattern and Insect and*

Disease Hazards. Portland, Oregon: U.S. Department of Agriculture, Forest Service, Pacific Northwest Research Station, 1994.

Lemon, James. *The Best Poor Man's Country: A Geographical Study of Early Southeastern Pennsylvania*, 1972.

Lewis, Ronald L. *Transforming the Appalachian Countryside: Railroads, Deforestation, and Social Change in West Virginia, 1880–1920*. Chapel Hill: University of North Carolina Press, 1998.

Little, Charles E. *The New Oregon Trail: An Account of the Development and Passage of State Land-Use Legislation in Oregon*. Washington, D.C.: Conservation Foundation, 1974.

———. *The Dying of the Trees: The Pandemic in America's Forests*. New York: Viking, 1995.

Loope, Lloyd L., and Arthur C. Medeiros. "Strategies for Long-Term Protection of Biological Diversity in Rainforests of Haleakala National Park and East Maui, Hawaii." *Endangered Species Update* 12(6) (June 1995): 1–6.

Lutz, Harold J. "Original Forest Composition in Northwestern Pennsylvania as Indicated by Early Land Survey Notes." *Journal of Forestry* 28 (December 1930): 1098–1103.

Lyon-Jenness, Cheryl. "A Telling Tirade: What Was the Controversy Surrounding Nineteenth-Century Midwestern Tree Agents Really All About?" *Agricultural History* 72 (fall 1998): 675–707.

MacCleery, Douglas W. "America's Forests in the Early Days." *Forests and People* 43 (1993): 19–20.

Magruder, John. "Early Wood Uses." *Virginia Forests* 50 (winter 1995): 35–37.

Manson, Marsden. *Observations on the Denudation of Vegetation — A Suggested Remedy for California*. San Francisco: n.p., 1899.

Martin, Calvin. "Fire and Forest Structure in Aboriginal Eastern Forest." *Indian Historian* 6 (summer 1973): 23–26.

Maunder, Elwood R. "The History of Land Use in the Housatonic Valley." *Connecticut Woodlands* 33 (spring 1968): 10–14.

McBride, Joe, and Diana Jacobs. *The Ecology of Redwood and the Impact of Man's Use of the Redwood Forest for Recreational Activities*. Berkeley: University of California Press, 1977.

McCullough, Robert. *The Landscape of Community: A History of Communal Forests in New England*. Hanover, N.H.: University Press of New England, 1995.

McDevitt, Robert, ed. *From Sawmills to Villages*. Marion, Wisc.: Marion Advertiser, 1992.

McGregor, Robert Kuhn. "Changing Technologies and Forest Consumption in the Upper Delaware Valley, 1790–1880." *Journal of Forest History* 32 (April 1988): 69–81.

Mecozzi, Maureen. "It Started with Fire." *Wisconsin Natural Resources* 18 (February 1994): 1–16.

Meeker, Ezra. *Washington Territory West of the Cascade Mountains, Containing a Description of Puget Sound, and Rivers Emptying Into It, the Lower Columbia, Shoalwater Bay, Gray's Harbor, Timber, Lands, Climate, Fisheries.* Olympia, Wash.: Printed at the Transcript Office, 1870.

Mehls, Steven F. *The New Empire of the Rockies: A History of Northeast Colorado.* Cultural Resources Series, no. 16. Denver: Bureau of Land Management, 1984.

Mei, Mary A. *Timber Resources of Northwest Oregon.* Portland, Ore.: Department of Agriculture, Forest Service, Pacific Northwest Forest and Range Experiment Station, 1979.

Meinecke, Emilio Pepe Michael. *A Report upon the Effect of Excessive Tourist Travel on the California Redwood Parks.* Sacramento: California State Printing Office, 1929.

Middleton, Harry, and Nick Lyons. "Metallic Mountain." *Audubon* 95 (November/December 1993): 56–58.

Miller, Char. "Sawdust Memories: Pinchot and the Making of Forest History." *Journal of Forestry* 92 (February 1994): 8–12.

———. *American Forests: Nature, Culture, and Politics.* Lawrence: University Press of Kansas, 1998.

Miller, Herman Lunden. *Lumbering in Early Twentieth Century Michigan, The Kneeland-Bigelow Company Experience.* Lewiston, Mich.: Walnut Hill Press, 1995.

Miller, Perry. *Errand Into the Wilderness.* Cambridge: Harvard University Press, 1956.

Miller, Robin. "Reforestation: A Louisiana Success Story." *Forests and People* 45(2) (1995): 4–5, 7.

Minore, Don, Joseph N. Graham, and Edward W. Murray. *Environment, Vegetation, and Regeneration After Timber Harvest in the Applegate Area of Southwestern Oregon.* Portland, Ore.: U.S. Department of Agriculture, Forest Service, Pacific Northwest Forest and Range Experiment Station, 1982.

Muir, John. "The American Forests." *Atlantic Monthly* 80 (August 1897): 145–57.

———. "The Wild Parks and Forest Reservations of the West." *Atlantic Monthly* 81 (January 1898): 15–28.

———. *Our National Parks.* Boston: Houghton Mifflin, 1901.

Murphy, Thomas Dowler. *Oregon, the Picturesque: A Book of Rambles in the Oregon Country and in the Wilds of Northern California …* Boston: Page Company, 1917.

Napier, John Hawkins, III. "The Gilded Pearl: From Settlers to Sawmill Hands." *Gulf Coast Historical Review* 10 (fall 1994): 111–21.

Newman, Cathy. "The Lure of the Catskills." *National Geographic* 182(5) (November 1992): 108–31.

Nodvin, Stephen C., and Thomas A. Waldrop, eds. *Fire and the Environment: Ecological and Cultural Perspectives.* Asheville, N.C.: Southeastern Forest Experiment Station, 1990.

Northern Pacific Railroad Company. *The Northern Pacific Railroad: Its Route, Resources, Progress and Business: The New Northwest and Its Great Thoroughfare.* Philadelphia: J. Cooke, 1872.

Nowak, David J. "Historical Vegetation Change in Oakland and Its Implications for Urban Forest Management." *Journal of Arboriculture* 19 (September 1993): 313–19.

Ogden, J. Gordon. "Forest History of Martha's Vineyard, Massachusetts. I. Modern and Pre-Colonial Forests." *American Midland Naturalist* 66 (October 1961): 417–30.

Osborn, John. "Clearcuts and Railroads: Reforming the 1864 Northern Pacific Land Grant." *Wild Forest Review* (February 1994): 49–52.

Paananen, Donna M., Richard F. Fowler, and Louis F. Wilson. "The Aerial War Against Eastern Region Forest Insects, 1921–86." *Journal of Forest History* 31 (October 1987): 173–186.

Patterson, Rich. "Fire in the Oaks." *American Forests* 98 (November/December 1992): 58–59.

Pennsylvania Heritage 20 (winter 1994): 17–23.

Pfaff, Tim. *Settlement and Survival: Building Towns in the Chippewa Valley, 1850–1925.* Eau Claire, Wisc.: Chippewa Valley Museum Press, 1994.

Pierce, Arthur Dudley. *Iron in the Pines: The Story of New Jersey's Ghost Towns and Bog Iron.* New Brunswick, N.J.: Rutgers University Press, 1957.

Pierson, George, and George Zimmerman. "Restoring Jersey's Atlantic White-cedar." *Allegheny News* 3 (winter 1994): 17–19.

Pinchot, Gifford. *Biltmore Forest, the Property of Mr. George W. Wanderbiltf...* Chicago: R. R. Donnelley, 1893.

——. *The Adirondack Spruce: A Study of the Forest in Ne-Ha-Sa-Ne Park.* New York: Arno, 1898.

——. "The Blazed Trail of Forest Depletion." *American Forestry* 29 (June 1923): 323–28, 374.

Pinchot, Gifford, and W. W. Ashe. *Timber Trees and Forests of North Carolina.* Winston, N.C.: M. I. and J. C. Steward, Public Printers, 1897.

Pratt, George DuPont. "The Use of the New York State Forests for Public Recreation." *Proceedings of the Society of American Foresters* 11 (July 1916): 281–85.

Pyne, Stephen J. *Fire in America: A Cultural History of Wildland and Rural Fire.* Princeton: Princeton University Press, 1982.

——. "Keeper of the Flame: A Survey of Anthropogenic Fire." In *Fire in the Environment: The Ecological, Atmospheric, and Climatic Importance of Vegetation Fires,* edited by P. J. Crutzen and J. G. Goldammer. New York: Wiley, 1993, pp. 245–66

Pytte, Alyson. "Timber, Spotted Owl Interests Find Middle Ground Elusive." *Congressional Quarterly Weekly Report* 48(39) (29 September 1990): 3104–7

Rajala, Richard A. *Clearcutting the Pacific Rain Forest: Production, Science, and Regulation.* Seattle: University of Washington Press, 1988.

——. "The Forest as Factory: Technological Change and Worker Control in the West Coast Logging Industry, 1880–1930." *Labour/Le Travail* 32 (fall 1993): 73–104.

Robbins, William G. *American Forestry: A History of National, State, and Private Cooperation.* Lincoln and London: University of Nebraska Press, 1985.

——. "Landscape and Environment: Ecological Change in the Intermontane Northwest." *Pacific Northwest Quarterly* 84 (October 1993): 140–49.

Robertson, Janet. *The Magnificent Mountain Women: Adventures in the Colorado Rockies.* Lincoln: University of Nebraska Press, 1990.

Rothman, Hal K., ed. *"I'll Never Fight with My Bare Hands Again": Recollections of the First Forest Rangers of the Inland Northwest.* Lawrence: University Press of Kansas, 1994.

Runte, Alfred. *Trains of Discovery: Western Railroads and the National Parks.* Flagstaff, Ariz.: Northland Press, 1990.

Rupp, Alfred E. "History of Land Purchase in Pennsylvania." *Journal of Forestry* 22 (May 1924): 490–497.

Ryan, J. C. "Who Logged Here?" *Timber Bulletin* 49 (October/November 1994): 18–21.

Sarvis, Will. "The Mount Rogers National Recreation Area and the Rise of Public Involvement in Forest Service Planning." *Environmental History Review* 18 (summer 1994): 40–65.

Schama, Simon. *Landscape and Memory.* New York: Knopf, 1995.

Sieber, Ellen, and Cheryl Ann Munson. *Looking at History: Indiana's Hoosier National Forest Region, 1600–1950.* Washington, D.C.: Government Printing Office, 1992.

Silver, Timothy. *A New Face on the Countryside: Indians, Colonists, and Slaves in the South Atlantic Forests, 1500–1800.* New York: Cambridge University Press, 1990.

Silversides, C. Ross. *Broadaxe to Flying Shear: The Mechanization of Forest Harvesting East of the Rockies.* Ottawa: National Museum of Science and Technology, 1997.

Smith, Kenneth L. *Sawmill: The Story of Cutting the Last Great Virgin Forest East of the Rockies.* Fayetteville: University of Arkansas Press, 1986.

Spurr, Stephen H., and Burton V. Barnes. "The American Forest Since 1600." In *Forest Ecology.* New York: Ronald Press, 1964, pp. 557–71.

Stanger, Frank M. *Sawmills in the Redwoods: Logging on the San Francisco Peninsula, 1849–1967.* San Mateo, Calif.: San Mateo County Historical Society, 1967.

Stevens, Sylvester K. "When Timber Was King in Pennsylvania." *Pennsylvania History* 19 (October 1952): 391–395.

Stillgoe, John R. *Common Landscape of America, 1580–1845.* New Haven: Yale University Press, 1982.

Strong, Douglas H. *Tahoe: An Environmental History.* Lincoln: University of Nebraska Press, 1984.

Tamura, Linda. "Railroads, Stumps, and Sawmills: Japanese Settlers of the Hood River Valley." *Oregon Historical Quarterly* 94 (winter 1993–94): 368–98.

Theiss, Lewis Edwin. "The Pioneer and the Forest." *Pennsylvania History* 23 (October 1956): 487–503.

Thompson, Jeanne Porges. "Stehekin, Washington: An Analysis of National Park Service Land Acquisition and Management 25 Years After Establishment of the Lake Chelan National Recreation Area." *Journal of Environmental Law and Litigation* 9(1) (1994): 215–47.

Thoreau, Henry David. *Walden*. Boston: Ticknor and Fields, 1854.

———. *The Maine Woods*. Boston: Ticknor and Fields, 1864.

Thoreau, Henry D., John Burroughs, John Muir, Bradford Torrey, Dallas Lore Sharp, and Olive Thorne Miller. *In American Fields and Forests*. New York: Houghton Mifflin, 1909.

Thrower, Nina Gale. "Medicinal Value of Plants in Alabama's Forests." *Alabama's Treasured Forests* 13 (fall 1994): 10–11.

U.S. Department of Agriculture, Forest Service, Boise National Forest. *Snapshot in Time: Repeat Photography on the Boise National Forest, 1870–1992*. Washington, D.C.: Government Printing Office for U.S. Department of Agriculture, Forest Service, Boise National Forest, 1993.

Vale, Thomas R., and Geraldine R. Vale. *Time and the Tuolumne Landscape: Continuity and Change in the Yosemite High Country*. Salt Lake City: University of Utah Press, 1994.

Van Duers, George. "The Russians Logged the Redwoods First." *American Forests* 65 (January 1959): 29–31, 61–62.

Vogel, John N. *Great Lakes Lumber on the Great Plains: The Laird, Norton Lumber Company in South Dakota*. Iowa City: University of Iowa Press, 1992.

Walker, Laurence C. *The North American Forests: Geography, Ecology, and Silviculture*. New York: CRC Press, 1998.

Warren, Viola Lockhart. "The Eucalyptus Crusade." *Historical Society of Southern California Quarterly* 44 (March 1962): 31–42.

Webb, Walter Prescott. *The Great Frontier*. Boston: Houghton Mifflin, 1952.

Weidensaul, Scott. *Mountains of the Heart: A Natural History of the Appalachians*. Golden, Colo.: Fulcrum, 1994.

White, Peter S. "Conserving Biodiversity: Lessons from the Smokies." *Forum for Applied Research and Public Policy* 10 (summer 1995): 116–20.

White, Richard. *Land Use, Environment, and Social Change: The Shaping of Island County, Washington*. Seattle: University of Washington Press, 1980.

Whitney, Gordon G. *From Coastal Wilderness to Fruited Plain: A History of Environmental Change in Temperate North America from 1500 to the Present*. New York: Cambridge University Press, 1994.

Wilkerson, Hugh, and John Van der Zee. *Life in the Peace Zone: An American Company Town*. New York: Macmillan, 1971.

Williams, Michael. *Americans and Their Forests: A Historical Geography*. New York: Cambridge University Press, 1989.

Williams, Ted. "Whose Woods Are These? Saving the Northern Forest Will Take More Than Task Forces and Studies." *Audubon* 96(3) (May/June 1994): 26–34.

Winter, Kate H. *The Woman in the Mountain: Reconstructions of Self and Land by Adirondack Women Writers.* Albany: State University of New York Press, 1989.

Wood, Richard G. *A History of Lumbering in Maine, 1820–1861.* Orono, Maine: University Press, 1935.

Woolsey, Nathan. "Simpson and Company: Victorian Sawmill of the Gulf Coastal Plain, 1865–1882." *Gulf Coast Historical Review* 10 (fall 1994): 122–39.

Wrinn, Jim. "Rails Through the Hills." *Wildlife in North Carolina* 58 (January 1994): 10–17.

Wyckoff, William, and Lary M. Dilsaver, eds. *The Mountainous West: Explorations in Historical Geography.* Lincoln: University of Nebraska Press, 1995.

Young, Janet. "Mason: A Lumbermill Town." *Historical Happenings* 15 (spring 1994): 1–2, 8.

Zedler, Paul H., Clayton R. Gautier, and Gregory S. McMaster. "Vegetation Change in Response to Extreme Events: The Effect of a Short Interval Between Fires in California Chaparral and Coastal Scrub." *Ecology* 64(4) (1983): 809–18.

Zeisler-Vralsted, Dorothy. "Reclaiming the Arid West: The Role of the Northern Pacific Railway in Irrigating Kennewick, Washington." *Pacific Northwest Quarterly* 84 (October 1993): 130–39.

15. Mining History

Amundson, Michael A. "Home on the Range No More: The Boom and Bust of a Wyoming Uranium Mining Town, 1957–1988." *Western Historical Quarterly* 26(4) (winter 1995): 483–506.

Anon. *Colorado's Gold Fields; America's Most Famous Gold Mining Camps.* Denver: Calhoun, 1897.

Bining, Arthur Cecil. *Pennsylvania Iron Manufacture in the Eighteenth Century.* Harrisburg: Pennsylvania Historical Commission, 1938.

Black, Brian. "Oil Creek as Industrial Apparatus: Re-creating the Industrial Process Through the Landscape of Pennsylvania's Oil Boom." *Environmental History* 3(2) (April 1998): 210–29.

Boehnert, Caryl. "Juneau After the Gold Rush: Fighting Dangerous Development." *Multinational Monitor* 14(3) (March 1993): 9–13.

Bushnell, H.K. *Butte City, Montana, the Largest Mining Camp on Earth.* Helena, Mont.: H.K. Bushnell, 189-?.

Campbell, Lawrence James, et al. *Skagway: A Legacy of Gold.* Anchorage: Alaska Geographic, 1992.

Canfield, John G. *Mines and Mining Men of Colorado, Historical, Descriptive and Pictorial; An Account of the Principal Producing Mines of Gold and Silver.* Denver, Colo.: J.G. Canfield, 1893.

Caughey, J.W. *The California Gold Rush* [1948]. Berkeley: University of California Press, 1975.

Chilson, Peter. "Coal Miners' Story." *Audubon* 96(2) (March/April 1994): 50–65.

Clappe, Louise Amelia. *The Shirley Letters, Being Letters Written in 1851–1852 from the California Mines.* Salt Lake City: Peregrine Smith Books, 1985.

Coates, Peter A. *The Trans-Alaska Pipeline Controversy: Technology, Conservation, and the Frontier.* Bethlehem, Penn.: Lehigh University Press, 1991.

Cohen, Stan. *Gold Rush Gateway, Skagway and Dyea, Alaska.* Missoula, Mont.: Pictorial Histories, 1986.

Cole, Terrence. *Crooked Past: The History of a Frontier Mining Camp, Fairbanks, Alaska.* Fairbanks: University of Alaska Press, 1991.

Comstock, David A. "Proper Women at the Mines: Life at Nevada City in the 1850's." *Pacific Historian* 28(3) (1984): 65–73.

Davis, Joseph A. "The Wasting of Nevada: Yucca Mountain as a Repository for High-Level Nuclear Waste." *Sierra* 73(4) (July/August 1988).

Derickson, Alan. *Black Lung: Anatomy of a Public Health Disaster.* Ithaca, N.Y.: Cornell University Press, 1998.

Dillon, Richard, ed. "Mother Lode Memoir: Reminiscences of George A. Marshall." *Journal of the West* 3(3) (1964): 355–68.

Fahy, Neil E. *San Mateo County Sourdoughs in the Klondike and Nome Gold Rushes of 1896–1900.* San Mateo, Calif.: San Mateo County Historical Association, 1988.

Fairfax, Sally K., and Carolyn Yale. *Federal Lands: A Guide to Planning, Management, and State Revenues.* Washington, D.C.: Island Press, 1987.

Finn, Janet L. *Tracing the Veins: Of Copper, Culture, and Community from Butte to Chuquicamata.* Berkeley: University of California Press, 1998.

Freeman, Harry Campbell. *A Brief History of Butte, Montana, the World's Greatest Mining Camp; Including a Story of the Extraction and Treatment of Ores from Its Gigantic Copper Properties.* Chicago: H. O. Shepard Company, Printers, 1900.

Gates, Michael. *Gold at Fortymile Creek: Early Days in the Yukon.* Vancouver: UBC Press, 1994.

Gharst, Michael, ed. "The Klondike Journal of William Mann." *Pacific Northwest Forum* 8 (winter/spring 1995): 5–28.

Gutiérrez, Ramón, and Richard J. Orsi, eds. *Contested Eden: California Before the Gold Rush.* Berkeley: University of California Press, 1998.

Harts, William Wright. *The Control of Hydraulic Mining in California by the Federal Government.* New York: n.p., 1906.

Hevly, Bruce and John M. Findlay, eds. *The Atomic West.* Seattle: Center for the Study of the Pacific Northwest in association with the University of Washington Press, 1998.

Hurtado, Albert L. "Sex, Gender, Culture, and a Great Event: The California Gold Rush." *Pacific Historical Review* 68 (November 1999): 1–20.

Jackson, W. Turrentine. *The History of Mining in the Plumas Eureka State Park Area, 1851–1890.* Sacramento, Calif.: Division of Beaches and Parks, October 1960.

Jones, Fayette Alexander. *Old Mining Camps of New Mexico, 1854–1904.* Santa Fe, N.M.: Stagecoach Press, 1964.

Kelley, Robert L. "Forgotten Giant: The Hydraulic Gold Mining Industry in California." *Pacific Historical Review* (November 1954): 343–56.

———. *Gold vs. Grain: The Hydraulic Mining Controversy in California's Sacramento Valley,* 1959.

Kirchhoff, M. J. *Historic McCarthy: The Town that Copper Built.* Juneau: Alaska Cedar Press, 1993.

Klyza, Christopher McGrory. *Who Controls Public Lands?: Mining, Forestry, and Grazing Policies, 1870–1990.* Chapel Hill: University of North Carolina Press, 1996.

Limerick, Patricia. *The Legacy of Conquest.* New York: Norton, 1987.

Manning, Richard. "Going for the Gold: Under an 1872 Law, Miners Tore Up the West." *Audubon* 96(1) (January/February 1994): 68–77.

Meals, Hank, and Dennis Stevens. *Tailings and Time Travel — The Plot "Slickens": A History of Placer Mining Methods and Communities Along a Portion of the North Yuba River, 1848–1942.* N.p.: U.S. Department of Agriculture, Forest Service, 1993.

Paddison, Joshua, ed. *A World Transformed: Firsthand Accounts of California Before the Gold Rush.* Berkeley: Heyday Books, 1999.

Quinn, M. L. "Industry and Environment in the Appalachian Copper Basin, 1890–1930." *Technology and Culture* 34(3) (July 1993): 575–613.

Rawls, James J., and Richard J. Orsi, eds. *A Golden State: Mining and Economic Development in Gold Rush California*. Berkeley: University of California Press, 1999.

Robbins, William G. *Colony and Empire: The Capitalist Transformation of the American West*. Lawrence: University Press of Kansas, 1994.

Roppel, Patricia. *Fortunes from the Earth: An History of the Base and Industrial Minerals of Southeast Alaska*. Manhattan, Kans.: Sunflower University Press, 1991.

Royce, Sarah. *A Frontier Lady: Recollections of the Gold Rush and Early California* [1932]. Lincoln: University of Nebraska Press, 1977.

Smith, Duane A. *Rocky Mountain Mining Camps: The Urban Frontier*. Bloomington: Indiana University Press, 1967.

———. *Mining America: The Industry and the Environment, 1800–1980*. Lawrence: University Press of Kansas, 1987.

Stanton, Robert Brewster. *The Hoskaninni Papers: Mining in Glen Canyon, 1897–1902*. Edited by C. Gregory Crampton and Dwight L. Smith. Salt Lake City: University of Utah Press, 1961.

Tonopah Promotion Association. *Nye County, Nevada and the Mineral Resources of Her Fifty Mining Camps*. Topnopah, Nev.: Tonopah Promotion Association, 1909.

Trexler, Harrison A. *Flour and Wheat in the Montana Gold Camps, 1862–1870*. Missoula, Mont.: Dunstan Printing and Stationery, 1918.

Turnipseed, Donna. "Skim Digging for Dignity: A Look at Placer Mining in Idaho County During the Great Depression." *Idaho Yesterdays* 42 (spring 1998): 18–24.

Vischer, Edward. "A Trip to the Mining Regions in the Spring of 1859." *California Historical Society Quarterly* 11(4) (1932): 321–38.

Wellock, Thomas Raymond. *Critical Masses: Opposition to Nuclear Power in California, 1958–1978*. Madison: The University of Wisconsin Press, 1998.

Wells, Merle W. *Gold Camps and Silver Cities: Nineteenth Century Mining in Central and Southern Idaho*. Moscow, Idaho: Idaho Department of Lands, Bureau of Mines and Geology, 1983.

West, Elliott. *The Contested Plains: Indians, Goldseekers, and the Rush to Colorado*. Lawrence: University of Kansas Press, 1998.

Weston, Silas. *Four Months in the Mines of California: or, Life in the Mountains*. Providence, R.I.: B. T. Albro, printer, 1854.

Whitney, J. D. *The Metallic Wealth of the United States*. Philadelphia: Lippincott, Grambo, 1854.

Williams, Ted. "Alaska's Rush for the Gold." *Audubon* 95(6) (November/ December 1993): 50–55.

Willis, George F. *"It Is a Hard Country, Though": Historic Resource Study, Bering Land Bridge National Preserve*. Washington, D.C.: National Park Service, U.S. Department. of the Interior, 1986.

Wyckoff, William. *Creating Colorado: The Making of a Western American Landscape, 1860–1940*. New Haven: Yale University Press, 1999.

Woods, Thomas J. "Alaskan Oil and Gas Prospects: Boom or Bust?" *Oil and Gas Journal* 92(7) (14 February 1994): 91–95.

16. Pollution

Ayres, Robert U., and Samuel R. Rod. "Patterns of Pollution in the Hudson-Raritan Basin: Reconstructing an Environmental History." *Environment* 28 (May 1986): 14–20, 39–43.

Bay Area Air Pollution Control District. *Air Pollution and the San Francisco Bay Area*. 8th ed. San Francisco: Bay Area Air Pollution Control District, 1973.

Brown, Michael H. *Laying Waste: The Poisoning of America by Toxic Chemicals*. New York: Pantheon, 1980.

Clark, John G. *Energy and the Federal Government: Fossil Fuel Policies, 1900–1946*. Champaign: University of Illinois Press, 1987.

Commoner, Barry. *The Closing Circle*. New York: Knopf, 1972.

Davis, Lewis G. *The Problem of Air Pollution in the San Francisco Bay Area*. Berkeley: University of California, 1971.

Dunlap, Thomas R. *DDT: Scientists, Citizens, and Public Policy*. Princeton: Princeton University Press, 1981.

Eichstaedt, Peter H. *If You Poison Us: Uranium and Native Americans*. Santa Fe, N.M.: Red Crane Books, 1994.

Ellis, William, and Jim Richardson. "The Mississippi: River Under Siege." *National Geographic* (special edition: "Water: The Power, Promise, and Turmoil of North America's Fresh Water") 184(n5A) (November 1993): 90–104.

Fabry, Judith. "Enlightened Selfishness: Great Falls and the Sun River Project." *Montana* 44 (winter 1994): 14–27.

Forstenzer, Martin. "Mono Lake's Deadly Dust." *Audubon* 95(5) (September/October 1993): 29–34.

Gilliam, Harold. *Between the Devil and the Deep Blue Bay: The Struggle to Save San Francisco Bay.* San Francisco: Chronicle Books, 1969.

Green, Rick. "Another Kesterson?" *Sierra* 70(5) (September 1985): 29–34.

Harris, Glenn, and Seth Wilson. "Water Pollution in the Adirondack Mountains: Scientific Research and Governmental Response, 1890–1930." *Environmental History Review* 17 (winter 1993): 47–71.

Harris, Tom. "The Kesterson Syndrome: The Federal Irrigation Projects That Made the Deserts Bloom Are Now Killing Wildlife Throughout the West." *The Amicus Journal* 11(4) (fall 1989): 4–7.

Holker, Douglas L. *Effects on the Los Angeles River Due to Urbanization.* 1982.

Kuletz, Valerie. *The Tainted Desert: Environmental Ruin in the American West.* New York: Routledge, 1998.

Krier, James, and Edmund Ersin. *Pollution and Policy: A Case Essay on California and the Federal Experience with Motor Vehicle Air Pollution.* Berkeley: University of California Press, 1977.

Lear, Linda J. "Bombshell in Beltsville: The USDA and the Challenge of "Silent Spring." *Agricultural History* 66(2) (spring 1992): 151–71.

Lee, Douglas B. "Tragedy in Alaska Waters." *National Geographic* 176(2) (August 1989): 260–64.

Matsen, Brad. "The Once and Future Spill: In the Wake of 1989's Exxon Valdez Oil Spill Disaster, Has Anything Really Changed?" *Audubon* 96(4) (July/August 1994): 116.

McCarty, James C. *Water Pollution and San Francisco Bay.* San Francisco: Federal Water Pollution Control Administration, U.S. Department of the Interior, 1966.

Melosi, Martin V. *Garbage in the Cities: Refuse, Reform, and the Environment, 1880–1980.* College Station: Texas A&M University Press, 1981.

———. *The Sanitary City: Urban Infrastructure in America from Colonial Times to the Present.* Baltimore: Johns Hopkins University Press, 2000.

———. *Effluent America: Cities, Industry, Energy, and the Environment.* Pittsburgh: University of Pittsburgh Press, 2001.

———, ed. *Pollution and Reform in American Cities, 1870–1930.* Austin: University of Texas Press, 1980.

Nash, Gerald D. *United States Oil Policy, 1890–1964: Business and Government in Twentieth Century America.* Pittsburgh: University of Pittsburgh Press, 1968.

O'Neill, Dan. "Project Chariot: How Alaska Escaped Nuclear Excavation." *Bulletin of the Atomic Scientists* 45(10) (December 1989): 28–38.

Richards, Ellen Swallow. *Sanitation in Daily Life.* Boston: Whitcomb and Barrows, 1907.

Riis, Jacob. *How the Other Half Lives.* New York: Scribner's, 1890.

Sinclair, Upton. *The Jungle.* New York, Doubleday, Page, 1906.

Solnit, Rebecca. "Dust, or Erasing the Future: The Nevada Test Site" in *Savage Dreams.* San Francisco: Sierra Club Books, 1994.

Steinberg, Theodore. *Nature Incorporated: Industrialization and the Waters of New England.* New York: Cambridge University Press, 1991.

Tarr, Joel A. "Out of Sight, Out of Mind: A Brief History of Sewage Disposal in the United States." *American History Illustrated* 10 (1976): 40–47.

——. "Historical Turning Points in Municipal Water Supply and Wastewater Disposal, 1850–1932." *Civil Engineering* 47 (1977): 82–91.

——. *The Search for the Ultimate Sink: Urban Pollution in Historical Perspective.* Akron, Ohio: University of Akron press, 1996.

Turner, R. Eugene, and Nancy N. Rabalais. "Changes in Mississippi River Water Quality this Century." *BioScience* 41(3) (March 1991): 140–48.

Vollers, Maryanne. "Everyone Has Got to Breathe" (Air Pollution in Chester, Pennsylvania). *Audubon* 97(2) (March/April 1995): 64–74.

Wargo, John. *Our Children's Toxic Legacy: How Science and Law Fail to Protect Us From Pesticides.* New Haven: Yale University Press, 1996.

Williams, Ted. "Death in a Black Desert: in California's Fields, Toxic Runoff is Poisoning the Land." *Audubon* 96(1) (January/February 1994): 24–30.

Wirt, Laurie. *Radioactivity in the Environment: A Case Study of the Puerco and Little Colorado River Basins, Arizona and New Mexico.* Tucson: U.S. Geological Survey, U.S. Department of the Interior, 1994.

Wirth, John. *Smelter Smoke in North America: The Politics of Transborder Pollution.* Lawrence: University Press of Kansas, 2000.

Worldwatch Institute. *Nuclear Waste: The Problem that Won't Go Away.* Washington, D.C.: Worldwatch Institute, December 1991.

17. Range History

Bozell, John R. "Culture, Environment, and Bison Populations on the Late Prehistoric and Early Historic Central Plains." *Plains Anthropologist* 40(152) (May 1995): 145–64.

Branch, Edward Douglas. *The Hunting of the Buffalo*. Introduction by J. Frank Dobie. 1st ed. 1929. Lincoln: University of Nebraska Press, 1997.

Brandt, C. A., and W. H. Richard. "Alien Taxa in the North American Shrub-Steppe Four Decades After Cessation of Livestock Grazing and Cultivation Agriculture." *Biological Conservation* 68(2) (1994): 95–105.

Branson, Farrel Allen. *Vegetation Changes on Western Rangelands*. Denver: Society for Range Management, 1985.

Chadwick, Douglas H. "Sagebrush Country: America's Outback." *National Geographic* 175(1) (January 1989): 52–80.

———. "What Good Is a Prairie?" *Audubon* 97(6) (November/December 1995): 36–50.

Chadwick, Douglas H., and Jim Brandenburg. "The American Prairie: Roots of the Sky." *National Geographic* 184(4) (October 1993): 90–120.

Clawson, Marion. *The Western Range Livestock Industry*. New York: McGraw-Hill, 1950.

Cleland, R. G. *The Place Called Sespe: A History of a California Ranch*. Alhambra, Calif.: private printing, 1940.

Cleland, Robert G. *The Cattle on a Thousand Hills: Southern California, 1850–1880*. 2d ed. San Marino, Calif.: Huntington Library, 1951.

Cook, C. Wayne, and Edward F. Redente. "Development of the Ranching Industry in Colorado." *Rangelands* 15 (October 1993): 204–207.

Crampton, B. *Grasses in California*. Berkeley: University of California Press, 1974.

Cronise, Titus Fey. *The Natural Wealth of California*. San Francisco and New York: H. H. Bancroft, 1868.

Dana, Samuel T., and Sally K. Fairfax. *Forest and Range Policy, Its Development in the United States*. 2d ed. New York: McGraw-Hill, 1980.

Dixon, Joseph Scattergood. *The Common Hawks and Owls of California from the Standpoint of the Rancher*. Berkeley: University of California, College of Agriculture, Agricultural Experiment Station, 1922.

Duram, Leslie Aileen. "The National Grasslands: Past, Present and Future." *Rangelands* 17 (April 1995): 36–42.

Faragher, John Mack. *Women and Men on the Overland Trail*. New Haven: Yale University Press, 1979.

Fleharty, Eugene D. *Wild Animals and Settlers on the Great Plains*. Norman: University of Oklahoma Press, 1995.

Flores, Dan. "Bison Ecology and Bison Diplomacy: The Southern Plains from 1800 to 1850." *Journal of American History* 78 (September 1991): 465–85.

Harrington, John A., Jr., and Jay R. Harman. "Climate and Vegetation in Central North America: Natural Patterns and Human Alterations." *Great Plains Quarterly* (spring 1991): 103–112.

Hodgson, Bryan. "Buffalo: Back Home on the Range." *National Geographic* 186(5) (November 1994): 64–90.

Igler, David. "Industrial Cowboys: Corporate Ranching in Late Nineteenth-Century California." *Agricultural History* 69(2) (spring 1995): 201–16.

Isenberg, Andrew. *The Destruction of the Bison: Social and Ecological Changes in the Great Plains, 1750–1920.* New York: Cambridge University Press, 2000.

Klyza, Christopher McGrory. *Who Controls Public Lands?: Mining, Forestry, and Grazing Policies, 1870–1990.* Chapel Hill: University of North Carolina Press, 1996.

Lindgren, H. Elaine. *Land in Her Own Name: Women as Homesteaders in North Dakota.* Fargo: North Dakota Institute for Regional Studies, 1991.

Luoma, Jon R. "Back Home on the Range" (Dubious Future for the American Bison's Habitat). *Audubon* 95(2) (March/April 1993): 46–53.

Madson, John. *Where the Sky Began: Land of the Tallgrass Prairie.* Revised ed. Ames: Iowa State University Press, 1995. Original edition, Boston: Houghton Mifflin, 1982.

Malin, James C. *The Grassland of North America: Prolegomena to Its History.* Lawrence, Kans.: James C. Malin (Ann Arbor, Mich.: Edwards Brothers), 1947.

Manning, Richard. *Grasslands: The History, Biology, Politics, and Promise of the American Prairie.* New York: Viking, 1995.

Matthews, Anne. *Where the Buffalo Roam.* New York: Grove Weidenfeld, 1992.

Myres, Sandra L. *Westering Women and the Frontier Experience, 1800–1915.* Albuquerque : University of New Mexico Press, 1982.

O'Brien, Michael J., et al. *Grassland, Forest, and Historical Settlement: An Analysis of Dynamics in Northeast Missouri.* Lincoln: University of Nebraska Press, 1984.

Olson, Steven. *The Prairie in Nineteenth-Century American Poetry.* Norman: University of Oklahoma Press, 1994.

Parkman, Francis. *The Oregon Trail.* New York, Chicago: Scribner's, 1924.

Patterson-Black, Sheryll. "Women Homesteaders on the Great Plains Frontier." *Frontiers* 1(2) (spring 1976): 67–88.

Popper, Frank J., and Deborah E. Popper. "The Reinvention of the American Frontier." *Amicus* (summer 1991): 4–7.

Rosenberg, Norman J. "Climate of the Great Plains Region of the United States." *Great Plains Quarterly* 7(1) (1987): 22–32.

Rowley, William D. *U.S. Forest Service Grazing and Rangelands: A History.* College Station: Texas A&M University Press, 1985.

Shirk, David Lawson. *The Cattle Drives of David Shirk from Texas to the Idaho Mines, 1871 and 1873: Reminiscences of David L. Shirk.* Portland, Ore.: Champoeg Press, 1956.

Smits, David D. "The Frontier Army and the Destruction of the Buffalo: 1865–1883." *Western Historical Quarterly* 25 (autumn 1994): 313–38.

Starrs, Paul F. *Let the Cowboy Ride: Cattle Ranching in the American West.* 1st ed. Baltimore: Johns Hopkins University Press, 1998.

Taniguchi, N. J. "Land, Laws, and Women: Decisions of the General Land Office, 1881–1920: A Preliminary Report." *Great Plains Quarterly* 13(4) (fall 1993): 223–36.

Thomas, Heather Smith. "History of Public Land Grazing." *Rangelands* 16 (December 1994): 50–55.

Webb, Walter Prescott. *The Great Plains.* Boston: Ginn and Company, 1931.

——. *The Great Frontier.* Boston: Houghton Mifflin, 1952.

Wood, W. Raymond. "An Introduction to the History of the Fur Trade on the Northern Plains." *North Dakota History* 61 (summer 1994): 2–6.

18. Water and Irrigation History

Abruzzi, William S. *Dam That River! Ecology and Mormon Settlement in the Little Colorado River Basin.* Lanham, Maryland: University Press of America, 1993.

Adams, Frank. *Investigations of the Economical Duty of Water for Alfalfa in Sacramento Valley, California, 1910–1915.* Sacramento: California State Printing Office, 1917.

Adams, Henry A. *The Story of Water in San Diego and What the Southern California Mountain Water Company Has Done to Solve the Problem.* Chula Vista, Calif.: Denrich Press, 1916.

Adams, John, Jr. *Damming the Colorado: The Rise of the Lower Colorado River Authority, 1933–1939*. College Station: Texas A&M University Press, 1990.

Adams, William Yewdale. *Ninety Years of Glen Canyon Archaeology, 1869–1959*. Flagstaff, Ariz.: Northern Arizona Society of Science and Art, 1960.

Alexander, Thomas G. "Stewardship and Enterprise: The LDS Church and the Wasatch Oasis Environment, 1847–1930." *Western Historical Quarterly* 25 (autumn 1994): 341–364.

Aton, James M., and Robert S. McPherson. *River Flowing from the Sunrise: An Environmental History of the Lower San Juan*. Logan, Utah: Utah State University Press, 2000.

August, Jack L., Jr. *Vision in the Desert: Carl Hayden and Hydropolitics in the American Southwest*. Fort Worth: Texas Christian University Press, 1999.

Austin, Mary H. *The Colorado River Controversy*. New York, 1927.

Babbitt, Bruce. "Age-Old Challenge: Water and the West." *National Geographic* 179(6) (June 1991): 2–4.

Barker, F. C. *Irrigation in Mesilla Valley, New Mexico*. Washington, D.C.: Government Printing Office, 1898.

Barry, John M. *Rising Tide: The Great Mississippi Flood of 1927 and How It Changed America*. New York: Simon and Schuster, 1997.

Bates, Sarah F., et al. *Searching Out the Headwaters: Change and Rediscovery in Western Water Policy*. Washington, D.C.: Island Press for the Natural Resources Law Center, University of Colorado School of Law, 1993.

Blake, Nelson Manfred. *Land into Water — Water into Land: A History of Water Management in Florida*. Tallahassee: University of Florida Press, 1980.

Bogue, Margaret Beattie. *Fishing the Great Lakes: An Environmental History, 1783–1933*. Madison: University of Wisconsin Press, 2000.

Brigham, Jay L. *Empowering the West: Electrical Politics Before FDR*. Lawrence: University Press of Kansas, 1998.

Brossard, Edgar Bernard. *Some Types of Irrigation Farming in Utah*. Logan: Utah Agricultural College Experiment Station, 1920.

Brough, Charles Hillman. *Irrigation in Utah*. Baltimore: The Johns Hopkins Press, 1898.

Buffum, Burt C. *The Use of Water in Irrigation in Wyoming and Its Relation to the Ownership and Distribution of the Natural Supply*. Washington, D.C.: Government Printing Office, 1900.

Carrels, Peter. *Uphill Against Water: The Great Dakota Water Wars*. Lincoln: University of Nebraska Press, 1999.

Carrier, Jim. "The Colorado: A River Drained Dry." *National Geographic* 179(6) (June 1991): 4–36.

Clark, Ira G. *Water in New Mexico: A History of its Management and Use*. Albuquerque: University of New Mexico Press, 1987.

Clements, Kendrick A. *Hoover, Conservation, and Consumerism: Engineering the Good Life*. Lawrence: University Press of Kansas, 2000.

Coate, Charles. "'The Biggest Water Fight in American History': Stewart Udall and the Central Arizona Project." *Journal of the Southwest* 37 (spring 1995): 79–101.

Conniff, Richard, and Rick Rickman. "California: Desert in Disguise." *National Geographic* (special edition: "Water: The Power, Promise, and Turmoil of North America's Fresh Water") 184(n5A) (November 1993): 38–54.

Cooke, Ronald U., and Richard W. Reeves. *Arroyos and Environmental Change in the American South-West*. London: Clarendon Press, 1976.

Crampton, C. Gregory. *The Complete Las Vegas: Including Hoover Dam and the Desert Water World of Lake Meade and Lake Mohave, Together with a Peep at Death Valley and a Visit to Zion National Park*. Salt Lake City: Peregrine Smith, 1976.

———. *Ghosts of Glen Canyon: History Beneath Lake Powell*. St. George, Utah: Publishers Place, 1986.

Cromwell, Larry, and James D. Goodridge. *Mono Lake, California, Water Balance*, 1979.

Culliney, John L. *Islands in a Far Sea: Nature and Man in Hawaii*. San Francisco: Sierra Club Books, 1988.

Davis, Arthur Powell. *Irrigation near Phoenix, Arizona*. Washington, D.C.: Government Printing Office, 1897.

———. *Report on the Irrigation Investigation: For the Benefit of the Pima and Other Indians on the Gila River Indian Reservation, Arizona*. Washington, D.C.: Government Printing Office, 1897.

deBuys, William. *Salt Dreams: Land and Water in Low-Down California*. Albuquerque: The University of New Mexico Press, 1999.

de Roos, Robert. *The Thirsty Land: The Story of the Central Valley Project*. Stanford, Calif.: Stanford University Press, 1948.

Dietrich, William. *Northwest Passage: The Great Columbia River*. Seattle: University of Washington Press, 1996.

Downey, Tom. "Riparian Rights and Manufacturing in Antebellum South Carolina: William Gregg and the Origins of the 'Industrial Mind.'" *Journal of Southern History* 65 (February 1999): 77–108.

Duplaix, Nicole. "Paying the Price" (South Florida Water Supplies). *National Geographic* 178(1) (July 1990): 88–114.

Elkind, Sarah S. "Industry and Water Distribution in California: The East Bay Municipal Utility District, 1920–1930." *Environmental History Review* 18 (winter 1994): 63–88.

——. *Bay Cities and Water Politics: The Battle for Resources in Boston and Oakland.* Maps by Aaron J. Weier. Lawrence: University Press of Kansas, 1998.

Espeland, Wendy Nelson. *The Struggle for Water: Politics, Rationality, and Identity in the American Southwest.* Chicago: University of Chicago Press, 1998.

Fiege, Mark. *Irrigated Eden: The Making of an Agricultural Landscape in the American West.* Foreword by William Cronon. Seattle: University of Washington Press, 1999.

Forbes, Robert Humphrey. *Irrigation in Arizona.* Washington, D.C.: Government Printing Office, 1911.

Fortier, Samuel, and Arthur A. Young. *Irrigation Requirements of the Arid and Semiarid Lands of the Southwest.* Washington, D.C.: U.S. Department of Agriculture, 1930.

Gammon, James R. *The Wabash River Ecosystem.* Plainfield, Ind.: Cinergy Corporation, 1998.

Gephard, Stephen R. "A History of Chapman's Pond and Its Preservation: Part II." *Connecticut Woodlands* 48 (spring 1983): 15–16.

Gerstell, Richard. *American Shad: A Three-Hundred-Year History in the Susquehanna River Basin.* University Park: Pennsylvania State University Press, 1998.

Gill, Garmulch, Edward Gray, and David Seckler. "The California Water Plan and Its Critics." In *California Water,* edited by David Seckler. Berkeley: University of California Press, 1971, pp. 3–27.

Gomez, Gay M. *A Wetland Biography: Seasons on Louisiana's Chenier Plain.* Austin: University of Texas Press, 1998.

Greely, A. W. *Report on the Climate of Arizona, with Particular Reference to Questions of Irrigation and Water Storage in the Arid Region.* Washington, D.C.: Government Printing Office, 1891.

Gregor, H. "Water and the California Paradox." In *California Revolution,* edited by Cary McWilliams. New York: Grossman, 1968, pp. 159–84.

Hanson, Elden G. *Irrigation Requirements for Cotton in Southern New Mexico*. New Mexico: New Mexico Agricultural Experiment Station, 1956.

Harden, Blaine. *A River Lost: The Life and Death of the Columbia*. New York: Norton, 1996.

Harris, Karl, et al. *Cotton Irrigation in the Southwest*: Agricultural Research Service, U.S. Department of Agriculture, 1959.

Hastings, Stephen H. *Irrigation and Related Cultural Practices with Cotton in the Salt River Valley of Arizona*. Washington, D.C.: U.S. Department of Agriculture, 1932.

Hem, John David. *Quality of Water of the Gila River Basin Above Coolidge Dam, Arizona*. (Series: U.S. Geological Survey, Water-supply paper 1104.) Washington, D.C.: Government Printing Office, 1950.

Hoffman, Abraham. "Joseph Barlow Lippincott and the Owens Valley Controversy: Time for Revision." *Southern California Quarterly* 54(3) (1972): 239–54.

———. *Vision or Villainy: Origins of the Owens Valley-Los Angeles Water Controversy*. College Station: Texas A&M University Press, 1981.

Hundley, Norris, Jr. *Property, Proprietorship and Politics: Law and the Structure of Strategic Opportunities in the California Water Industry*. Berkeley: University of California Press, 1963.

———. "The Politics of Water and Geography: California and the Mexican-American Treaty of 1944." *Pacific Historical Review* 36(2) (1967): 209–26.

———. *Water and the West: The Colorado River Compact and the Politics of Water in the American West*. Berkeley: University of California Press, 1975.

———. "The West Against Itself: The Colorado River — An Institutional History." In *New Courses for the Colorado River: Major Issues for the Next Century*, edited by Gary D. Weatherford and F. Lee Brown. Albuquerque: University of New Mexico Press, 1986, pp. 9–49.

———. *The Great Thirst: Californians and Water, 1770s–1990s*. Berkeley: University of California Press, 1992.

Hurt, R. Douglas. "Irrigation in the West." *Journal of the West* 30 (April 1991): 63–77.

Igler, David. "When Is a River Not a River?" *Environmental History* 1(2) (April 1996): 52–69.

Introcaso, David M. *Coolidge Dam, Pinal County, Arizona*. San Francisco: Historic American Building Survey, National Park Service, Western Region, U.S. Department of the Interior, 1986.

———. "The Politics of Technology: The 'Unpleasant Truth About Pleasant Dam.'" *Western Historical Quarterly* 26 (autumn 1995): 333–52.

Jackson, William T., and Stephen D. Mikesell. *The Stanislaus River Drainage Basin and the New Melones Dam: Historical Evolution of Water Use Priorities.* Davis: University of California Press, 1979.

Jackson, William T., and Donald J. Pisani. *Lake Tahoe Water: A Chronicle of Conflict Affecting the Environment.* Davis: Institute of Governmental Affairs, University of California, Davis, 1972.

Janfinson, John O. "The Secret History of the Mississippi's Earliest Locks and Dams." *Minnesota History* 54 (summer 1995): 254–67.

Kahrl, William L. *Water and Power: The Conflict over Los Angeles' Water Supply in the Owens Valley.* Berkeley: University of California Press, 1982.

Kelso, Maurice M., et al. *Water Supplies and Economic Growth in an Arid Environment: An Arizona Case Study.* Tucson: University of Arizona Press, 1973.

Kendrick, Gregory D., ed. *Beyond the Wasatch: The History of Irrigation in the Uinta Basin and Upper Provo River Area of Utah.* Washington, D.C.: Government Printing Office, 1989.

Kepfield, S. S. "The Liquid Gold Rush: Groundwater Irrigation and Law in Nebraska, 1900–1993." *Great Plains Quarterly* 13(4) (fall 1993): 237–50.

———. "El Dorado on the Platte: The Development of Agricultural Irrigation and Water Law in Nebraska, 1860–1895." *Nebraska History* 75 (fall 1994): 232–43.

———. "'They Were in Far Too Great Want': Federal Drought Relief to the Great Plains, 1887–1895." *South Dakota History* 28 (winter 1998): 244–70.

Koppes, Clayton R. "Public Water, Private Land: Origins of the Acreage Limitation Controversy, 1933–1953." *Pacific Historical Review* 47(4) (1978): 607–36.

Leopold, Luna B. "Vegetation of Southwestern Watersheds in the Nineteenth Century." *Geographical Review* 41 (1951): 295–316.

Limerick, Patricia Nelson. *Desert Passages: Encounters with the American Deserts.* Albuquerque: University of New Mexico Press, 1985.

Little, Charles E. "The Great American Aquifer." *Wilderness* 51 (fall 1987): 43–46.

Long, David R. "Pipe Dreams: Hetch Hetchy, the Urban West, and the Hydraulic Society Revisited." *Journal of the West* (July 1995): 19–31.

Love, Frank. *Mining Camps and Ghost Towns; A History of Mining in Arizona and California Along the Lower Colorado.* Los Angeles: Westernlore Press, 1974.

McCool, Daniel. *Command of the Waters: Iron Triangles, Federal Water Development, and Indian Water.* Berkeley: University of California Press, 1987.

McKibben, Bill. "A Refuge Without Borders (Connecticut River)." *Audubon* 97(1) (January/February 1995): 58–70.

McNamee, Gregory. *Gila: The Life and Death of an American River.* New York: Orion Books, 1994.

Melcher, Nick B., and Charles Parrett. "1993 Upper Mississippi River Floods." *Geotimes* 38(12) (December 1993): 15–18.

Meyer, Michael C. *Water in the Hispanic Southwest: A Social and Legal History.* Tucson: University of Arizona Press, 1984.

Miller, Char. *Water in the West: A High Country News Reader.* Corvallis, Ore.: Oregon State University Press, 2000.

Miller, Char, ed. *Fluid Agruments: Five Centuries of Western Water Conflict.* Tucson: University of Arizona Press, 2001.

Miller, M. Catherine. "Who Owns the Water?: Law, Property, and the Price of Irrigation." *Journal of the West* 29 (October 1990): 35–41.

Momatuk, Yva, and John Eastcott. "Liquid Land: Louisiana's Atchafalaya Swamp." *Audubon* 97(5) (September/October 1995): 48–59.

Murray, John A., ed. *A Republic of Rivers: Three Centuries of Nature Writing from Alaska and the Yukon.* New York: Oxford University Press, 1990.

Nash, Gerald. "Problems and Projects in the History of Nineteenth-Century California Land Policy." *Arizona and the West* 2 (winter 1960): 327–40.

Neel, Susan Rhoades. "Newton Drury and the Echo Park Dam Controversy." *Forest and Conservation History* 38(2) (April 1994): 56–66.

———. "A Place of Extremes: Nature, History, and the American West." *Western Historical Quarterly* 25 (winter 1994): 489–505.

Nichols, John. *The Milagro Beanfield War.* New York: Ballantine, 1974.

Opie, John. *Ogallala: Water for a Dry Land.* Lincoln: University of Nebraska Press, 1993.

Owens, Harry P. *Steamboats and the Cotton Economy: River Trade in the Yazoo-Mississippi Delta.* Jackson: University Press of Mississippi, 1990.

Perkins, Priscilla C. *Scientific Information in the Decision to Dam Glen Canyon.* Los Angeles: University of California, 1975.

Petersen, Keith C. *River of Life, Channel of Death: Fish and Dams on the Lower Snake.* Lewiston, Idaho: Confluence Press, 1995.

Peterson, Charles S. "Headgates and Conquest: The Limits of Irrigation on the Navajo Reservation, 1880–1950." *New Mexico Historical Review* 68 (July 1993): 269–90.

Pisani, Donald J. "The Origins of Reclamation in the Arid West: William Ralston's Canal and the Federal Irrigation Commission of 1873." *Journal of the West* 22(2) (1983): 9–19.

———. *From the Family Farm to Agribusiness: The Irrigation Crusade in California and the West, 1850–1931.* Berkeley: University of California Press, 1984.

———. "Land Monopoly in Nineteenth-Century California." *Agricultural History* 65(4)(fall 1991): 15–37.

———. *To Reclaim a Divided West: Water, Law, and Public Policy, 1848–1902.* Albuquerque: University of New Mexico Press, 1992.

Powell, John Wesley. *Report on the Arid Region of the United States.* Edited by Wallace Stegner. Cambridge: Harvard University Press, 1962.

Prince, Hugh. *Wetlands of the American Midwest: A Historical Geography of Changing Attitudes.* Chicago: University of Chicago Press, 1997.

Reid, Robert L, ed. *Always a River: The Ohio River and the American Experience.* Bloomington: Indiana University Press, 1991.

Reisner, Marc. *Cadillac Desert: The American West and Its Disappearing Water.* New York: Viking Penguin, 1986.

Reuss, Martin. *Designing the Bayous: The Control of Water in the Atchafalaya Basin, 1800–1995.* Alexandria, Va.: U.S. Army Corps of Engineers, 1998.

———. "The Art of Scientific Precision: River Research in the United States Army Corps of Engineers to 1945." *Technology and Culture* 40(2) (April 1999): 292–323.

Rivera, José A. *Acequia Culture: Water, Land, and Community in the Southwest.* Albuquerque: University of New Mexico Press, 1998.

Rogge, A. E., et al. *Raising Arizona's Dams: Daily Life, Danger, and Discrimination in the Dam Construction Camps of Central Arizona, 1890s–1940s.* Tucson: University of Arizona Press, 1995.

Scarpino, Philip. *Great River: An Environmental History of the Upper Mississippi, 1890–1950.* Columbia: University of Missouri Press, 1985.

Schneiders, Robert Kelley. *Unruly River: Two Centuries of Change Along the Missouri.* Lawrence: University Press of Kansas, 1999.

Sirey, Joseph. *Marshes of the Ocean Shore: Development of an Ecological Ethic.* College Station: Texas A&M University Press, 1984.

Smith, Karen L. *The Magnificent Experiment: Building the Salt River Reclamation Project, 1890–1917.* Tucson: University of Arizona Press, 1986.

Stegner, Wallace. *Beyond the Hundredth Meridian: John Wesley Powell and the Second Opening of the West.* Boston: Houghton Mifflin, 1954.

Steinberg, Theodore. *Nature Incorporated: Industrialization and the Waters of New England.* New York: Cambridge University Press, 1991.

Stine, Jeffrey K. *Mixing the Waters: Environment, Politics, and the Building of the Tennessee-Tombigbee Waterway.* Akron, Ohio: University of Akron Press, 1993.

Storper, Michael, and Richard Walker. *The Price of Water: Surplus and Subsidy in the California State Water Project.* Berkeley: Institute of Governmental Studies, University of California, Berkeley, 1984.

Sullivan, Vernon L. *Irrigation in New Mexico.* Washington, D.C.: Government Printing Office, 1909.

Teisch, Jessica. "The Drowning of Big Meadows: Nature's Managers in Progressive-Era California." *Environmental History* 4(1) (January 1999): 32–53.

Turner, Raymond M., and Martin M. Karpiscak. *Recent Vegetation Changes Along the Colorado River Between Glen Canyon Dam and Lake Mead, Arizona.* Washington, D.C.: Government Printing Office, 1980.

U.S. Bureau of Reclamation. *The Colorado River: A Comprehensive Report on the Development of the Water Resources of the Colorado River Basin for Irrigation, Power Production, and Other Beneficial Uses in Arizona, California, Colorado, Nevada, New Mexico, Utah, and Wyoming, by the United States Department of the Interior.* Washington, D.C.: U.S. Bureau of Reclamation, 1946.

——. *Navajo Indian Irrigation Project: New Mexico, San Juan County.* Washington, D.C.: U.S. Bureau of Reclamation, 1983.

U.S. Congress. *An Act to Provide for the Settlement of Certain Claims of the Papago Tribe of Arizona Arising from the Construction of Tat Momolikot Dam.* Washington, D.C.: Government Printing Office, 1986.

U. S. Congress. House. *Gila River and Tributaries Below Gillespie Dam, Arizona: Letter from the Secretary of the Army, Transmitting ... a Report ... for a Preliminary Examination and Survey of Gila River, in Arizona, from Gillespie Dam Downstream to a Point Near Wellton.* 81st Cong., 1st sess., H. Doc. 331. Washington, D.C.: U.S. Government Printing Office, 1950.

Valentine, Rodney J. "Pioneer Settlers' Abuse of Land Laws in the Nineteenth Century: The Case of the Boise River Valley, Idaho." *Agricultural History* 67 (summer 1993): 47–65.

Van Dyke, T.S. "Irrigation in Southern California." *Independent* 45(4) (May 1983): 8.

Vileisis, Ann. *The Unknown Landscape: A History of America's Wetlands.* Washington, D.C.: Island Press, 1997.

Walker, Richard. *The California Water System: Another Round of Expansion.* Berkeley: Institute of Governmental Studies, University of California, 1979.

Wallace, Henry A. *Irrigation Frontier: On the Trail of the Corn-Belt Farmer.* Norman: University of Oklahoma, 1909.

Wells, Andrew Jackson. *Government Irrigation and the Settler: California, Oregon, Nevada and Arizona, Including a Description of the Imperial Valley Project.* San Francisco: Passenger Department [of] Southern Pacific, 1910.

Whayne, Jeannie, and Willard B. Gatewood, eds. *The Arkansas Delta: Land of Paradox.* Fayetteville: University of Arkansas Press, 1993.

Wilkinson, Charles, F. *Crossing the Next Meridian: Land, Water, and the Future of the West.* Washington, D.C.: Island Press, 1992.

Worster, Donald. "Hydraulic Society in California: An Ecological Interpretation." *Agricultural History* 56 (July 1982): 503–15.

———. *Rivers of Empire: Water, Aridity, and the Growth of the American West.* New York: Pantheon, 1985.

———. *A River Running West: The Life of John Wesley Powell.* New York: Oxford, 2001.

Wright-Peterson, Ralph. "Benny Ambrose: Life in the Boundary Waters." *Minnesota History* 54 (fall 1994): 124–37.

Zwingle, Erla. "Ogallala Aquifer: Wellspring of the High Plains." *National Geographic* 183(3) (March 1993): 54–80.

19. Wilderness Preservation

Abbey, Edward. *Desert Solitaire.* Salt Lake City: Peregrine Smith, 1981.

Albright, Horace M., and Marian Albright Schenck. *Creating the National Park Service: The Missing Years.* Norman: University of Oklahoma Press, 1999.

Allin, Craig W. *The Politics of Wilderness Preservation.* Westport, Conn.: Greenwood Press, 1982.

Austin, Mary H. *The Colorado River Controversy.* New York, 1927.

Backes, David. *Canoe Country: An Embattled Wilderness.* Minocqua, Wisc.: NorthWord Press, 1991.

Benton, Lisa M. *The Presidio: From Army Post to National Park*. Boston: Northeastern University Press, 1998.

Bogue, Margaret Beattie. "To Save the Fish: Canada, the United States, the Great Lakes, and the Join Commission of 1892." *Journal of American History* 79(4) (March 1993): 1429–55.

———. *Fishing the Great Lakes: An Environmental History, 1783–1933*. Madison: University of Wisconsin Press, 2000.

Brown, Margaret Lynn. *The Wild East: A Biography of the Great Smoky Mountains*. Gainesville, Fla.: University Press of Florida, 2000.

Buell, Lawrence. *The Environmental Imagination: Thoreau, Nature Writing, and the Formation of American Culture*. Cambridge: Belknap Press of Harvard University Press, 1995.

Burnham, Philip. *Indian Country, God's Country: Native Americans and the National Parks*. Washington, D.C. Island Press, 2000.

Callicott, J. Baird, and Michael P. Nelson, eds. *The Great New Wilderness Debate: An Expansive Collection of Writings Defining Wilderness from John Muir to Gary Snyder*. Athens: University of Georgia Press, 1998.

Carroll, Peter. *Puritanism and the Wilderness: The Intellectual Significance of the New England Frontier, 1629–1700*. New York: Columbia University Press, 1969.

Catton, Theodore. *Inhabited Wilderness: Indians, Eskimos, and National Parks in Alaska*. Albuquerque: University of New Mexico Press, 1997.

Chase, Alston. *Playing God in Yellowstone*. San Diego: Harcourt Brace Jovanovich, 1986.

Cohen, Michael. *The Pathless Way: John Muir and American Wilderness*. Madison: University of Wisconsin Press, 1984.

Cook, Gregory F. "The Public Trust Doctrine in Alaska." *Journal of Environmental Law and Litigation* 8 (1993): 1–49.

Cronon, William. "Kennecott Journey: The Paths Out of Town." In *Under an Open Sky: Rethinking America's Western Past*, edited by William Cronon, George Miles, and Jay Gitlin. New York, Norton, 1992, pp. 28–51.

Davis, Mike. "The Case for Letting Malibu Burn." *Environmental History Review* 19 (summer 1995): 1–36.

Devall, Bill, ed. *Clearcut: The Tragedy of Industrial Forestry*. San Francisco: Sierra Club Books / Earth Island Press, 1993.

Dilsaver, Lary M., and Douglas H. Strong. "Sequoia and Kings Canyon National Parks: One Hundred Years of Preservation and Resource Management." *California History* 69 (summer 1990): 98–117.

Dilsaver, Lary M., ed. *America's National Park System: The Critical Documents*. Lanham, Maryland: Rowman and Littlefield, 1994.

Drache, Hiram M. *Taming the Wilderness: The Northern Border Country, 1910–1939*. Danville, Ill.: Interstate, 1992.

Fox, Steven. *John Muir and His Legacy: The American Conservation Movement*. Boston: Little, Brown, 1981.

Frome, Michael. *Regreening the National Parks*. Tucson: University of Arizona Press, 1992.

Gottlieb, Robert. *Thirst for Growth: Water Agencies as Hidden Government in California*. Tucson: University of Arizona Press, 1991.

Grant, Madison. *Saving the Redwoods, An Account of the Movement during 1919 to Preserve the Redwoods of California*. New York: New York Zoological Society, 1919.

Grant, Richard A. "The Fight for the California Desert: Conserve or Destroy?" *Cry California* 8(1) (1972–73): 4–12.

Harvey, Mark W. T. "Echo Park, Glen Canyon, and the Postwar Wilderness Movement." *Pacific Historical Review* 60(1) (February 1991): 43–68.

———. *A Symbol of Wilderness: Echo Park and the American Conservation Movement*. Albuquerque: University of New Mexico Press, 1994.

———. "Battle for Dinosaur: Echo Park Dam and the Birth of the Modern Wilderness Movement." *Montana* 45 (winter 1995): 32–45.

Hoff, Jeff. *The Legal Battle Over Mono Lake*. San Francisco: State Bar of California, 1982.

Huth, Hans. "Yosemite: The Story of an Idea." *Sierra Club Bulletin* 33 (March 1948): 47–78.

Jacoby, Karl. *Crimes Against Nature: Squatters, Poachers, Thieves, and the Hidden History of American Conservation*. Berkeley: University of California Press, 2001.

Jaehn, Thomas. *The Environment in the Twentieth-Century American West: A Bibliography*. Albuquerque: Center for the American West, Department of History, University of New Mexico, 1990.

Jones, Holway R. "Mysterious Origin of the Yosemite Park Bill." *Sierra Club Bulletin* 48 (December 1963): 69–79.

———. *John Muir and the Sierra Club: The Battle for Yosemite*. San Francisco: Sierra Club, 1965.

Kaufman, Polly Welts. *National Parks and the Woman's Voice: A History*. 1st ed. Albuquerque: University of New Mexico Press, 1997.

Keller, Robert H., and Michael F. Turek. *American Indians and National Parks*. Tucson: University of Arizona Press, 1998.

Kittredge, William. "Second Change at Paradise: After Decades of Conquering the Land, Can We at Last Learn to Live in the West?" *Audubon* 96(3) (May/June 1994): 68–73.

Knobloch, Frieda. *The Culture of Wilderness: Agriculture as Colonization in the American West.* Chapel Hill: University of North Carolina Press, 1996.

Kundell, James E., et al. *Land-use Policy and the Protection of Georgia's Environment.* Athens: Carl Vinson Institute of Government, University of Georgia, 1989.

Langford, Nathaniel Pitt. *The Discovery of Yellowstone Park; Journal of the Washburn Expedition to the Yellowstone and Firehole Rivers in the Year 1870.* Foreword by Aubrey L. Haines. 1st ed. 1905. Reprint, Lincoln: University of Nebraska Press, 1972.

Limbaugh, Ronald H. *John Muir's "Stickeen" and the Lessons of Nature.* Fairbanks: University of Alaska Press, 1996.

Linn, Amy. "Treaty in the Tallgrass" (Preservation of Kansas Prairie). *Audubon* 92(2) (March/April 1995): 118–23.

Machlis, Gary E., and Donald R. Field, eds. *National Parks and Rural Development: Practice and Policy in the United States.* Washington, D.C.: Island Press, 2000.

McCally, David. *The Everglades: An Environmental History.* Gainesville: University Press of Florida, 1999.

McCarthy, George Michael. *Hour of Trial: The Conservation Conflict in Colorado and the West, 1891–1907.* Norman: University of Oklahoma Press, 1977.

McClelland, Linda Flint. *Building the National Parks: Historic Landscape Design and Construction.* Baltimore: Johns Hopkins Univesity Press, 1998.

McCloskey, Michael, and Albert Hill. *Mineral King: Mass Recreation Versus Park Protection in the Sierra.* San Francisco: Sierra Club, 1971.

McKibben, Bill. *The End of Nature.* New York: Random House, 1989.

McPhee, John. *The Control of Nature.* New York: Farrar, Straus, Giroux, 1989.

Merrill, Marlene. *The Discovery of Yellowstone Park: Journals, Letters, and Images from the 1871 Hayden Expedition.* Lincoln: University of Nebraska Press, 1999.

Meyer, Judith L. *The Spirit of Yellowstone: The Cultural Evolution of a National Park.* Lanham, Md.: Rowman and Littlefield (distributed by National Book Network), 1996.

Muir, John. *Our National Parks.* Boston: Houghton Mifflin, 1901.

———. "The Hetch-Hetchy Valley: A National Question." *Sierra Club Bulletin* 16(5) (1910): 263–69.

———. "The Creation of Yosemite National Park." *Sierra Club Bulletin* 29 (October 1944): 49–60.

———. *Letters from Alaska.* Edited by Robert Engberg and Bruce Merrell. Madison: University of Wisconsin, 1993.

Nash, Roderick. "John Muir, William Kent, and the Conservation Schism." *Pacific Historical Review* 36 (November 1967): 423–33.

———. "The American Invention of National Parks." *American Quarterly* 22 (fall 1970): 726–35.

———. *Wilderness and the American Mind.* Rev. ed., 1973. 3d ed., 1982. New Haven: Yale University Press.

———. *Creating the West: Historical Interpretations, 1890–1990.* Albuquerque: University of New Mexico Press, 1991.

———, ed. *American Environmentalism: Readings in Conservation History.* New York: McGraw-Hill, 1990.

Neumann, Mark. *On the Rim: Looking for the Grand Canyon.* Minneapolis: University of Minnesota Press, 1999.

Novak, Barbara. *Nature and Culture: American Landscape Painting, 1825–1875.* New York: Oxford University Press, 1980.

O'Brien, Bob. *Our National Parks and the Search for Sustainability.* Austin: University of Texas Press, 1999.

Oelschlaeger, Max. *The Idea of Wilderness from Prehistory to the Age of Ecology.* New Haven: Yale University Press, 1991.

———. *The Wilderness Condition: Essays on Environment and Civilization.* San Francisco: Sierra Club Books, 1992.

Pearson, Byron E. "Salvation for Grand Canyon: Congress, the Sierra Club, and the Dam Controversy of 1966–1968." *Journal of the Southwest* 36 (summer 1994): 159–75.

Pinkett, Harold T. "Sources of American Forest and Conservation History." *Journal of Forest History* 25 (October 1981): 210–12.

Porter, Eliot. *The Place No One Knew: Glen Canyon on the Colorado.* Edited by David Brower. San Francisco: Sierra Club, 1963.

Pyne, Stephen J. *How the Canyon Became Grand: A Short History.* New York: Viking, 1998.

Richardson, Elmo R. *Dams, Parks, and Politics: Resource Development and Preservation in the Truman-Eisenhower Era.* Lexington: University Press of Kentucky, 1973.

Riley, Glenda. *Women and Nature: Saving the "Wild" West.* Lincoln : University of Nebraska Press, 1999.

Robbins, Jim. *Last Refuge: The Environmental Showdown in Yellowstone and the American West.* New York: Morrow, 1993.

Rohman, Hal K. "Stumbling Toward the Millennium: Tourism, the Post-industrial World and the Transformation of the American West." *California History* 77 (fall 1998): 140–55.

Runte, Alfred. *Yosemite: The Embattled Wilderness.* Lincoln: University of Nebraska Press, 1990.

——. *National Parks: The American Experience.* 3rd ed. Lincoln: University of Nebraska Press, 1997.

Schmitt, Peter J. *Back to Nature: The Arcadian Myth in Urban America.* New York: Oxford University Press, 1969.

Schrepfer, Susan. *The Fight to Save the Redwoods: A History of Environmental Reform, 1917–1978.* Madison: University of Wisconsin Press, 1983.

Senft, Dennis, and William J. McGinnies. "Conserving the Great Plains for All." *Agricultural Research* 40(8) (August 1992): 4–10.

Smiley, Jane. "So Shall We Reap." *Sierra* 79 (March/April 1994): 74–82, 140–141.

Solnit, Rebecca. *Savage Dreams: A Journey into the Hidden Wars of the American West.* San Francisco: Sierra Club Books, 1994.

Steely, James Wright. *Parks for Texas: Enduring Landscapes of the New Deal.* Austin: University of Texas Press, 1999.

Stegner, Wallace. *The American West as Living Space.* Ann Arbor: The University of Michigan Press, 1987.

Strong, Douglas H. "The Sierra Forest Preserve: The Movement to Preserve the San Joaquin Valley Watershed." *California Historical Society Quarterly* 46(1) (1967): 3–17.

Terrie, Philip G. *Forever Wild: A Cultural History of Wilderness in the Adirondacks.* Syracuse, New York: Syracuse University Press, 1994.

Utley, Dan K., and James Steely. *Guided with a Steady Hand: The Cultural Landscape of a Rural Texas Park.* Waco, Tex.: Baylor University Press, 1998.

Webb, Melody. *The Last Frontier: A History of the Yukon Basin of Canada and Alaska.* Albuquerque: University of New Mexico Press, 1985.

West, Patrick, and Steven R. Brechin, eds. *Resident Peoples and National Parks: Social Dilemmas and Strategies in International Conservation.* Tucson: University of Arizona Press, 1991.

Wheeler, Ray. "Two Weeks to Wander: Walking Alone from Desert Bad-
    lands to Frost-Shattered Summits, a Hiker Finds Otherworldly Re-
    wards — and Disturbing Signs of Change — in Southern Utah." *Sierra*
    75(1) (January/February 1990): 32–43.
Whitehead, John. "Hawaii: The First and Last Far West?" *Western Historical*
    Quarterly 23(2) (May 1992): 152–77.
Wise, Kenneth, and Ron Petersen. A *Natural History of Mount Le Conte*.
    Knoxville: University of Tennessee Press, 1998.
Worster, Donald. *Under Western Skies: Nature and History in the American*
    West. New York: Oxford University Press, 1992.
——, ed. *American Environmentalism: The Formative Period, 1860–1915*.
    New York: Wiley, 1973.
Wright, John B. *Rocky Mountain Divide: Selling and Saving the West*.
    Austin: University of Texas Press, 1993.

20. Wildlife

Ackerman, Diane, and Bill Curtsinger. "Last Refuge of the Monk Seal." *Na-*
    tional Geographic 181(1) (January 1992): 128–45.
Ager, Thomas Alan. *Late Quaternary Environmental History of the Tanana*
    Valley, Alaska. Columbus: Reseaerch Foundation, Ohio State Univer-
    sity, 1975.
Alaska Geographic. "Arctic National Wildlife Refuge." *Alaska Geographic*
    20 (1993): 1–95.
Amos, William H. "Hawaii's Volcanic Cradle of Life." *National Geographic*
    178(1) (July 1990): 70–88.
Asquith, Adam. "Alien Species and the Extinction Crisis of Hawaii's Inverte-
    brates." *Endangered Species Update* 12(6) (June 1995): 6–12.
Baker, Rollin H. "Texas Wildlife Conservation — Historical Notes." *East*
    Texas Historical Journal 33(1) (1995): 59–72.
Barrow, Mark V., Jr. A *Passion for Birds: American Ornithology After*
    Audubon. Princeton: Princeton University Press, 1998.
Beans, Bruce E. *Eagle's Plume: The Struggle to Preserve the Life and Haunts*
    of America's Bald Eagle. Lincoln: University of Nebraska Press 1997.
Bellomy, M.D. "The Unconquered Gypsy." *American Forests* 61 (June
    1955): 26–27, 56–59.
Belsky, Joy. "Cattle and Sheep — The Forgotten Pathogens." *Inner Voice* 7
    (July/August 1995): 6–7.

Blair, James P. "A Portrait of the Missouri Botanical Garden: The Plant Hunters." *National Geographic* 178(2) (August 1990): 124–41.

Brodhead, Michael J. "The United States Army Signal Service and Natural History in Alaska, 1874–1883." *Pacific Northwest Quarterly* 86 (spring 1995): 72–82.

Brown, Bruce. *Mountain in the Clouds: A Search for Wild Salmon.* Seattle: University of Washington Press, 1982, first illustrated edition, 1995.

Bryan, Edwin H., Jr., et al. *The Natural and Cultural History of Honaunau, Kona, Hawai'i.* Honolulu: Department of Anthropology, Bernice Pauahi Bishop Museum, 1986.

Busch, Briton Cooper. *The War Against the Seals: A History of the North American Seal Fishery.* Toronto: McGill-Queen's University Press, 1985.

Caldwell, Francis E. *Land of the Ocean Mists: The Wild Ocean Coast West of Glacier Bay.* Edmonds, Wash.: Alaska Northwest, 1986.

Campbell, Louis W. *The Marshes of Southwestern Lake Erie.* Athens: Ohio University Press, 1995.

Catton, Theodore, and Lisa Mighetto. *The Fish and Wildlife Job on the National Forests: A Century of Game and Fish Conservation, Habitat Protection, and Ecosystem Management.* Missoula, Mont.: Historical Research Associates, Inc., for the U.S. Department of Agriculture, Forest Service, 1999.

Chadwick, Douglas H. "Denali: Alaska's Wild Heart." *National Geographic* 182(2) (August 1992): 62–86.

Chapple, Steve. "Far Away from It All: Waipio Valley, on Hawaii's Big Island, Is a Lost Place Where Wilderness Thrives." *Audubon* 96(3) (May/June 1994): 34–36.

Cone, Joseph, and Sandy Ridlington, eds. *The Northwest Salmon Crisis: A Documentary History.* Corvallis: Oregon State University Press, 1996.

Conniff, Richard. "Blackwater Country." *National Geographic* 181(4) (April 1992): 34–64.

Cowdrey, Albert E. *This Land, This South: An Environmental History.* Lexington: University Press of Kentucky, 1983.

Crosby, Alfred W. *The Columbian Exchange: Biological and Cultural Consequences of 1492.* Westport, Conn.: Greenwood Press, 1972.

———. *Ecological Imperialism: The Biological Expansion of Europe.* New York: Cambridge University Press, 1986.

———. *Germs, Seeds, and Animals: Studies in Ecological History.* New York: M.E. Sharpe, 1994.

Dalke, Paul D., A. Starker Leopold, and David L. Spencer. *The Ecology and Management of the Wild Turkey in Missouri.* Jefferson City, Mo.: Conservation Commission, Federal Aid–Wildlife Program, State of Missouri, 1946.

Davis, Lance E., Robert E. Gallman, and Teresa D. Hutchins. "The Decline of U.S. Whaling: Was the Stock of Whales Running Out?" *Business History Review* 62 (winter 1988): 569–95.

Derr, Mark. "Redeeming the Everglades." *Audubon* 95(5) (September/October 1993): 48–61.

Doughty, Robin W. "San Francisco's Nineteenth-Century Egg Basket: The Farallons." *Geographical Review* 61(4) (1971): 544–72.

Dunbar, Kurt, and Chris Friday. "Salmon, Seals, and Science: The *Albatross* and Conservation in Alaska, 1888–1914." *Journal of the West* 33 (October 1994): 6–13.

Dunlap, Thomas R. *Saving America's Wildlife.* Princeton: Princeton University Press, 1988.

Elias, Scott A. *The Ice Age History of Alaskan National Parks.* Washington, D.C.: Smithsonian Institution Press, 1995.

Evenden, Matthew D. "The Laborers of Nature: Economic Ornithology and the Role of Birds as Agents of Biological Pest Control in North American Agriculture, ca. 1880–1930." *Forest and Conservation History* 39 (October 1995): 172–183.

Fields, Leslie Leyland. *The Entangling Net: Alaska's Commercial Fishing Women Tell Their Lives.* Urbana: University of Illinois Press, 1997.

Foster, Michael S. *The Ecology of Giant Kelp Forests in California: A Community Profile.* Washington, D.C.: U.S. Fish and Wildlife Service, 1985.

Gay, James Thomas. *American Fur Seal Diplomacy: The Alaskan Fur Seal Controversy.* New York: P. Lang, 1987.

Gerstell, Richard. *American Shad: A Three-Hundred-Year History in the Susquehanna River Basin.* University Park: Pennsylvania State University Press, 1998.

Gildrie, Richard P. "Towards an Environmental History of Tennessee." *Tennessee Historical Quarterly* 53 (spring 1994): 42–53.

Gosse, Philip Henry. *Letters from Alabama, (U.S.) Chiefly Relating to Natural History.* Tuscaloosa: University of Alabama Press, 1993.

Graham, Frank, Jr. "Building on Adversity: Near an Industrial Waste Ground, Sanctuaries Now Thrive on Tampa Bay." *Audubon* 96(4) (July/August 1994): 106–109.

Harris, Tom. *Death in the Marsh (Kesterson National Wildlife Refuge).* Washington, D.C.: Island Press, 1991.

Hoover, Herbert T. *Wildlife on the Cheyenne River and Lower Brule Sioux Reservations: A History of Use and Jurisdiction.* Vermillion: University of South Dakota Press, 1992.

Hornocker, Maurice G., and George F. Mobley. "Learning to Live with Mountain Lions." *National Geographic* 182(1) (July 1992): 52–66.

Horton, Tom. "Chesapeake Bay: Hanging in the Balance." *National Geographic* 183(6) (June 1993): 2–36.

Huth, Paul C., and Robi Josephson. "Slabsides Centennial: 'Wildlife About My Cabin.'" *Conservationist* 49 (April 1995): 16–19.

Johnson, Charles Eugene. "The Beaver in the Adirondacks: Its Economics and Natural History." *Roosevelt Wild Life Bulletin* 4 (July 1927): 501–641.

Kay, E. Alison, ed. *A Natural History of the Hawaiian Islands: Selected Readings II.* Honolulu: University of Hawaii Press, 1994.

Kenna, Michael. "Eastern Wildlife: Bittersweet Success." *National Geographic* 181(2) (February 1992): 70–90.

Klinkenborg, Verlyn. "The Mustang Myth: America's Wild Horses Have Inspired One of the Nation's Most Passionate Conservation Battles." *Audubon* 96(1) (January/February 1994): 34–44.

Laycock, George. "How to Kill a Wolf: Laws to the Contrary, 'Mechanical Hawks' Decimate Alaskan Packs." *Audubon* 92(6) (November 1990): 44–48.

Leffler, Marilyn, and Nancy Matthews. "Women in Wildlife." *Women in Natural Resources* 20 (fall 1998): 9–12.

Leopold, A. Starker, and Tupper Ansel Blake. *Wild California: Vanishing Lands, Vanishing Wildlife.* Berkeley: University of California Press, 1985.

Leopold, Aldo. *Game Management.* New York: Scribner's, 1933.

Lord, Philip, Jr. "Of Eels and the River." *Conservationist* 50 (August 1995): 16–19.

Lowe, Charles H. *Arizona's Natural Environment: Landscapes and Habitats.* Tucson: University of Arizona Press, 1964.

Lyman, R. Lee. *White Goats, White Lies: The Misuse of Science in Olympic National Park.* Salt Lake City: University of Utah Press, 1998.

MacCameron, Robert. "Environmental Change in Colonial New Mexico." *Environmental History Review* 18 (summer 1994): 17–39.

Mahan, John, and Ann Mahan. *Wild Lake Michigan.* Stillwater, Minn.: Voyageur Press, 1991.

Mairson, Alan. "The Everglades: Dying for Help." *National Geographic* 185(4) (April 1994): 2–26.

Maxwell, Jessica. "Swimming with Salmon." *Natural History* 104 (September 1995): 26–39.

McClaran, Mitchel P., and Ward W. Brady. "Arizona's Diverse Vegetation and Contributions to Plant Ecology." *Rangelands* 16 (October 1994): 208–17.

McCullough, Dale R. *The Tule Elk: Its History, Behavior, and Ecology.* Berkeley: University of California Press, 1971.

McIntyre, Rick. *A Society of Wolves: National Parks and the Battle over the Wolf.* Stillwater, Minn.: Voyageur Press, 1993.

McKee, Edwin Dinwiddie. *The Environment and History of the Toroweap and Kaibab Formations of Northern Arizona and Southern Utah.* Washington, D.C.: Carnegie Institution of Washington, 1938.

Merriam, C. Hart. *Results of a Biological Survey of Mount Shasta, California, by C. Hart Merriam, Chief of Division of Biological Survey.* Washington, D.C.: Government Printing Office, 1899.

Mighetto, Lisa. *Wild Animals and American Environmental Ethics.* Tucson: University of Arizona Press, 1991.

———. *Saving the Salmon: A History of the U.S. Army Corps of Engineers' Efforts to Protect Anadromous Fish on the Columbia and Snake Rivers.* Seattle, Wash.: Historical Research Associates, 1994.

———, ed. *Muir Among the Animals: The Wildlife Writings of John Muir.* San Francisco: Sierra Club Books, 1986.

Mitchell, John G. "Our Disappearing Wetlands." *National Geographic* 182(4) (October 1992): 2–46.

Mitman, Gregg. *Reel Nature: America's Romance with Wildlife on Film.* Cambridge: Harvard University Press, 1999.

Muir, John. *My First Summer in the Sierra.* Boston: Houghton Mifflin, 1911.

———. *The Yosemite.* New York: Century, 1912.

———. *Northwest Passages: From the Pen of John Muir in California, Oregon, Washington, and Alaska.* Edited by Scott Lankford. Palo Alto, Calif.: Tioga, 1988.

Norwood, Vera. *Made From this Earth: American Women and Nature.* Chapel Hill: University of North Carolina Press, 1993.

O'Clair, Rita M. *The Nature of Southeast Alaska: A Guide to Plants, Animals, and Habitats.* Anchorage: Alaska Northwest Books, 1992.

Orsi, Jared. "From Horicon to Hamburgers and Back Again: Ecology, Ideology, and Wildfowl Management, 1917–1935." *Environmental History Review* 18 (winter 1994): 19–40.

Petersen, David. *Ghost Grizzlies.* New York: Holt, 1995.

Postel, Mitchell. *Vigil on the Golden Gate: The Environmental History of the San Francisco Bay Since 1850.* N.p., 1977.

Preston, William. *Vanishing Landscapes: Land and Life in the Tulare Lake Basin.* Berkeley: University of California Press, 1981.

Pritchard, James. "Charles C. Adams and Early Ecological Rationales for Yellowstone National Park, 1916–1941." *George Wright Forum* 15(4) (1998): 27–35.

Quammen, David. "Island of the Bears (Grizzly Bears of Yellowstone National Park)." *Audubon* 97(2) (March/April 1995): 82–90.

Raventon, Edward. *Island in the Plains: A Black Hills Natural History.* Boulder: Johnson Books, 1994.

Reisner, Marc. "California's Vanishing Wetlands." *The Amicus* 9(1) (winter 1987): 8–15.

Rieger, John. "Western Riparian and Wetland Ecosystems." *Restoration and Management Notes* 10 (summer 1992): 52–55.

Roppel, Patricia. *Salmon from Kodiak: An History of the Salmon Fishery of Kodiak Island, Alaska.* Anchorage: Alaska Historical Commission, 1986.

Rowell, Galen. "Falcon Rescue." *National Geographic* 179(4) (April 1991): 106–16.

Royte, Elizabeth. "On the Brink: Hawaii's Vanishing Species." *National Geographic* 188(3) (September 1995): 2–32.

Russell, Emily W. B. "Mt. Tabor, New Jersey: An Environmental History." *New Jersey History* 95 (autumn 1977): 157–69.

Scott, Susan. *Plants and Animals of Hawai'i.* Hololulu: Bess Press, 1991.

Seton, Ernest Thompson. *The Biography of a Grizzly.* New York: Grosset and Dunlap, 1900.

———. *Animals Worth Knowing: Selected from "Life Histories of Northern Animals."* Garden City, N.Y.: Doubleday, Page, 1925.

———. *Ernest Thompson Seton's America: Selections from the Writings of the Artist-Naturalist.* Edited by Farida A. Wiley. New York: Devin-Adair, 1954.

Sheldon, Charles. *The Wilderness of Desert Bighorns and Seri Indians: The Southwestern Journals of Charles Sheldon.* Phoenix: Arizona Desert Bighorn Sheep Society, 1979.

Skinner, J. E. *An Historical Review of the Fish and Wildlife Resources of the San Francisco Bay*. Sacramento: Water Project Branch Report; California Department of Fish and Game, 1962.

Sorensen, Victor. "The Wasters and Destroyers: Community-sponsored Predator Control in Early Utah Territory." *Utah Historical Quarterly* 62 (winter 1994): 26–41.

Stap, Don. "Florida's Ancient Shores (Apalachicola River)." *Audubon* 97(3) (May/June 1995): 36–38.

Stilgoe, John R. "A New England Coastal Wilderness." *Geographical Review* 71 (January 1981): 33–50.

Stoker, Sam W., et al. *Biological Conditions in Prince William Sound, Alaska: Following the Valdez Oil Spill, 1989–1992*. Anchorage, Alaska: Woodward-Clyde Consultants, 1992.

Stone, Charles P., and Linda W. Pratt. *Hawai'i's Plants and Animals: Biological Sketches of Hawaii Volcanoes National Park*. Honolulu: Hawaii Natural History Association, National Park Service, 1994.

Strong, Douglas H. "Lassen Volcanic National Park's Manzanita Lake: A Brief History." *Pacific Historian* 15 (fall 1971): 68–82.

Tamkins, Theresa. "Hog Wilds in Hawaii: The Islands' Big Pig Problem." *Audubon* 96(1) (January/February 1994): 17–19.

Taylor, David, ed. *South Carolina Naturalists: An Anthology, 1700–1860*. Columbia: University of South Carolina Press, 1998.

Taylor, Joseph E., III. "Burning the Candle at Both Ends: Historicizing Overfishing in Oregon's Nineteenth-Century Salmon Fisheries." *Environmental History* 4(1) (January 1999): 54–79.

———. *Making Salmon: An Environmental History of the Northwest Fisheries Crisis*. Seattle: University of Washington Press, 1999.

Taylor, Walter K. *Wild Shores: Exploring the Wilderness Areas of Eastern North Carolina*. Asheboro, North Carolina: Down Home Press, 1993.

Thiel, Richard P. *The Timber Wolf in Wisconsin: The Death and Life of a Majestic Predator*. Madison: University of Wisconsin Press, 1993.

Vencil, Betsy. "The Migratory Bird Treaty Act: Protecting Wildlife on Our National Refuges—California's Kesterson Reservoir, a Case in Point." *Natural Resources Journal* 26(3) (July 1986): 609–27.

Vorren, Ornulv. *Saami, Reindeer, and Gold in Alaska: The Emigration of Saami from Norway to Alaska*. Prospect Heights, Ill.: Waveland Press, 1994.

Walters, Nick. "Living With Beaver." *Alabama's Treasured Forests* 13 (spring 1994): 12–14.

Ward, Fred. "Florida's Coral Reefs Are Imperiled." *National Geographic* 178(1) (July 1990): 114–33.

Warren, Louis S. *The Hunter's Game: Poachers and Conservationists in Twentieth-Century America.* New Haven: Yale University Press, 1997.

Warrick, Sheridan F., and Elizabeth D. Wilcox, eds. *Big River: The Natural History of an Endangered Northern California Estuary.* Santa Cruz, Calif.: University of California Environmental Field Program, 1981.

Wassink, Jan L. *Mammals of the Central Rockies.* Missoula, Mont.: Mountain Press, 1993.

Webb, Robert Lloyd. *On the Northwest: Commercial Whaling in the Pacific Northwest, 1790–1967.* Pacific Maritime Studies, vol. 6. Vancouver: University of British Columbia Press, 1988.

Welch, Margaret. *The Book of Nature: Natural History in the United States, 1825–1875.* Boston: Northeastern University Press, 1998.

Wilcove, David S. *The Condor's Shadow: The Loss and Recovery of Wildlife in America.* New York: W. H. Freeman, 1999.

Wilkinson, Todd. "Back to the Badlands: Black-footed Ferrets, Once Considered Extinct, Are Being Reintroduced…" *National Parks* 68(11–12) (November/December 1994): 38–43.

Williams, Ted. "Torch of the Yankee Salmon" (Restoring Atlantic Salmon to New England). *Audubon* 97(3) (May/June 1995): 28–34

——. "Finding Safe Harbor (Habitat Conservation in North Carolina)." *Audubon* 98(1) (January/February 1996): 26–32.

Wolch, Jennifer, and Jody Emel, eds. *Animal Geographies: Place, Politics, and Identity in the Nature-Culture Borderlands.* New York: Verso, 1998.

Woodbury, Angus M. *Ecological Studies of Flora and Fauna in Glen Canyon.* Salt Lake City: University of Utah Press, 1959.

Zontek, Ken. "Idaho Big-Game Management and Harvesting During the Great Depression." *Idaho Yesterdays* 42 (spring 1998): 3–13.

21. The Urban Environment

Adams, Robert. *Los Angeles Spring.* New York: Aperture, 1986.

Addams, Jane. *Twenty Years at Hull-House.* New York: The Macmillan Company, 1938.

Bolles, Albert S. *Industrial History of the United States: From the Earliest Settlements to the Present Times: Being a Complete Survey of American Industries …* Norwich, Conn.: Henry Bill, 1878.

Bronson, William. *How to Kill a Golden State.* Garden City, N.Y.: Double-day, 1968.

Bullard, Robert D., Glenn S. Johnson, and Angel O. Torres. *Sprawl City: Race, Politics and Planning in Atlanta.* Washington, D.C. Island Press, 2000.

California Governor's Office of Planning and Research. *CEQA: California Environmental Quality Act: Statutes and Guidelines.* Sacramento, Calif.: Governor's Office of Planning and Research, 1992.

Cicchetti, Charles J. "The Route Not Taken: The Decision to Build the Trans-Alaska Pipeline and the Aftermath." *American Enterprise* 4(5) (September/October 1993): 38–46.

Clark, Claudia. *Radium Girls: Women and Industrial Health Reform, 1910–1935.* Chapel Hill: University of North Carolina Press, 1997.

Colton, Craig, ed. *Transforming New Orleans and Its Environs: Centuries of Change.* Pittsburgh: University of Pittsburgh Press, 2000.

Cooper, Gail. *Air-Conditioning America: Engineers and the Controlled Environment, 1900–1960.* Baltimore: Johns Hopkins University Press, 1998.

Commoner, Barry. *The Closing Circle.* New York: Knopf, 1972.

Dawson, Robert, and Gray Brechin. *Farewell, Promised Land: Waking from the California Dream.* Berkeley: University of California Press, 1999.

Davis, Mike. "The Case for Letting Malibu Burn." *Environmental History Review* 19 (summer 1995): 1–36.

Davis, Mike. *City of Quartz.* New York: Vintage, 1992.

——. *Ecology of Fear: Los Angeles and the Imagination of Disaster.* New York: Holt, 1998.

Davis, Susan G. *Spectacular Nature: Corporate Culture and the Sea World Experience.* Berkeley: University of California Press, 1997.

——. "Landscapes of Imagination: Tourism in Southern California." *Pacific Historical Review* 68(2) (May 1999): 173–92.

Deverell, William, Greg Hise, and David C. Sloane, "Orange Empires: Comparing Miami and Los Angeles." *Pacific Historical Review* 68(2) (May 1999): 145–52.

Domosh, Mona. *Invented Cities: The Creation of Landscape in Nineteenth-century New York and Boston.* New Haven: Yale University Press, 1996.

Dowall, David E. *Land-use Policies in the San Francisco Bay Area: A Survey of Local Governments.* Berkeley: Institute of Urban and Regional Development, University of California, 1981.

——. *The Suburban Squeeze: Land Conversion and Regulation in the San Francisco Bay Area.* Berkeley: University of California Press, 1984.

Elkind, Sarah S. *Bay Cities and Water Politics: The Battle for Resources in Boston and Oakland.* Lawrence: University Press of Kansas, 1998.

Flink, James J. *The Car Culture.* Cambridge: MIT Press, 1975.

Gallagher, Thomas J., and Anthony F. Gasbarro. "The Battles for Alaska: Planning in America's Last Wilderness." *Journal of the American Planning Association* 55(4) (autumn 1989): 433–45.

Gómez, Arthur R. *Quest for the Golden Circle: The Four Corners and the Metropolitan West, 1945–1970.* Lawrence: University Press of Kansas, 1994.

Hahn, Steven, and Jonathan Prude, eds. *The Countryside in the Age of Capitalist Transformation.* Chapel Hill: University of North Carolina Press, 1985.

Hays, Samuel P., ed. *City at the Point: Essays on the Social History of Pittsburgh.* Pittsburgh: University of Pittsburgh Press, 1989.

Havlick, Spenser W. *The Urban Organism: The City's Natural Resources from an Environmental Perspective.* New York: Macmillan, 1974.

Herzog, Lawrence A. *Where North Meets South: Cities, Space, and Politics on the U.S.-Mexico Border.* Austin: Center for Mexican American Studies, University of Texas at Austin, 1990.

Issel, William. "Land Values, Human Values, and the Preservation of the City's Treasured Appearance: Environmentalism, Politics, and the San Francisco Freeway Revolt." *Pacific Historical Review* 68 (November 1999): 611–46.

Jackson, W. Turrentine, and Donald J. Pisani. *From Resort Area to Urban Recreation Center: Themes in the Development of Lake Tahoe, 1946–1956.* Davis: Institute of Governmental Affairs, University of California, 1973.

Keeble, John. "A Parable of Oil and Water: Revisiting Prince William Sound, Four Years After." *Amicus Journal* 15(1) (spring 1993): 35–43.

Klein, Maury. *Unfinished Business: The Railroad in American Life.* Hanover: University of Rhode Island, 1994.

Kowsky, Francis R. *Country, Park and City: The Architecture and Life of Calvert Vaux.* New York: Oxford University Press, 1998.

Krakauer, Jon. "Ice, Mosquitoes and Muskeg — Building the Road to Alaska." *Smithsonian* 23(4) (July 1992): 102–12.

Kulikoff, Alan. "The Transition to Capitalism in Rural America." *William and Mary Quarterly* 46(1) (1989): 120–44.

Landau, Sarah Bradford and Carl W. Condit. *Rise of the New York Skyscraper, 1865–1913.* New Haven: Yale University Press, 1996.

Machor, James. *Pastoral Cities: Urban Ideals and the Symbolic Landscape of America*. Madison: University of Wisconsin Press, 1987.

McGowan, "Fault-lines: Seismic Safety and the Changing Political Economy of California's Transit System." *California History* (summer 1993): 170–193.

McPhee, John. "Los Angeles Against the Mountains." In *The Control of Nature*. New York: Farrar, Straus, Giroux, 1989.

Melosi, Martin V. *Garbage in the Cities: Refuse, Reform, and the Environment, 1880–1980*. College Station: Texas A&M University Press, 1981.

——. *Coping with Abundance: Energy and Environment in Industrial America*. New York: Knopf, 1985.

——. *The Sanitary City: Urban Infrastructure in America from Colonial Times to the Present*. Baltimore: Johns Hopkins University Press, 2000.

——. *Effluent America: Cities, Industry, Energy, and the Environment*. Pittsburgh: University of Pittsburgh Press, 2001.

Melosi, Martin V., ed. *Pollution and Reform in American Cities, 1870–1930*. Austin: University of Texas Press, 1980.

Miller, Angela. *The Empire of the Eye: Landscape Representation and American Cultural Politics, 1825–1875*. Ithaca, N.Y.: Cornell University Press, 1993

Naske, Claus-M. "Some Attention, Little Action: Vacillating Federal Efforts to Provide Territorial Alaska with an Economic Base." *Western Historical Quarterly* 26 (spring 1995): 37–68.

Norris, Frank. *The Octopus: A Story of California*. Garden City, N.Y.: Doubleday, 1948.

Pincetl, Stephanie S. *Transforming California: A Political History of Land Use and Development*. Baltimore and London: The Johns Hopkins University Press, 1999.

Price, John A. *The Urbanization of Baja California*. San Diego, Calif.: J. A. Price, 1968.

Quam-Wickham, Nancy. "'Cities Sacrificed on the Altar of Oil': Popular Opposition to Oil Development in 1920s Los Angeles." *Environmental History* 3(2) (April 1998): 189–209.

Robbins, William G. *Colony and Empire: The Capitalist Transformation of the American West*. Lawrence: University Press of Kansas, 1994.

Rome, Adam. *The Bulldozer in the Countryside: Suburban Sprawl and the Rise of American Environmentalism*. New York: Cambridge University Press, 2001.

Ross, C. George. *The Urbanization of Rural California: A Study of Land Use and Markets for Recreational Land Developments in State, 1964–1974.* Berkeley: Center for Real Estate and Urban Economics, University of California, 1975.

Roth, Matthew W. "Mulholland Highway and the Engineering Culture of Los Angeles in the 1920s." *Technology and Culture* 40(3) (July 1999): 545–75.

Saunders, Richard. "'Rags! Rags!! Rags!!!': Beginnings of the Paper Industry in the Salt Lake Valley, 1849–58." *Utah Historical Quarterly* 62 (winter 1994): 42–52.

Schama, Simon. *Landscape and Memory.* New York: Knopf, 1995.

Sellers, Christopher C. *Hazards of the Job: From Industrial Disease to Environmental Health Science.* Chapel Hill: University of North Carolina Press, 1997.

Steinberg, Theodore. *Nature Incorporated: Industrialization and the Waters of New England.* New York: Cambridge University Press, 1991.

Stephenson, R. Bruce. *Visions of Eden: Environmentalism, Urban Planning, and City Building in St. Petersburg, Florida, 1900–1995.* Columbus: Ohio State University Press, 1997.

Still, Bayrd. *Urban America: A History with Documents.* Boston: Little, Brown, 1974.

Stillgoe, John R. *Common Landscape of America, 1580–1845.* New Haven: Yale University Press, 1982.

Storper, Michael, and Richard Walker. *The Capitalist Imperative: Territory, Technology, and Industrial Growth.* Oxford: Blackwell, 1989.

Swerdlow, Joel L., and Bob Sacha. "Erie Canal: Living Link to Our Past." *National Geographic* 178(5) (November 1990): 38–66.

Tarr, Joel A. *The Search for the Ultimate Sink: Urban Pollution in Historical Perspective.* Akron, Ohio: University of Akron Press, 1996.

Williams, Raymond. *The Country and the City.* New York: Oxford University Press, 1976.

Wilson, Alexander. *The Culture of Nature: North American Landscape from Disney to the Exxon Valdez.* Cambridge, Mass.: Blackwell, 1992.

Index